Spatio-temporal design

STATISTICS IN PRACTICE

Statistics in Practice is an important international series of texts which provide detailed coverage of statistical concepts, methods and worked case studies in specific fields of investigation and study.

With sound motivation and many worked practical examples, the books show in down-to-earth terms how to select and use an appropriate range of statistical techniques in a particular practical field within each title's special topic area.

The books provide statistical support for professionals and research workers across a range of employment fields and research environments. Subject areas covered include medicine and pharmaceutics; industry, finance and commerce; public services; the earth and environmental sciences, and so on.

The books also provide support to students studying statistical courses applied to the above areas. The demand for graduates to be equipped for the work environment has led to such courses becoming increasingly prevalent at universities and colleges.

It is our aim to present judiciously chosen and well-written workbooks to meet everyday practical needs. Feedback of views from readers will be most valuable to monitor the success of this aim.

A complete list of titles in this series appears at the end of the volume.

Spatio-temporal design

Advances in efficient data acquisition

Edited by

Jorge Mateu

Department of Mathematics of the University
Jaume I of Castellon, Spain

Werner G. Müller

Department of Applied Statistics
Johannes Kepler University Linz, Austria

WILEY

A John Wiley & Sons, Ltd., Publication

This edition first published 2013

Registered office
John Wiley & Sons Ltd, The Atrium, Southern Gate, Chichester, West Sussex, PO19 8SQ, United Kingdom

For details of our global editorial offices, for customer services and for information about how to apply for permission to reuse the copyright material in this book please see our website at www.wiley.com.

Library of Congress Cataloging-in-Publication Data

Spatio-temporal design : advances in efficient data acquisition / edited by Jorge Mateu, Department of Mathematics of the University Jaume I of Castellon, Spain, Werner G. Müller, Department of Applied Statistics, Johannes Kepler University Linz, Austria.
pages cm. – (Statistics in practice)
ISBN 978-0-470-97429-2 (hardback)
1. Sampling (Statistics) 2. Spatial analysis (Statistics) I. Mateu, Jorge, editor of compilation. II. Müller, W. G. (Werner G.), editor of compilation.
QA276.6.S63 2013
001.4′33–dc23

2012027161

A catalogue record for this book is available from the British Library.

ISBN: 978-0-470-97429-2

Set in 10/12pt Times by Laserwords Private Limited, Chennai, India.

To Eva and Evelyn

Contents

Contributors xv

Foreword xix

1 Collecting spatio-temporal data 1
Jorge Mateu and Werner G. Müller
1.1 Introduction 1
1.2 Paradigms in spatio-temporal design 2
1.3 Paradigms in spatio-temporal modeling 3
1.4 Geostatistics and spatio-temporal random functions 4
1.4.1 Relevant spatio-temporal concepts 4
1.4.2 Properties of the spatio-temporal covariance and variogram functions 6
1.4.3 Spatio-temporal kriging 8
1.4.4 Spatio-temporal covariance models 10
1.4.5 Parametric estimation of spatio-temporal covariograms 11
1.5 Types of design criteria and numerical optimization 13
1.6 The problem set: Upper Austria 17
1.6.1 Climatic data 17
1.6.2 Grassland usage 18
1.7 The chapters 23
Acknowledgments 28
References 28

2 Model-based frequentist design for univariate and multivariate geostatistics 37
Dale L. Zimmerman and Jie Li
2.1 Introduction 37
2.2 Design for univariate geostatistics 38
2.2.1 Data-model framework 38
2.2.2 Design criteria 38
2.2.3 Algorithms 42
2.2.4 Toy example 42

2.3 Design for multivariate geostatistics 45
2.3.1 Data-model framework 45
2.3.2 Design criteria 47
2.3.3 Toy example 48
2.4 Application: Austrian precipitation data network 50
2.5 Conclusions 52
References 53

3 Model-based criteria heuristics for second-phase spatial sampling 54
Eric M. Delmelle
3.1 Introduction 54
3.2 Geometric and geostatistical designs 56
3.2.1 Efficiency of spatial sampling designs 56
3.2.2 Sampling spatial variables in a geostatistical context 57
3.2.3 Sampling designs minimizing the kriging variance 58
3.3 Augmented designs: Second-phase sampling 59
3.3.1 Additional sampling schemes to maximize change in the kriging variance 59
3.3.2 A weighted kriging variance approach 60
3.4 A simulated annealing approach 63
3.5 Illustration 65
3.5.1 Initial sampling designs 66
3.5.2 Augmented designs 68
3.6 Discussion 68
References 69

4 Spatial sampling design by means of spectral approximations to the error process 72
Gunter Spöck and Jürgen Pilz
4.1 Introduction 72
4.2 A brief review on spatial sampling design 75
4.3 The spatial mixed linear model 76
4.4 Classical Bayesian experimental design problem 77
4.5 The Smith and Zhu design criterion 79
4.6 Spatial sampling design for trans-Gaussian kriging 81
4.7 The spatDesign toolbox 82
4.7.1 Covariance estimation and variography software 83
4.7.2 Spatial interpolation and kriging software 84
4.7.3 Spatial sampling design software 85
4.8 An example session 89
4.8.1 Preparatory calculations 89
4.8.2 Optimal design for the BSLM 93
4.8.3 Design for the trans-Gaussian kriging 94
4.9 Conclusions 98
References 99

5 Entropy-based network design using hierarchical Bayesian kriging **103**
Baisuo Jin, Yuehua Wu and Baiqi Miao
5.1 Introduction 103
5.2 Entropy-based network design using hierarchical Bayesian kriging 105
5.3 The data 107
5.4 Spatio-temporal modeling 107
5.5 Obtaining a staircase data structure 111
5.6 Estimating the hyperparameters $\mathcal{H}_g$ and the spatial correlations between gauge stations 113
5.7 Spatial predictive distribution over the 445 areas located in the 18 districts of Upper Austria 117
5.8 Adding gauge stations over the 445 areas located in the 18 districts of Upper Austria 120
5.9 Closing down an existing gauge station 122
5.10 Model evaluation 124
Appendix 5.1: Hierarchical Bayesian spatio-temporal modeling (or kriging) 124
Appendix 5.2: Some estimated parameters 128
Acknowledgments 129
References 129

6 Accounting for design in the analysis of spatial data **131**
Brian J. Reich and Montserrat Fuentes
6.1 Introduction 131
6.2 Modeling approaches 134
6.2.1 Informative missingness 134
6.2.2 Informative sampling 135
6.2.3 A two-stage approach for informative sampling 136
6.3 Analysis of the Austrian precipitation data 137
6.4 Discussion 139
References 141

7 Spatial design for knot selection in knot-based dimension reduction models **142**
Alan E. Gelfand, Sudipto Banerjee and Andrew O. Finley
7.1 Introduction 142
7.2 Handling large spatial datasets 145
7.3 Dimension reduction approaches 146
7.3.1 Basic properties of low rank models 146
7.3.2 Predictive process models: A brief review 148
7.4 Some basic knot design ideas 149
7.4.1 A brief review of spatial design 149
7.4.2 A strategy for selecting knots 151

7.5 Illustrations 153
7.5.1 A simulation example 153
7.5.2 A simulation example using the two-step analysis 159
7.5.3 Tree height and diameter analysis 160
7.5.4 Austria precipitation analysis 162
7.6 Discussion and future work 165
References 166

8 Exploratory designs for assessing spatial dependence 170
Agnes Fussl, Werner G. Müller and Juan Rodríguez-Díaz
8.1 Introduction 170
8.1.1 The dataset and its visualization 172
8.2 Spatial links 174
8.2.1 Spatial neighbors 175
8.2.2 Spatial weights 176
8.3 Measures of spatial dependence 178
8.4 Models for areal data 180
8.4.1 H_0: A spaceless regression model 181
8.4.2 H_0: Spatial regression models 185
8.5 Design considerations 190
8.5.1 A design criterion 192
8.5.2 Example 194
8.6 Discussion 195
Appendix 8.1: R code 198
Acknowledgments 202
References 203

9 Sampling design optimization for space-time kriging 207
Gerard B.M. Heuvelink, Daniel A. Griffith, Tomislav Hengl and Stephanie J. Melles
9.1 Introduction 207
9.2 Methodology 209
9.2.1 Space-time universal kriging 209
9.2.2 Sampling design optimization with spatial simulated annealing 211
9.3 Upper Austria case study 212
9.3.1 Descriptive statistics 212
9.3.2 Estimation of the space-time model and universal kriging 215
9.3.3 Optimal design scenario 1 218
9.3.4 Optimal design scenario 2 219
9.3.5 Optimal design scenario 3 219
9.4 Discussion and conclusions 221
Appendix 9.1: R code 222

Acknowledgment 227
References 228

10 Space-time adaptive sampling and data transformations **231**
José M. Angulo, María C. Bueso and Francisco J. Alonso
10.1 Introduction 231
10.2 Adaptive sampling network design 233
10.2.1 A simulated illustration 235
10.3 Predictive information based on data transformations 238
10.4 Application to Upper Austria temperature data 242
10.5 Summary 246
Acknowledgments 247
References 247

11 Adaptive sampling design for spatio-temporal prediction **249**
Thomas R. Fanshawe and Peter J. Diggle
11.1 Introduction 249
11.2 Review of spatial and spatio-temporal adaptive designs 251
11.3 The stationary Gaussian model 253
11.3.1 Model specification 253
11.3.2 Theoretically optimal designs 254
11.3.3 A comparison of design strategies 254
11.4 The dynamic process convolution model 257
11.4.1 Model specification 257
11.4.2 A comparison of design strategies 258
11.5 Upper Austria rainfall data example 262
11.6 Discussion 264
Appendix 11.1 266
References 267

12 Semiparametric dynamic design of monitoring networks for non-Gaussian spatio-temporal data **269**
Scott H. Holan and Christopher K. Wikle
12.1 Introduction 269
12.2 Semiparametric non-Gaussian space-time dynamic design 271
12.2.1 Semiparametric spatio-temporal dynamic Gamma model 271
12.2.2 Simulation-based dynamic design 274
12.2.3 Extended Kalman filter for dynamic gamma models 275
12.2.4 Extended Kalman filter design algorithm 277
12.3 Application: Upper Austria precipitation 278
12.4 Discussion 282
Acknowledgments 282
References 283

13 Active learning for monitoring network optimization **285**
Devis Tuia, Alexei Pozdnoukhov, Loris Foresti and Mikhail Kanevski
13.1 Introduction 285
13.2 Statistical learning from data 287
13.2.1 Algorithmic approaches to learning 288
13.2.2 Over-fitting and model selection 288
13.3 Support vector machines and kernel methods 289
13.3.1 Classification: SVMs 290
13.3.2 Density estimation: One-class SVM 292
13.3.3 Regression: Kernel ridge regression 293
13.3.4 Regression: SVR 294
13.4 Active learning 294
13.4.1 A general framework 295
13.4.2 First steps in active learning: Reducing output variance 296
13.4.3 Exploration–exploitation strategies: Towards mixed approaches 297
13.5 Active learning with SVMs 297
13.5.1 Margin sampling 297
13.5.2 Diversity of batches of samples 299
13.5.3 Committees of models 299
13.6 Case studies 300
13.6.1 Austrian climatological data 300
13.6.2 Cesium-137 concentration after Chernobyl 304
13.6.3 Wind power plants sites evaluation 307
13.7 Conclusions 312
Acknowledgments 314
References 314

14 Stationary sampling designs based on plume simulations **319**
Kristina B. Helle and Edzer Pebesma
14.1 Introduction 319
14.2 Plumes: From random fields to simulations 320
14.3 Cost functions 324
14.3.1 Detecting plumes 324
14.3.2 Mapping and characterising plumes 325
14.3.3 Combined cost functions 325
14.4 Optimisation 326
14.4.1 Greedy search 326
14.4.2 Spatial simulated annealing 328
14.4.3 Genetic algorithms 329
14.4.4 Other methods 331
14.4.5 Evaluation and sensitivity 331
14.4.6 Use case: Combination and comparison of optimisation algorithms 332

14.5 Results 334
14.5.1 Simulations 334
14.5.2 Greedy search 335
14.5.3 Sensitivity of greedy search to the plume simulations 336
14.5.4 Comparison of optimisation algorithms 337
14.6 Discussion 340
Acknowledgments 341
References 341

Index 345

Contributors

Francisco J. Alonso
Department of Statistics
University of Granada
Spain

José M. Angulo
Department of Statistics
University of Granada
Spain

Sudipto Banerjee
Division of Biostatistics
School of Public Health
University of Minnesota
Minneapolis, USA

María C. Bueso
Department of Applied Mathematics and Statistics
Technical University of Cartagena
Murcia, Spain

Eric M. Delmelle
Geography and Earth Sciences
University of North Carolina at Charlotte, USA

Peter J. Diggle
Lancaster Medical School
Lancaster University, UK
and
Institute of Infection and Global Health
University of Liverpool, UK

Thomas R. Fanshawe
Lancaster Medical School
Lancaster University, UK

Andrew O. Finley
Department of Geography and
Department of Forestry
Michigan State University
East Lansing, USA

Loris Foresti
Institute of Geomatics and Analysis of Risk (IGAR)
University of Lausanne
Switzerland

Montserrat Fuentes
Department of Statistics
North Carolina State University
USA

Agnes Fussl
Department of Applied Statistics
Johannes Kepler University Linz
Austria

Alan E. Gelfand
Department of Statistical Science
Duke University
Durham, USA

Daniel A. Griffith
School of Economic, Political and Policy Sciences
University of Texas at Dallas
USA

Kristina B. Helle
Institute for Geoinformatics (IFGI)
University of Muenster
Germany

Tomislav Hengl
ISRIC – World Soil Information
Wageningen
The Netherlands

Gerard B.M. Heuvelink
Department of Environmental Sciences
Wageningen University
The Netherlands

Scott H. Holan
Department of Statistics
University of Missouri
Columbia, USA

Baisuo Jin
School of Management
University of Science and Technology of China, Hefei
People's Republic of China

Mikhail Kanevski
Institute of Geomatics and Analysis of Risk (IGAR)
University of Lausanne
Switzerland

Jie Li
Department of Statistics
Virginia Tech University
USA

Jorge Mateu
Department of Mathematics
University of Jaume I of Castellon
Spain

Stephanie J. Melles
Biology Department
Trent University
Ontario, Canada

Baiqi Miao
School of Management
University of Science and Technology of China, Hefei
People's Republic of China

Werner G. Müller
Department of Applied Statistics
Johannes Kepler University Linz
Austria

Edzer Pebesma
Institute for Geoinformatics (IFGI)
University of Muenster
Germany

Jürgen Pilz
Department of Statistics
University of Klagenfurt
Austria

Alexei Pozdnoukhov
National Centre for Geocomputation
National University of Ireland
Maynooth, Ireland

Brian J. Reich
Department of Statistics
North Carolina State University
USA

Juan Rodríguez-Díaz
Faculty of Science
University of Salamanca
Spain

Gunter Spöck
Department of Statistics
University of Klagenfurt
Austria

Devis Tuia
Image Processing Laboratory
University of Valencia
Spain
and
Laboratory of Geographic
Information Systems
Lausanne Institute of Technology
EPFL
Switzerland

Christopher K. Wikle
Department of Statistics
University of Missouri
Columbia, USA

Yuehua Wu
Department of Mathematics and
Statistics
York University
Toronto, Canada

Dale L. Zimmerman
Department of Statistics and Actuarial
Science
University of Iowa, USA

Foreword

Imagine driving a car that has not been built yet. Design the car: look at principles of combustion, stability, and comfort; consider the manufacturing tools available, including the individuals who will build the car; and keep it within budget. Spatio-temporal sampling design has some of the same features: the principles of stratification, replication, and randomization are important; measuring instruments have to be bought or built and possibly moved around the spatio-temporal domain by field teams; and there is still a bottom line to adhere to.

In this edited volume of chapters on spatio-temporal sampling design, by and large the long-term design focus has been on 'driving the car,' that is, data analysis and inference are very much on the minds of the designers. There, it is the spatio-temporal variability that is the coin of the realm. Controlling this variability allows for more precise inferences and a greater likelihood of detecting 'signals' in the data. Importantly, relating this benefit to a cost, allows an efficient allocation of a study's resources.

Readers of the book's chapters will find a myriad of techniques for linking the design with the analysis, and by far the majority of the authors concentrate on model-based designs. The models are statistical and require some knowledge of the underlying spatio-temporal variability, presumably from a pilot study. (Build a prototype and test drive it!) Statistical analyses require assumptions, and an important one in spatio-temporal design is that the spatio-temporal sampling inadequacies not be confounded with sources of variability due to the process being studied. A small amount of randomization in the design can be very prudent, like putting a spare tire in the back of the car. So too can a sampling protocol that includes samples very close together in space and time to disentangle measurement error from microscale variation.

One of the strengths of the book is that the editors asked the authors to use a common dataset, namely rainfall, temperature, and grassland-usage measurements in Upper Austria. Readers can see how different chapters' design criteria relate to this dataset. While it is not large in size, there are many scientific applications where sampling is limited (e.g., computer experiments, and wellington-boots-on-the-ground field studies).

To see which design approaches scale up to massive datasets, it is usually better to simulate first from a known process and determine what can be recovered from noisy, incompletely sampled, but massive data. In the geosciences, these are sometimes called Observing System Simulation Experiments

(OSSEs), and they are used to mimic the data explosion coming from satellite remote-sensing instruments. This spatio-temporal data explosion is also coming from the mobile devices we carry around as we move through our space-time continuum; crowd-sourcing of this sort offers new statistical sampling challenges to build an accurate information base, and then a knowledge base for making important societal decisions.

The authors of the chapters in this book are eminent in their field, and the editors have meticulously framed a state-of-the-art snapshot for 2012. This is a fertile area for future research, but we should not forget the bottom line, that sampling is costly.

Noel Cressie
University of Wollongong, Australia
and The Ohio State University, USA

1

Collecting spatio-temporal data

Jorge Mateu[1] and Werner G. Müller[2]

[1]*Department of Mathematics, University of Jaume I of Castellon, Spain*

[2]*Department of Applied Statistics, Johannes Kepler University Linz, Austria*

1.1 Introduction

In this volume we intend to provide a comprehensive state-of-the-art presentation combining both classical and modern treatments of network design and planning for spatial and spatio-temporal data acquisition. A common problem set is interwoven throughout the chapters, providing various perspectives to illustrate a complete insight to the problem at hand. Motivated by the high demand for statistical analysis of data that takes spatial and spatio-temporal information into account, this book incorporates ideas from the areas of time series, spatial statistics and stochastic processes, and combines them to discuss optimum spatio-temporal sampling design.

The past has seen, perhaps initiated by Gribik *et al.* (1976), a great number of statistical papers devoted to the purely spatial aspect of sampling design mainly in the context of monitoring networks. Other early papers include those by Caselton and Zidek (1984), Olea (1984) and Fedorov and Müller (1988); book-length treatments are given by Müller (1998, 2007) and de Gruijter *et al.*

Spatio-temporal design: Advances in efficient data acquisition, First Edition.
Edited by Jorge Mateu and Werner G. Müller.

(2006). An excellent recent overview over this literature is provided by Zidek and Zimmerman (2010) and we will take the liberty in this introduction to draw heavily from their structure and exposition, albeit complementing it with aspects that enter due to the additional temporal component. Another excellent review paper is that by Dobbie *et al.* (2008), who provide some material on spatio-temporal aspects. As those two texts are comprehensive we can restrict ourselves to a brief exposition with the goal and emphasis to lead the readers into the more substantial subsequent chapters.

1.2 Paradigms in spatio-temporal design

An important clarification that needs to be made before thinking about spati(o-tempor)al design is whether we assume the randomness of the observations to stem from stochastic disturbances or from the sampling process itself. This leads to the distinction of so-called model-based and design-based (rather than this common but confusing expression we prefer to call them probability-based) inferences and their respective design procedures.

The probability-based paradigm is rooted in classical sampling theory and assumes the ability of defining a population explicitly and a respective randomness in the design. These methods aim at restoring unobserved observations and more importantly general attributes of the spatial population, such as total means (cf. de Gruijter and ter Braak 1990) and variances (cf. Fewster 2011). Probability-based inferences of these attributes are bias-free and allow uncertainty assessments under mild assumptions. The corresponding design techniques reach from the benchmark random sampling to stratified, two-stage, cluster or sequential random sampling with a multitude of variants (Stehman 1999) that all lend themselves to straightforward extensions into incorporating a temporal dimension (Brus and de Gruijter 2011). An excellent account of the latter can be found in Part IV of de Gruijter *et al.* (2006), which can in general be considered the most definitive text for the probability-based design paradigm.

The model-based paradigm on the other hand requires a statistical model to describe the data-generating spatio-temporal process. Here we typically assume that observations stem from a random field generally given by

$$Z(x, \mathbf{s}, t) = \eta(x(\mathbf{s}, t), \mathbf{s}, t, \beta) + \epsilon(x, \mathbf{s}, t), \quad \mathbf{s} \in D, t \in T, \tag{1.1}$$

where $\mathbf{s}$ denotes a spatial location, t a time point, x some potentially space and time dependent regressors, and η a parametrized trend model (a nearly encyclopedic reference for these type of processes is Cressie and Wikle 2011). Note that the random element here is the error ϵ rather than the design mechanism. This allows to assign meaning to purely geometric designs, such as regular grids or space-filling lattices, that are common in applications. Another advantage here is that by borrowing inference strength from the model we can make meaningful inferences from rather small samples and for very specific aspects derived from the model parameters, such as threshold exceedances, times of trend-reversals, local outliers, etc.

We believe that unless a reasonable modeling is out of reach the model-based approach offers more flexibility and statistical power, which is why most of the contributions in this volume will fall into this category. However, this issue has been the subject of considerable debate in the literature and further details are provided in Papritz and Webster (1995), Brus and de Gruijter (1997), Stevens (2006), and an overview is given in Table 1.1 of Dobbie *et al.* (2008). A general discussion that goes beyond the spatio-temporal realm can be found in Thompson (2002). Recently, there also have been attempts to fuse the two paradigms, Brus and de Gruijter (2012) for instance employ probability-based sampling for the spatial coordinates, whereas they build a time series model and use a respective design for the temporal trend. How to include probability-based design in a hierarchical statistical modeling framework is surveyed in Cressie *et al.* (2009).

1.3 Paradigms in spatio-temporal modeling

Another dichotomy clearly shows when one examines the spati(o-tempor)al modeling literature. Historically, two schools have somewhat independently developed, one based on discrete time series model analogies and the other one derived from generalizations of stochastic process methodologies. The former is much used by geographers and economists particularly in the advent of what was termed 'new economic geography' and was consequently referred to as 'spatial econometrics' (for perhaps the earliest full exposition see Anselin 1988; a recent account of the history of the field thereafter by the same author can be found in Anselin 2010). The latter school stems from the theory of regionalized variables developed among mining scientists and geologists and has consequently been named 'geostatistics' (a book-length treatment is provided by Chilès and Delfiner 1999). Comparative discussions on these two paradigms can be found in the encyclopedic Cressie (1993) and more recently in Griffith and Paelinck (2007), Hae-Ryoung *et al.* (2008) and Haining *et al.* (2010).

Both of these modeling views are encompassed by the random field (1.1) and can be solely distinguished by the nature of the indexing variables s and t. In spatial econometrics spatial econometrics we usually assume the s's to form a discrete geographic lattice and their relationships are usually described in the form of a so-called spatial weight or link matrix W. Various types of dependence structures can be modeled by assigning particular forms of η and covariances of ϵ employing W, such as the common simultaneous and conditionally autoregressive regression models (SAR and CAR), the latter being spatial manifestations of Gaussian Markov random fields (GMRF; see Rue and Held 2005 for a definitive text).

In geostatistics the locations $\mathbf{s}$ are assumed to vary continuously in D and again the implied models differ by the choice of η and the error dependence usually determined by the so-called variogram $\gamma(\mathbf{s}, \mathbf{s}') = E(|Z(x, \mathbf{s}, t) - Z(x, \mathbf{s}', t)|^2)$. Under normality assumptions for the errors these processes also come under the notion of Gaussian random fields (GRF) and they are also much in use in other contexts such as machine learning and computer simulation experiments (cf. Rasmussen and Williams 2005). Though more flexible the corresponding models are usually much more estimation intensive than those for GMRF, which can typically be employed for much larger spatial datasets.

Despite this divide there have lately been successful attempts to merge those two spatio-temporal modeling paradigms. While previously only results for regular sampling schemes were available (cf. Griffith and Csillag 1993), Lindgren *et al.* (2011) provide an explicit link for arbitrary lattices, thus opening the issue for the question of sampling design. In a discussion to this article Müller and Waldl (2011) indeed uncover relationships between the respective designs that will allow to exploit properties from both paradigms.

A great number of spatio-temporal extensions of these models exist particularly for GRF; see Cressie and Wikle (2011) for an extensive review and Baxevani *et al.* (2011) for a particular representation using velocity fields. GMRF are usually extended by modeling them in discrete time, so-called spatial panel models (see e.g., Elhorst 2012 for a recent survey); a continuous time extension of spatial panels is given in Oud *et al.* (2012).

1.4 Geostatistics and spatio-temporal random functions

Geostatistical research has typically analyzed random fields, in which every spatio-temporal location can be seen as a point on $\mathbb{R}^d \times \mathbb{R}$. While from a mathematical point of view $\mathbb{R}^d \times \mathbb{R} = \mathbb{R}^{d+1}$, from a physical perspective it would make no sense to consider spatial and temporal aspects in the same way, due to the significant differences between the two axes of coordinates. Therefore, while the time axis is ordered intrinsically (as it exists in the past, present and future), the same does not occur with the spatial coordinates.

Recalling (1.1), assume that observations stem from a random field (r.f.) given by $Z(x, \mathbf{s}, t) = \eta(x(\mathbf{s}, t), \mathbf{s}, t, \beta) + \epsilon(x, \mathbf{s}, t), \mathbf{s} \in D, t \in T$, where $\mathbf{s}$ denotes a spatial location, t a time point, x some potentially space and time dependent regressors, η a parametrized trend model, $D \subset \mathbb{R}^d$ (very often $d = 2$), and $T \subset \mathbb{R}$. For ease of notation, we remove the term in the covariates x, and write $Z(\mathbf{s}, t)$, assuming whenever necessary that any trend coming from a set of covariates has already been removed.

1.4.1 Relevant spatio-temporal concepts

A spatio-temporal r.f. $Z(\mathbf{s}, t)$ is said to be *Gaussian* if the random vector $\mathbf{Z} = (Z(\mathbf{s}_1, t_1), ..., Z(\mathbf{s}_n, t_n))'$ for any set of spatio-temporal locations

$\{(\mathbf{s}_1, t_1), \ldots, (\mathbf{s}_n, t_n)\}$ follows a multivariate normal distribution. When not stated explicitly, the indexes i and j will go from 1 to n.

The spatio-temporal r.f. $Z(\mathbf{s}, t)$ is said to have a *spatially stationary covariance* function if, for any two pairs $(\mathbf{s}_i, t_i)$ and $(\mathbf{s}_j, t_j)$ on $\mathbb{R}^d \times \mathbb{R}$, the covariance $C((\mathbf{s}_i, t_i), (\mathbf{s}_j, t_j))$ only depends on the distance between the locations $\mathbf{s}_i$ and $\mathbf{s}_j$ and the times t_i and t_j. And the spatio-temporal r.f. $Z(\mathbf{s}, t)$ is said to have a *temporarily stationary covariance* function if, for any two pairs $(\mathbf{s}_i, t_i)$ and $(\mathbf{s}_j, t_j)$ on $\mathbb{R}^d \times \mathbb{R}$, the covariance $C((\mathbf{s}_i, t_i), (\mathbf{s}_j, t_j))$ only depends on the distance between the times t_i and t_j and the spatial locations $\mathbf{s}_i$ and $\mathbf{s}_j$. If the spatio-temporal r.f. $Z(\mathbf{s}, t)$ has a *stationary covariance function* in both spatial and temporal terms, then it is said to have a stationary covariance function. In this case, the covariance function can be expressed as

$$C((\mathbf{s}_i, t_i), (\mathbf{s}_j, t_j)) = C(\mathbf{h}, u) \tag{1.2}$$

with $\mathbf{h} = \mathbf{s}_i - \mathbf{s}_j$ and $u = t_i - t_j$ the distances in space and time, respectively.

A spatio-temporal r.f. $Z(\mathbf{s},t)$ has a *separable covariance function* if there is a purely spatial covariance function $C_s(\mathbf{s}_i, \mathbf{s}_j)$ and a purely temporal covariance function $C_t(t_i, t_j)$ such that

$$C((\mathbf{s}_i, t_i), (\mathbf{s}_j, t_j)) = C_s(\mathbf{s}_i, \mathbf{s}_j)C_t(t_i, t_j) \tag{1.3}$$

for any pair of spatio-temporal locations $(\mathbf{s}_i, t_i)$ and $(\mathbf{s}_j, t_j) \in \mathbb{R}^d \times \mathbb{R}$.

A spatio-temporal r.f. $Z(\mathbf{s},t)$ has a *fully symmetrical covariance function* if

$$C((\mathbf{s}_i, t_i), (\mathbf{s}_j,t_j)) = C_s(\mathbf{s}_i, t_j)C_t(\mathbf{s}_j, t_i) \tag{1.4}$$

for any pair of spatio-temporal locations $(\mathbf{s}_i, t_i)$ and $(\mathbf{s}_j, t_j) \in \mathbb{R}^d \times \mathbb{R}$.

Separability is a particular case of complete symmetry and, as such, any test to verify complete symmetry can be used to reject separability. In the case of stationary spatio-temporal covariance functions, the condition of full symmetry reduces to

$$C(\mathbf{h}, u) = C(\mathbf{h}, -u) = C(-\mathbf{h}, u) = C(-\mathbf{h}, -u), \ \forall(\mathbf{h}, u) \in \mathbb{R}^d \times \mathbb{R}. \tag{1.5}$$

A spatio-temporal r.f. has a *compactly supported covariance function* if, for any pair of spatio-temporal locations $(\mathbf{s}_i, t_i)$ and $(\mathbf{s}_j, t_j) \in \mathbb{R}^d \times \mathbb{R}$, the covariance function $C((\mathbf{s}_i, t_i), (\mathbf{s}_j, t_j))$ tends towards zero when the spatial or temporal distance is sufficiently large.

If $C(\mathbf{s}_i - \mathbf{s}_j, t_i - t_j)$ depends only on the distance between positions, that is, $\left(\|\mathbf{s}_i - \mathbf{s}_j\|, t_i - t_j\right)$, the r.f., apart from being stationary, is also *isotropic* in space and time. Note that if the covariance function of a stationary r.f. is isotropic in space and time, then it is fully symmetrical.

The spatio-temporal *variogram* is defined as the function

$$2\gamma((\mathbf{s}_i, t_i), (\mathbf{s}_j, t_j)) = V(Z(\mathbf{s}_i, t_i) - Z(\mathbf{s}_j, t_j)), \tag{1.6}$$

where V is the variance, and half this quantity is called a semivariogram.

In the case of a r.f. with a zero mean,

$$2\gamma((\mathbf{s}_i, t_i), (\mathbf{s}_j, t_j)) = E[(Z(\mathbf{s}_i, t_i) - Z(\mathbf{s}_j, t_j))^2]. \tag{1.7}$$

Whenever it is possible to define the covariance function and the variogram, they will be related by means of the following expression

$$2\gamma((\mathbf{s}_i, t_i), (\mathbf{s}_j, t_j)) = V(Z(\mathbf{s}_i, t_i)) + V(Z(\mathbf{s}_j, t_j)) - 2C((\mathbf{s}_i, t_i), (\mathbf{s}_j, t_j)). \tag{1.8}$$

If the spatio-temporal r.f. $Z(\mathbf{s}, t)$ has an intrinsically stationary variogram in both space and time, then it is said to have an intrinsically stationary variogram. In this case, the variogram can be expressed as

$$2\gamma((\mathbf{s}_i, t_i), (\mathbf{s}_j, t_j)) = 2\gamma(\mathbf{h}, u). \tag{1.9}$$

The marginals $2\gamma(\cdot, u)$ and $2\gamma(\mathbf{h}, \cdot)$ are called purely spatial and purely temporal variograms, respectively.

A r.f. $Z(\mathbf{s}, t)$ is *strictly stationary* if its probability distribution is translation invariant. Second-order stationarity is a less demanding condition than strict stationarity. A spatio-temporal r.f. $Z(\mathbf{s}, t)$ is *second-order stationary* in the broad sense or weakly stationary if it has a constant mean and the covariance function depends on $\mathbf{h}$ and u.

A spatio-temporal r.f. $Z(\mathbf{s}, t)$ is said to be intrinsically stationary if it has a constant mean and an intrinsically stationary variogram. Intrinsic stationarity is less restrictive than second-order stationarity. Another widely used function when modeling implicit spatio-temporal dependence in a stationary r.f. is the correlation function. Let $Z(\mathbf{s}, t)$ be a second-order stationary r.f. with a priori variance $\sigma^2 = C(\mathbf{0}, 0) > 0$. The autocorrelation function of this r.f. is defined as

$$\rho(\mathbf{h}, u) = \frac{C(\mathbf{h}, u)}{C(\mathbf{0}, 0)}. \tag{1.10}$$

If $\rho(\mathbf{h}, u)$ is a correlation function on $\mathbb{R}^d \times \mathbb{R}$, then its marginal functions $\rho(\mathbf{0}, u)$ and $\rho(\mathbf{h}, 0)$ will respectively be the spatial correlation function on $\mathbb{R}^d$ and the temporal correlation function on $\mathbb{R}$.

1.4.2 Properties of the spatio-temporal covariance and variogram functions

A function $C((\mathbf{s}_i, t_i), (\mathbf{s}_j, t_j))$ of real values, defined on $\mathbb{R}^d \times \mathbb{R}$ is a *covariance function* if it is symmetrical, $C((\mathbf{s}_i, t_i), (\mathbf{s}_j, t_j)) = C((\mathbf{s}_j, t_j), (\mathbf{s}_i, t_i))$ and positive-definite, that is,

$$\sum_{i=1}^{n} \sum_{j=1}^{n} a_i a_j C((\mathbf{s}_i, t_i), (\mathbf{s}_j, t_j)) \geq 0 \tag{1.11}$$

for any $n \in \mathbb{N}$, $(\mathbf{s}_i, t_i) \in \mathbb{R}^d \times \mathbb{R}$, and $a_i \in \mathbb{R}$, $i = 1, ..., n$. The condition (1.11) is *sufficient* if the covariance function can take complex values. Similarly, one *necessary and sufficient* condition for a non-negative function of real values $\gamma((\mathbf{s}_i, t_i), (\mathbf{s}_j, t_j))$ defined on $\mathbb{R}^d \times \mathbb{R}$ to be a *semivariogram* is that it is a symmetrical function and conditionally negative-definite, that is,

$$\sum_{i=1}^{n} \sum_{j=1}^{n} a_i a_j \gamma((\mathbf{s}_i, t_i), (\mathbf{s}_j, t_j)) \leq 0 \tag{1.12}$$

with $\sum_{i=1}^{n} a_i = 0$.

Schoenberg (1938) proved the following theorem characterizing the spatio-temporal semivariogram. Let $\gamma((\mathbf{s}_i, t_i), (\mathbf{s}_j, t_j))$ be a function defined on $\mathbb{R}^d \times \mathbb{R}$, with $\gamma((\mathbf{s}, t), (\mathbf{s}, t)) = 0$, $\forall(\mathbf{s}, t) \in \mathbb{R}^d \times \mathbb{R}$. Then the following statements are equivalent:

- $\gamma((\mathbf{s}_i, t_i), (\mathbf{s}_j, t_j))$ is a semivariogram on $\mathbb{R}^d \times \mathbb{R}$.
- $\exp\big(-\theta\gamma((\mathbf{s}_i, t_i), (\mathbf{s}_j, t_j))\big)$ is a covariance function on $\mathbb{R}^d \times \mathbb{R}$, for any $\theta > 0$.
- $C((\mathbf{s}_i, t_i), (\mathbf{s}_j, t_j)) = \gamma((\mathbf{s}_i, t_i), (\mathbf{0}, 0)) + \gamma((\mathbf{s}_j, t_j), (\mathbf{0}, 0)) - \gamma((\mathbf{s}_i, t_i), (\mathbf{s}_j, t_j))$ is a covariance function on $\mathbb{R}^d \times \mathbb{R}$.

In case of stationarity, the above results reduce to functions depending on spatial and temporal lags. Another seminal result that characterizes covariance functions is that given in Bochner (1933). A function $C(\mathbf{h}, u)$ defined on $\mathbb{R}^d \times \mathbb{R}$ is a stationary covariance function if, and only if, it has the following form

$$C(\mathbf{h}, u) = \int\int e^{i(\boldsymbol{\omega}'\mathbf{h}+\tau u)} dF(\boldsymbol{\omega}, \tau), \quad (\mathbf{h}, u) \in \mathbb{R}^d \times \mathbb{R} \tag{1.13}$$

where the function F is a non-negative distribution function with a finite mean defined on $\mathbb{R}^d \times \mathbb{R}$, which is known as a *spectral distribution function*. Therefore, the class of stationary spatio-temporal covariance functions on $\mathbb{R}^d \times \mathbb{R}$ is identical to the class of Fourier transforms of non-negative distribution functions with finite means on that domain. If the function C can also be integrated, then the spectral distribution function F is absolutely continuous and the representation (1.13) simplifies to

$$C(\mathbf{h}, u) = \int\int e^{i(\boldsymbol{\omega}'\mathbf{h}+\tau u)} f(\boldsymbol{\omega}, \tau) d\boldsymbol{\omega} d\tau, \quad (\mathbf{h}, u) \in \mathbb{R}^d \times \mathbb{R} \tag{1.14}$$

where f is a non-negative, continuous and integrable function that is known as a *spectral density function*. The covariance function C and the spectral density function f then form a pair of Fourier transforms , and

$$f(\boldsymbol{\omega}, \tau) = (2\pi)^{-d-1} \int\int e^{-i(\boldsymbol{\omega}'\mathbf{h}+\tau u)} C(\mathbf{h}, u) d\mathbf{h} du \tag{1.15}$$

The decomposition (1.13) can be specialized for fully symmetrical covariance functions. Let $C(\cdot,\cdot)$ be a continuous function defined on $\mathbb{R}^d \times \mathbb{R}$, then $C(\cdot,\cdot)$ is a fully symmetrical stationary covariance function if, and only if, the following decomposition is possible

$$C(\mathbf{h}, u) = \int\int \cos(\boldsymbol{\omega}'\mathbf{h}) \cos(\tau u) dF(\boldsymbol{\omega}, \tau), \quad (\mathbf{h}, u) \in \mathbb{R}^d \times \mathbb{R} \tag{1.16}$$

where F is the non-negative and symmetrical spectral distribution function defined on $\mathbb{R}^d \times \mathbb{R}$.

Cressie and Huang (1999) provide a theorem for characterizing the class of stationary spatio-temporal covariance functions under the additional hypothesis of integrability. Let $C(\cdot,\cdot)$ be a continuous, bounded, symmetrical and integrable function defined on $\mathbb{R}^d \times \mathbb{R}$, then $C(\cdot,\cdot)$ is a stationary covariance function if, and only if, in view of $u \in \mathbb{R}$,

$$C_{\boldsymbol{\omega}}(u) = \int e^{-i\boldsymbol{\omega}'\mathbf{h}} C(\mathbf{h}, u) d\mathbf{h}, \tag{1.17}$$

is a covariance function for every $\boldsymbol{\omega} \in \mathbb{R}^d$ except, at the most, in a set with a null Lebesgue mean. Gneiting (2002) generalizes this result for C defined on $\mathbb{R}^d \times \mathbb{R}^l$, from which the previous statement is a particular case for $l = 1$.

Both the covariance function and the spectral density function are important tools for characterizing random stationary spatio-temporal fields. Mathematically speaking, both functions are closely related as a pair of Fourier transforms. Furthermore, the spectral density function is particularly useful in situations where there is no explicit expression of the covariance function. Stein (2005) shows the benefit of using smooth covariance functions far from the origin, which can be tested by verifying whether their spectral densities have derivatives of certain orders.

1.4.3 Spatio-temporal kriging

Kriging is aimed at predicting an unknown point value $Z(\mathbf{s}_0, t_0)$ at a point $(\mathbf{s}_0, t_0)$ that does not belong to the sample. To do so, all the information available about the regionalized variable is used, either at the points in the entire domain or in a subset of the domain called the *neighborhood*.

Assume that the value of the r.f. has been observed on a set of n spatio-temporal locations $\{Z(\mathbf{s}_1, t_1), \ldots, Z(\mathbf{s}_n, t_n)\}$. We now want to predict the value of the r.f. on a new spatio-temporal location $(\mathbf{s}_0, t_0)$, for which we use the linear predictor

$$Z^*(\mathbf{s}_0, t_0) = \sum_{i=1}^{n} \lambda_i Z(\mathbf{s}_i, t_i) \tag{1.18}$$

constructed from the random variables $Z(\mathbf{s}_i, t_i)$. As in the spatial case, spatio-temporal kriging equations will depend on the degree of stationarity attributed to the r.f. that supposedly generates the observed realization. The most widely used kriging techniques in the spatio-temporal case are simple spatio-temporal kriging, ordinary spatio-temporal kriging, and universal spatio-temporal kriging. In the case of *simple spatio-temporal kriging* we assume that $Z(\mathbf{s}, t)$ is a second-order stationary spatio-temporal r.f., with a constant and known mean $\mu(\mathbf{s}, t)$, constant and known variance $C(\mathbf{0}, 0)$, and a known covariance function $C(\mathbf{h}, u)$. The kriging equations (n equations with n unknown elements) are of the form

$$\sum_{j=1}^{n} \lambda_j C\left(\mathbf{s}_i - \mathbf{s}_j, \mathrm{t}_i - \mathrm{t}_j\right) = C\left(\mathbf{s}_i - \mathbf{s}_0, \mathrm{t}_i - \mathrm{t}_0\right), \forall i = 1, ..., n \tag{1.19}$$

from which we obtain the values λ_i that minimize the prediction error variance, which is given by

$$V\left[Z^*(\mathbf{s}_0, t_0) - Z(\mathbf{s}_0, t_0)\right] = C(\mathbf{0}, 0) - \sum_{i=1}^{n} \lambda_i C(\mathbf{s}_i - \mathbf{s}_0, t_i - t_0) \tag{1.20}$$

In the case of *ordinary spatio-temporal kriging*, the constant mean $\mu(\mathbf{s}, t)$ is not known, and the covariance function $C(\mathbf{h}, \mathrm{u})$ is known, under second-order stationarity. In the case of an intrinsic r.f. the variance is unbounded. In these two cases, simple kriging cannot be performed as the mean cannot be subtracted. We must therefore impose a condition of unbiasedness. In these situations, ordinary spatio-temporal kriging equations can be expressed, in the first case, in terms of the covariance function, and in the second case, in terms of the semivariogram, as there is no covariance at the origin.

In the *universal kriging* approach, assume $Z(\mathbf{s}, t)$ is a r.f. with drift, and so the mean of the r.f. is not constant, but depends on the pairs $(\mathbf{s}, t)$. In this situation the so-called condition of unbiasedness is substantially affected. In this case, the r.f. can be disaggregated into two components: one deterministic $\mu(\mathbf{s}, \mathrm{t})$ and the other stochastic $e(\mathbf{s}, t)$ which can be treated as an intrinsically stationary r.f. with zero expectation, $E\left[e(\mathbf{s}, \mathrm{t})\right] = 0$

$$Z(\mathbf{s}, t) = \mu(\mathbf{s}, \mathrm{t}) + \mathrm{e}(\mathbf{s}, \mathrm{t}) \tag{1.21}$$

We can assume that the mean, even unknown, can be expressed locally by

$$\mu(\mathbf{s}, t) = \sum_{h=1}^{p} a_h f_h(\mathbf{s}, \mathrm{t}) \tag{1.22}$$

where $\left\{f_h(\mathbf{s}, \mathrm{t}),\ h = 1, ..., p\right\}$ are p known functions, a_h constant coefficients, and p the number of terms used in the approximation. It must be taken into account that this expression is only valid locally. In this case, the equations that yield the prediction of the weights are obtained from the prediction error conditions of zero expectation and minimum variance.

1.4.4 Spatio-temporal covariance models

One key stage in the spatio-temporal prediction procedure is choosing the covariance function (covariogram or semivariogram) that models the structure of the spatio-temporal dependence of the data. However, while the semivariogram is normally chosen for this purpose in the spatial case, in a spatio-temporal framework the covariance function is the most commonly chosen tool. By referring to a valid covariographic spatio-temporal model, we are implicitly stating that the covariance function must be positive-definite. The purely spatial and temporal covariance models have been widely studied and there is a long list of those which can be used to model spatial or spatio-temporal dependence that guarantee the (spatial or temporal) covariance function is positive-definite. However, this is not the case in the spatio-temporal scenario, in which constructing valid spatio-temporal covariance models is one of the main research activities. In addition, while it is difficult to demonstrate that a spatial or temporal function is positive-definite, it is even more so when seeking to determine valid spatio-temporal covariance models. For this reason, many authors began to study how to combine valid spatial and temporal models to obtain (valid) spatio-temporal covariance models.

By way of introduction, the first approximations to modeling spatio-temporal dependence using covariance functions were nothing more than generalizations of the stationary models used in the spatial scenario. In this sense, early studies often modeled the spatio-temporal covariance using metric models by defining a metric in space and time that allowed researchers to directly use isotropic models that are valid in the spatial case. Such metric models were characterized by being nonseparable, isotropic and stationary. The next step in this initial stage consisted of configuring spatio-temporal covariance functions by means of the sum or product of a spatial covariance and a temporal covariance, both of which were stationary, giving rise to separable, isotropic and stationary models. Later, realizing the limitations of the two procedures detailed above in terms of capturing the spatio-temporal dependence that really exists in the large majority of the phenomena studied, interest shifted towards including the interaction of space and time, in covariance models, giving rise to the so-called nonseparable models (while remaining isotropic and stationary). It is worth highlighting the nonseparable models developed by Jones and Zhang (1997), Cressie and Huang (1999), Brown *et al.* (2000), De Cesare *et al.* (2001a, b), De Iaco *et al.* (2001, 2002a, b, 2003), Gneiting (2002), Ma (2002, 2003a, c, 2005a, b, c), Fernández-Casal *et al.* (2003), Kolovos *et al.* (2004), and Stein (2005) among others.

Development continued with the search for nonseparable spatio-temporal, spatially anisotropic and/or temporally asymmetrical models such as those described in Fernández Casal *et al.* (2003), Porcu *et al.* (2006), and Mateu *et al.* (2007). Finally, we can cite some recent approaches to the problem of modeling nonstationary covariance functions, such as those made by Ma (2002, 2003b), Fuentes *et al.* (2005), Stein (2005), Chen *et al.* (2006), Porcu *et al.* (2006, 2007a, b, 2009), Mateu *et al.* (2007), Porcu and Mateu (2007) and Gregori *et al.* (2008).

1.4.5 Parametric estimation of spatio-temporal covariograms

The empirical determination of the covariance function or the variogram of a spatio-temporal process can be generalized naturally using the procedures for merely spatial processes. Let $Z(\cdot,\cdot)$ be an intrinsically stationary process observed on a set of n spatio-temporal pairs $\{(\mathbf{s}_1, t_1), ..., (\mathbf{s}_n, t_n)\}$. Two direct and popular alternatives to obtain an estimation of the variogram $2\gamma(\cdot,\cdot)$ [and its covariance function $C(\cdot,\cdot)$, if the process is also second-order stationary] are the classical estimator based on the method-of-moments (MoM), and the robust estimator proposed by Cressie and Hawkins (1980).

This MoM estimator for the variogram is given in its most general form by

$$2\hat{\gamma}(\mathbf{h}(l),u(k)) = \frac{1}{|N(\mathbf{h}(l),u(k))|} \sum_{(\mathbf{s}_i,t_i)(\mathbf{s}_j,t_j)\in N(\mathbf{h}(l),u(k))} \left(Z(\mathbf{s}_i,t_i) - Z(\mathbf{s}_j,t_j)\right)^2, \tag{1.23}$$

where $N(\mathbf{h}(l),u(k)) = \{(\mathbf{s}_i, t_i), (\mathbf{s}_j, t_j) : \mathbf{s}_i - \mathbf{s}_j \in N(\mathbf{h}(l)), t_i - t_j \in T(u(k))\}$, $T(\mathbf{h}(l))$ being an area of tolerance in $\mathbb{R}^d$ around $\mathbf{h}(l)$, $T(u(k))$ is a region of tolerance in $\mathbb{R}$ around $u(k)$, and $|N(\mathbf{h}(l),u(k))|$ is the number of different elements in $N(\mathbf{h}(l),u(k))$, with $l = 1, ..., L$ and $k = 1, ..., K$.

Although the classical estimation method has the advantage of being easy to calculate, it also has some practical drawbacks, such as not being robust in the case of extreme values. In order to avoid this problem, and following and extending Cressie and Hawkins (1980) we have the following variogram estimator for the spatio-temporal case

$$2\hat{\gamma}(\mathbf{h}(l),u(k)) = \left(\frac{1}{|N(\mathbf{h}(l),u(k))|} \sum_{(\mathbf{s}_i,t_i)(\mathbf{s}_j,t_j)\in N(\mathbf{h}(l),u(k))} |\, Z(\mathbf{s}_i,t_i) - Z(\mathbf{s}_j,t_j)\,|^{\frac{1}{2}}\right)^4 \times \left(0.457 + \frac{0.494}{|N(\mathbf{h}(l),u(k))|}\right)^{-1}$$

As in the spatial case, these estimators of the covariance function or variogram of the process do not generally fulfill the condition of being positive-definite or conditionally negative definite, respectively. For this reason, in practice we select a parametric model of covariance or variogram that we already know is valid, and estimate the parameters of the covariance function or variogram that best fits the values of the empirical estimator.

Any of the procedures for *least squares* (LS) estimation used in the spatial case can be easily generalized to the spatio-temporal case. Let us assume that the process being analyzed has been decomposed such that either the original process or the process of residuals after modeling its mean, is intrinsically stationary. In this case, we can estimate the parameters that define the semivariogram model chosen using ordinary (OLS), generalized (GLS) or weighted (WLS) least squares. But the OLS procedure does not take into account the behavior of the semivariogram (or covariance function) near the origin and, above all, does not consider the possibility of the values of the semivariogram being correlated, which could have an adverse effect on the estimations of the vector of parameters as the amount of data increases. This last problem is normally solved by resorting to GLS, but it is also true that in order to determine the covariance matrix, which is normally quite large, we should work in a Gaussian context and proceed using iterative methods (cf. Müller 1999). A compromise between efficiency and computation is provided by the WLS method, which is the most popular LS estimation method. The WLS method consists of (iteratively) minimizing the expression

$$\sum_{i=1}^{n} \omega_i (\hat{\gamma}(\mathbf{h}_i, u_i) - \gamma(\mathbf{h}_i, u_i; \boldsymbol{\theta}))^2 = (\hat{\gamma} - \gamma(\boldsymbol{\theta}))' W(\boldsymbol{\theta})^{-1} (\hat{\gamma} - \gamma(\boldsymbol{\theta})) \quad (1.24)$$

where $W(\boldsymbol{\theta})$ is a diagonal matrix, the elements of which, ω_i, are approximately the variances of $\hat{\gamma}$.

It is a well-known fact that the *maximum likelihood* (ML) method requires knowing the distribution of the r.f. underlying the phenomenon under study. However, only the case of Gaussian random functions has been developed in the literature (Mardia and Marshall 1984). With this in mind, if the spatio-temporal process is Gaussian, then it is possible to use the ML method, which is also more efficient than the least squares procedures when estimating the parameters of the valid covariance function (or semivariogram) chosen. ML estimators have desirable properties for spatial analysis, such as consistency and asymptotic normality in the field of extendible domains (Mardia and Marshall 1984). However, they also have one significant limitation: even in the merely spatial case they entail high computation costs as the number of parameters the model includes is high. The most critical part of the estimation process is calculating the determinant and the inverse of the covariance matrix. An alternative to ML is given by the restricted maximum likelihood (RML), in which the estimators are obtained by applying the ML method to the so-called 'error contrasts'.

As the problem with the ML procedure is computational, it is natural that research has focused on searching for approaches to the likelihood function that require less than $O\left(N^3\right)$ stages and which have desirable statistical properties. However, another line of research has resorted to exploiting the special structure of the covariance matrix in order to enhance computation. In this sense, Zimmerman (1989) indicates that in certain sampling procedures it is possible to compute ML more efficiently. Another way of simplifying the number of operations in the context of lattice data is to represent the process spectrally (cf. Whittle 1954). Kaufman *et al.* (2008) focus directly on the covariance matrix and propose calculating likelihood with a reduced version of the covariance matrix instead of the original matrix. Furrer *et al.* (2006) use this technique to interpolate observations rather than estimate the parameters.

The computational problem was once again addressed recently using likelihood approaches, such as composite likelihood. According to Varin and Vidoni (2005), there are two types of composite likelihood. The first includes the so-called 'subsetting methods' and is based on marginal functions or 'subset pieces' of full likelihood. The second is based on the so-called 'mission methods', obtaining composite likelihood by omitting components from the full likelihood. The latter intuitively yield computational benefits insofar as they do not consider certain 'pieces of likelihood' when computing composite likelihood. Three types of composite likelihood have been proposed in the literature to approach full likelihood when working with a large set of spatial data (see Vecchia 1988, Curriero and Lele 1999, Caragea and Smith 2006, and the modified version of Stein *et al.* 2004).

Vecchia (1988) proposes factorizing the full likelihood into a product of conditional densities to later reduce the size of the conditioning sets. Stein *et al.* (2004) suggest that it might be more efficient to carry out the above procedure in blocks. Stein (2005) extended this approach to the spatio-temporal field when the data are measured in monitoring stations at regular intervals of time.

The Weighted Composite Likelihood (WCL) technique (cf. Lindsay 1988) defines a general estimation method when working with large sets of observations and has been applied in a large number of fields of science in recent years (particularly in agronomy). It is easy to appreciate that the approximations in Vecchia (1988), Stein *et al.* (2004), Stein (2005), and Caragea and Smith (2006) are special cases of composite likelihood. This approach is the starting point of the procedure conceived by Bevilacqua *et al.* (2012). Using WCL to estimate the parameters of a valid covariographic or semivariographic model aims to obtain accurate estimates comparable with those obtained by ML, but with lower computation costs (cf. Bevilacqua *et al.* 2012). This issue is of vital importance in the spatio-temporal field, as there is usually a sufficient number of observations that make the ML prohibitive in terms of computation.

1.5 Types of design criteria and numerical optimization

When it comes to actually determining the spatio-temporal design we have a variety of choices and face a multitude of constraints. Perhaps it should be first stressed that many of the decisions we are going to make will have a considerable influence on the designs we yield but it is the decision to design in the first place, that will impact the quality of our inference the most. As Ver Hoef (2012) puts it: 'The main point [...] is that any search for a good design is better than relying on a randomly chosen one, [...]'. We (and he) argue for designs that are robust with respect to choice that need to be made like prior guesses of parameters and spatial correlations and insensitive to the constraints imposed, ideally even those stemming from cost considerations.

Nevertheless to make concrete progress it may be worthwhile to define clear-cut objectives that can be optimized by the sampling site determination, so-called design criteria. If we define the design as the finite collection of sites $\mathbf{s}_i$ and observation times t_i, the set $\xi = \{\mathbf{s}_1, t_1, ..., \mathbf{s}_n, t_n\}$ we can usually state the design problem as the minimization task

$$\xi^* = \arg\min_{\xi} \phi(\xi), \tag{1.25}$$

with ϕ being a suitable scalar function. The result ξ^* is referred to as a (ϕ)-optimal design. For spatio-temporal designs the time dimension is sometimes treated separately (cf. de Gruijter *et al.* 2006) since there are often specific restrictions as to whether and how temporal changes in the spatial configuration of measurement sites are allowed (see Wikle and Royle 1999 for an early description of the problem). Sometimes one even has to take into account measurement devices that move in space according to given trajectories (Ucinski and Chen 2005). In other cases one might want to exploit information gained from early stages of the data collecting process as time progresses leading to so-called sequential or adaptive designs (Berliner *et al* 1999).

The choice of a particular design criterion ϕ clearly reflects the purpose of a study. Is it (cf. Overton and Stehman 1996):

- exploratory in nature to gain a first understanding of the phenomenon under study?
- confirmatory aimed at estimating global or local characteristics or model parameters?
- solely used for most accurate prediction purposes?

For the first in this list one usually constructs designs that are filling the space (and sometimes period) available in some efficient way. One could now look for a design that is trying to get good coverage of every point in the space D. Let us for this purpose define a metric for the (inverse) distance between a given point $\mathbf{s}$ and a design ξ, i.e

$$d_p(\mathbf{s}, \xi) = \left(\sum_{i=1}^{n} \frac{1}{\|\mathbf{s} - \mathbf{s}_i\|^p} \right)^{\frac{1}{p}}, \quad p > 0,$$

which can be considered as a measure of (lack of) coverage of a design ξ for the point $\mathbf{s}$. Consequently we can use the L_q average over the space D

$$\phi_{[p,q]}(\xi) = \left(\int_{\mathbf{s} \in D} d_p^q(\mathbf{s}, \xi) d\mathbf{s} \right)^{\frac{1}{q}}, \quad q > 0,$$

which forms a design criterion to be minimized over ξ (cf. Royle and Nychka 1998; a probability-based concept with a similar purpose was devised in Stevens and Olsen 2004).

As a particular choice, it is natural to attempt to make the maximum distance from all the points in D to their closest points in ξ as small as possible. This is achieved by letting $p \to \infty$ and $q \to \infty$ giving

$$\phi_{[\infty,\infty]}(\xi) = \max_{\mathbf{s}\in D} \min_{\mathbf{s}_i} \|\mathbf{s} - \mathbf{s}_i\| = \phi_{mM}(\xi) \tag{1.26}$$

and we call the resulting design

$$\xi^*_{mM} = \arg\min_{\xi} \phi_{mM}(\xi)$$

a *minimax distance design*. If we instead seek for a design that wants to achieve a high spread solely amongst its support points, we can similarly to above define a criterion that L_q averages the (inverse) interdistances for maximization, i.e.

$$\phi^{-1}_{[q]}(\xi) = \left(\sum_{\mathbf{s}_i \neq \mathbf{s}_j \in D} \frac{1}{\|\mathbf{s}_i - \mathbf{s}_j\|^q} \right)^{\frac{1}{q}}.$$

Then a design that is seeking for a high spread amongst its support points within the design region must attempt to make the smallest distance between neighboring points in ξ as large as possible. That is ensured by the maximin design criterion which corresponds to letting $q \to \infty$,

$$\phi^{-1}_{[\infty]}(\xi) = \min_{\mathbf{s}_i \neq \mathbf{s}_j \in D} \|\mathbf{s}_i - \mathbf{s}_j\| = \phi^{-1}_{Mm}(\xi) \tag{1.27}$$

and we call the resulting design

$$\xi^*_{Mm} = \arg\max_{\xi} \phi^{-1}_{Mm}(\xi)$$

a *maximin distance design*, first introduced by Johnson *et al.* (1990). For more details on space-fillingness refer to Pronzato and Müller (2012).

For the second point in the above list, that is improving the quality of the estimation of the model parameters or derived characteristics it is customary to base the design criterion on the corresponding Fisher information matrix $M(\xi, \beta)$ for the parameters β. A large theory of optimal experimental design is built around those criteria (Atkinson *et al.* 2007), however most of it covers the case of independent errors ϵ, which is usually violated in spatial studies. Some attempts have been made to extend the theory into the correlated error case (cf. Fedorov 1996 and Müller and Pázman 2003), but the respective criteria are still popular for spatial design even without proper adaptations and despite lacking theoretical justification they seem to fare well. The most common choice for a criterion here is

$$\phi_D(\xi) = -\log\ \det M(\xi, \beta)$$

and resulting designs are termed *D-optimal*, although occasionally other scalar functions of $M(\xi, \beta)$ such as the trace (Ver Hoef 2012) are preferred for simplicity (see Nowak 2010 for a survey). Sometimes rather than estimating trend parameters the focus could be on fitting parametrized variograms $\gamma(\mathbf{s}, \mathbf{s}', \theta)$ and estimation of their parameters, which consequently leads to designs based upon information matrices $M'(\xi, \theta)$ (Müller and Zimmerman 1999). A compound version taking into account both purposes

$$\phi_{D\alpha} = -\alpha \log \det M(\xi, \beta) - (1 - \alpha) \log \det M'(\xi, \theta) \tag{1.28}$$

with a weighting factor $0 \leq \alpha \leq 1$ was proposed by Müller and Stehlík (2010). A Bayesian approach for the same problem was proposed in Diggle and Lophaven (2006) and extended in Nowak *et al.* (2010).

When it comes to providing most accurate prediction for unobserved locations and times it is naturally to base a design criterion on the so-called kriging variance $\text{Var}(\hat{Z}(\mathbf{s}, t))$, the mean squared prediction error of the best linear unbiased predictor of the random field (see McBratney and Webster 1981 for some early suggestions and Gao *et al.* 1996 for an efficient formulation). Most commonly the concrete design criterion employed for prediction is

$$\phi_G(\xi) = \max_{D,T} \text{Var}(\hat{Z}(\mathbf{s}, t)|\xi),$$

which amounts to minimizing the maximum prediction variance over the design space, which in design theory is often called *G-optimal*. Some authors prefer to use the average variance (cf. Brus and Heuvelink 2007) but all variants basically yield space-filling designs. It has, however, to be noted that the kriging variance undervalues the uncertainty of the prediction, when – as is common – the variogram parameters θ are estimated together with β from the same data (see also Marchant and Lark 2007 for a discussion of this issue). Therefore Zhu and Stein (2006) and Zimmerman (2006) instead calculate their optimal prediction designs based on the modified kriging variance

$$\phi_{\tilde{G}}(\xi) = \max_{D,T} \left\{ \text{Var}(\hat{Z}(\mathbf{s}, t)|\xi) + \text{tr}\left[M'^{-1}(\xi, \theta) \text{Var}(\partial \hat{Z}(\mathbf{s}, t)/\partial \theta) \right] \right\} \tag{1.29}$$

resulting in much less space-filling and more patchy optimal designs. Note that designs for criteria of type (1.29) are computationally much more difficult to obtain than those of (1.28) since they require evaluations at all points of the design space. However, for uncorrelated regression there exist a certain duality between D- and G-optimality via the celebrated general equivalence theorem by Kiefer and Wolfowitz (1960). This formed the basis of the promising investigations relating criteria $\phi_{D\alpha}$ and $\phi_{\tilde{G}}$ given in Müller *et al.* (2012).

Finally, particularly from a Bayesian viewpoint, it may be desirable to reduce the uncertainty associated to an (a posteriori) distribution. This naturally leads to the so-called entropy criterion, which is given a comprehensive exposition in Le and Zidek (2006) and offers a quite general unified approach at least in the Gaussian case. Note that for most of the above methods the application in the spatio-temporal setting will require adjustments for specific purposes, such as allowing for adapting sampling (Marchant and Lark 2006), ensuring robustness from model assumptions (Wiens 2005) or knowledge of variograms (Spöeck and Pilz 2010).

Finding exact optimum designs is obviously a challenging computational problem and most of the mentioned criteria are nonconvex functions of the design. Thus the algorithms known from the classical design theory (cf. Wynn 1970 and Fedorov 1971) either require some straightforward heuristic adaptions (cf. Brimkulov *et al.* 1980) or taylor-cut numerical routines. Recently, stochastic search algorithms such as spatial simulated annealing (van Groeningen and Stein 1998) and genetic algorithms (Ruiz-Cárdenas *et al.* 2010) or hybrids (Guedes *et al.* 2011) have been considered; a comparison of some of these methods with classic procedures can be found in Baume *et al.* (2010). An approach based on Markov chain Monte Carlo (MCMC) calculations first proposed by Müller (1999) seems to be particularly suitable for finding dynamic spatio-temporal designs (cf. Wikle and Royle 2005).

Available software implementation of the above discussed algorithms and methods is scarce, and we can only find something in R. Some indications can be found in Bivand *et al.* (2008); concrete useful packages are 'fields' by Furrer *et al.* (2011) (model-based) and 'spsurvey' by Kincaid and Olsen (2012) (design-based).

1.6 The problem set: Upper Austria

It was a cornerstone of this project to make the different approaches to design assembled in this volume comparable on an identical problem set. A natural choice was the region of Oberösterreich (Upper Austria), which has the city of Linz, where the university of one of the authors is located, as its capital and was also featured in Müller (1998, 2007). Since we wanted to encourage the use of methods for both discretely and continuously indexed random fields two complementary datasets were provided to the authors.

1.6.1 Climatic data

This dataset contains climatic data measured at 37 stations irregularly placed over the region provided from http://www.zamg.ac.at/fix/klima/oe71-00/klima2000 /klimadaten_oesterreich_1971_frame1.htm. Here, we have (incomplete) monthly data from 1994 to 2009 on average temperature and total rainfall. Due to some missing observations, however, not all of the stations could be effectively used. A map of the region with the respective locations of the measurement stations is displayed in Figure 1.1.

Both of these climatic indicators have been frequently analyzed by geostatistical methods in the past, a recent showcase analysis for rainfall is found in Grimes and Pardo-Igúzquiza (2010). A kriging variance based early study for the optimal design of a rain gauge network was undertaken by Papamichail and Metaxa (1996), whereas Zimmermann *et al.* (2010) provide a mixed probability and model-based approach for the related issue of throughfall

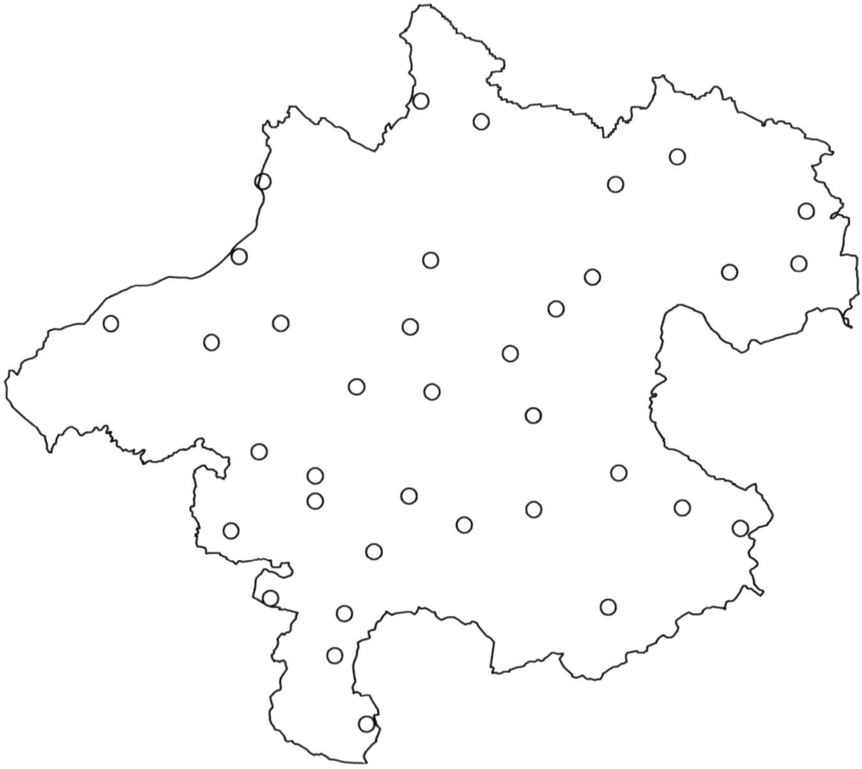

Figure 1.1 The sampling locations of the climatic dataset within the region of Upper Austria.

monitoring. For the current dataset (July 1994) the results of a straightforward kriging analysis employing an exponential semivariogram can be found in Figures 1.2 (temperature) and 1.3 (rainfall), respectively.

To get an impression of the temporal character of this dataset the rainfall measurements for Linz/Stadt from 1994 to 2009 are displayed in Figure 1.4. The seasonality is less pronounced here than in the temperature series, which require particular modeling.

1.6.2 Grassland usage

The second part of the dataset contains information on the grassland usage in the 444 municipal districts of Upper Austria over the years 1995, 1999, 2003, 2005, 2007 and 2008. Those data were processed from a set that was provided by the 'Abteilung Statistik' of the 'Amt der oberösterreichischen Landesregierung' (www.land-oberoesterreich.gv.at/statistik), a regionalized and extended version of what is provided by Statistics Austria (http://www.statistik.at/web_de/statistiken/land_und_forstwirtschaft/agrarstruktur_flaechen_ertraege/bodennutzung/index.html). The districts were represented by a lattice of local coordinates (easting and

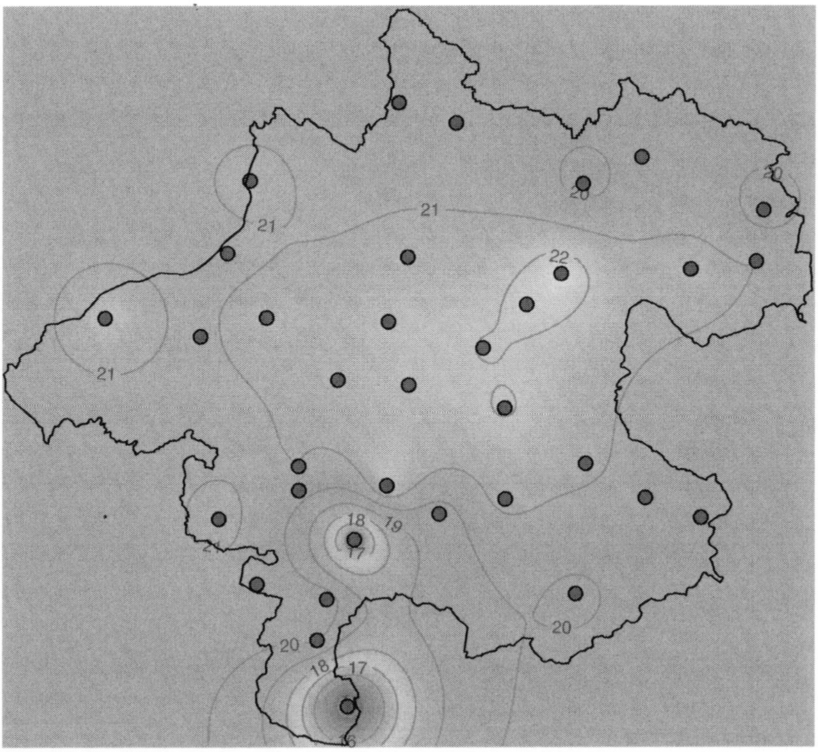

Figure 1.2 Average temperature map for July 1994 in Upper Austria.

northing) of the respective centroids and the boundary information is contained in shape-file (http://doris.ooe.gv.at/index.asp?MenuID = 4; see Figure 1.5).

Specifically we have provided the authors with the following variables: longitude, latitude (in local coordinates), LBBGG (identification number of municipality), BEZNR (identification number for district), FLKM2 (area of the region in square kilometers), ALTITUDE (elevation of the district's capital), and R95,R99,R03,R05,R07,R08 [log (area of arable land +1) − log (area of grassland +1)], i.e. practically the log of the ratio of the two areas for the periods 1995 (Figure 1.6) until 2008, making it scalefree. The latter variables should give some indication of the characteristic of interest to environmental scientists and policy makers, as to whether grassland is increasingly substituted by arable land leading to numerous economic and ecological impacts (see Vellinga *et al.* 2004 and Carlier *et al.* 2005 for a statistical analysis). The indicator is thus well suited for a spatio-temporal analysis and a positive temporal trend as well as some short range spatial autocorrelations can be expected (Yang *et al.* 2006). Furthermore some scientists postulated some interactions between climate change and land use (Thornley and Cannell 1997) thus allowing speculation about integrating some of our climatic variables into a potential spatio-temporal grassland usage model.

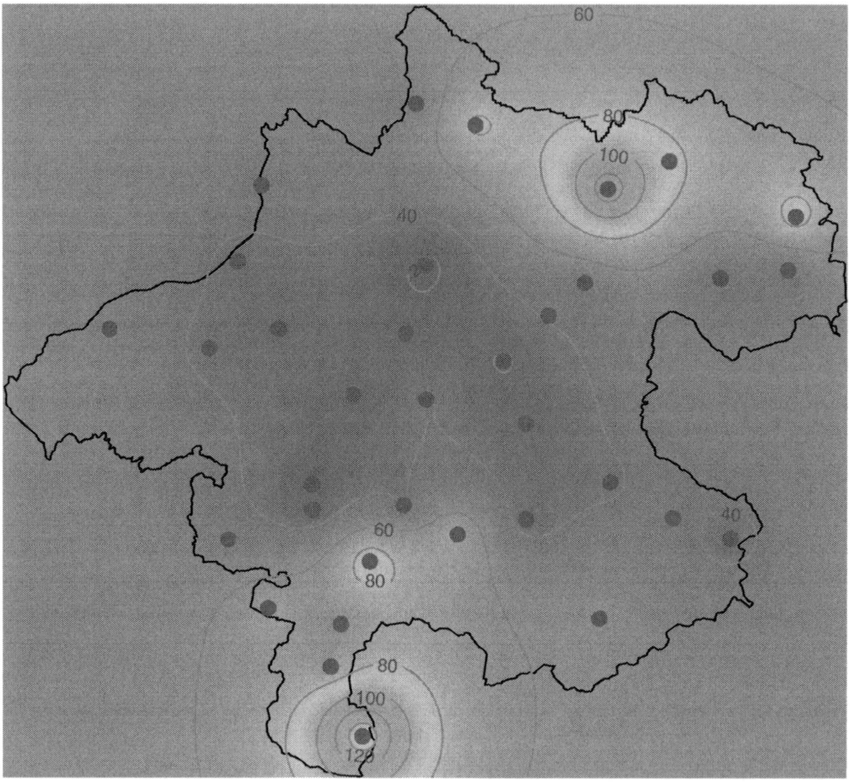

Figure 1.3 Average rainfall map for July 1994 in Upper Austria.

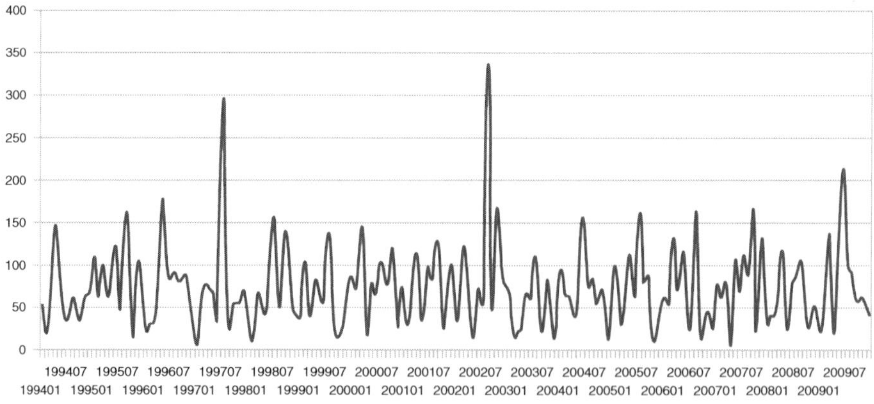

Figure 1.4 Monthly rainfall in Linz/Stadt.

Figure 1.5 The 444 municipal districts of Upper Austria.

The altitudes were not directly available, but were read off the respective Wikipedia pages (http://de.wikipedia.org/wiki/Liste_der_Gemeinden_in_Oberoe sterreich). Some authors have used a full digital elevation model such as provided by http://gdem.ersdac.jspacesystems.or.jp/; a corresponding plot can be found in Figure 1.7. As the local coordinate system (MGI Austria GK Central projection system based on Gauss-Krüger) was not sufficient for all purposes a transformation to the WGS84 geographical coordinates was performed (see http://spatial-analyst.net/book/uppera, where some more useful information on the case study can be found). A Google-Earth depiction of the area is presented in Figure 1.8.

Particularly in the environmental application areas we can naturally find most of the examples of spatio-temporal monitoring network design, see Vašát *et al.* (2009), Corwin *et al.* (2010), Brus and de Gruijter (2011), and Creelman and Risk (2011) for soil assessment, Reed and Minsker (2004) and Masoumi and Kerachian (2010) for groundwater and Murtojärvi *et al.* (2011) for sea water quality, Romary *et al.* (2011) and Wu and Bocquiet (2011) for air pollution, Abida *et al.* (2008) and Melles *et al.* (2011) for radioactive releases, Martínez *et al.* (2008) for solar radiation, Hooten *et al.* (2009) and Fewster (2011) for ecological surveys, Stehman (2009) for land cover evaluation and Barabesi *et al.* (2012) for forestry among the most recent. Note, however, that the problem of where and

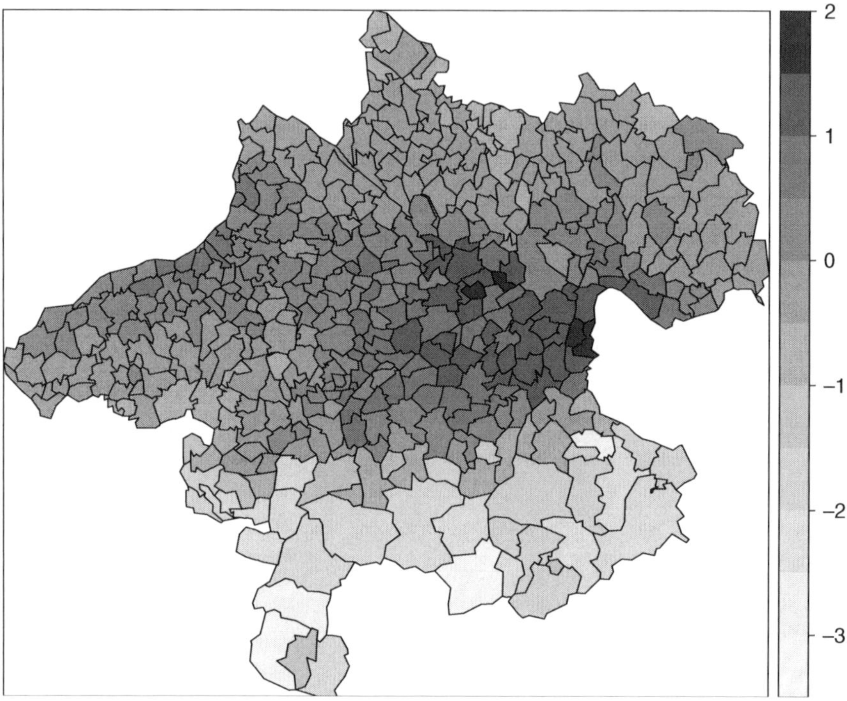

Figure 1.6 Grassland usage: log (area of arable land +1) − log (area of grassland +1) in Upper Austria 1995.

Figure 1.7 A perspective contour plot of Upper Austria.

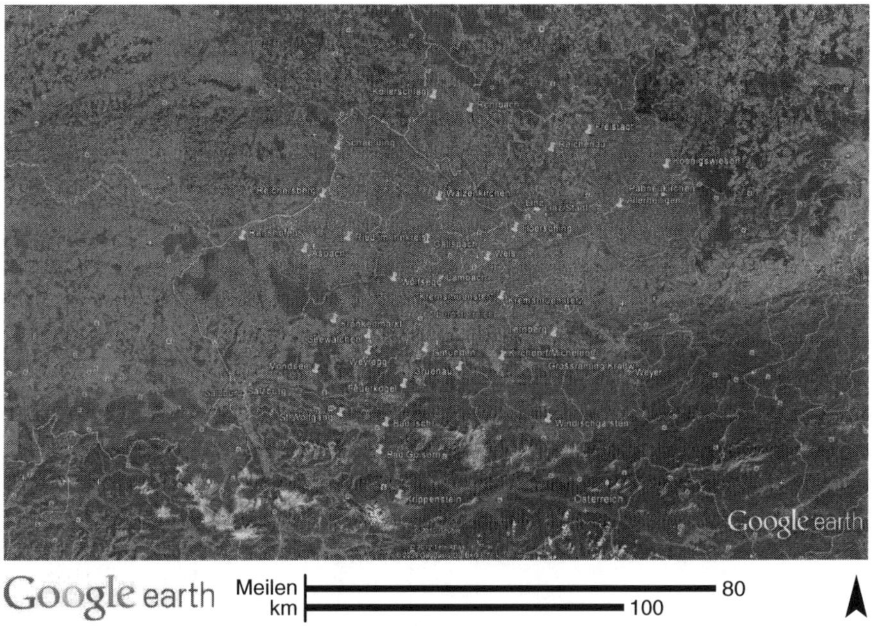

Figure 1.8 *A Google-Earth view of observation regions and sites. (Please see plate section for color version of the figure.)*

when to collect spatio-temporal data is also of relevance in many other fields, for example geophysics (Maurer *et al.* 2010), engineering (Ucinski and Patan 2010), industrial production (Borgoni *et al.* 2010), agriculture (Heuvelink and Egmond 2010), social sciences (Kumar 2007) and signal transmission (Rogerson *et al.* 2004). Furthermore the proposed techniques could even be fruitful for solving problems that are just loosely related to data collection (Guhaniyogi *et al.* (2011).

1.7 The chapters

Let us now give a quick overview of the contents of the book in terms of the individual chapters.

In the geostatistical context, the quality of inferences are affected substantially by the spatial configuration of the network of sites where measurements are taken. An extensive literature exists on spatial network design encompassing a variety of design objectives and statistical perspectives. Chapter 2 (*Model-based frequentist design for univariate and multivariate geostatistics*), written by Dale L. Zimmerman and Jie Li, considers spatial network design for four design objectives from a frequentist, parametric model-based perspective. The design objectives considered pertain to optimal estimation of model parameters and optimal spatial prediction, both for the univariate case (one spatially varying quantity of interest) and multivariate case (two or more such quantities).

Chapter 3 (*Model-based criteria heuristics for second-phase spatial sampling*), written by Eric M. Delmelle, discusses several objectives related to second-phase spatial sampling which is the process of collecting additional measurements of a spatial variable of interest. Different criteria exist to guide the location of these new measurements, and they generally are nonlinear. The author focuses on simulated annealing, a heuristic method which facilitates the process of finding a suitable sampling set among candidate locations.

As with any statistical prediction method, a general goal is to obtain as good predictions as possible for the complete area of investigation. One possibility to formalize this goal is to try to select the sampling locations in such a way that the sum of all kriging mean square errors of prediction becomes a minimum over the area of investigation. Notably, this is a very complicated optimization problem and becomes still more complicated by the fact that the covariance matrix enters the kriging mean square error of prediction in its inverse form. This design problem has been tackled in a number of recent research papers. Gunter Spöck and Jürgen Pilz in Chapter 4 (*Spatial sampling design by means of spectral approximations to the error process*) first consider the integrated kriging variance averaged over the area of investigation and then, as a combined design criterion, the averaged expected lengths of predictive intervals are considered. Finally it is shown how these design criteria can be generalized to spatial variables having skewed distributions. In such context a criterion for spatial sampling design with Box–Cox transformed spatial variables is proposed. Mathematically speaking, they approximate the investigated spatial random field by means of a large regression model consisting of cosine-sine Bessel surface harmonics with random amplitudes. This approximating regression model is a direct result of the polar spectral representation theorem for isotropic random fields. The authors show that kriging prediction in the original random field model is equivalent to trend prediction in this approximating Bayesian linear regression model.

Assessing the changes both spatially and temporally of data are highly important for monitoring purposes. Many spatio-temporal data are collected from monitoring stations. However, the selection of these stations is often influenced by administrative, political, and other pragmatic considerations. Thus it is important to evaluate if an existing station is statistically unnecessary. It is also crucial to find a location such that a new monitoring station upon it can greatly improve the data modeling. These are considered as environmental network design problems. Most of the approaches that tackle this type of problems can be classified into the following three categories: (1) geometry-based; (2) probability-based; (3) model-based. In Chapter 5 (*Entropy-based network design using hierarchical Bayesian kriging*), written by Baisuo Jin, Yuehua Wu, and Baiqi Miao, the attention is paid to the entropy-based design within Category (3). Although the authors cannot escape from the 'curse of dimensionality', they take advantage of recent increases in computational speed and numerical advances (e.g. MCMC) that allow the implementation of Bayesian space-time dynamical models in a hierarchical framework. Such specifications provide simple strategies for incorporating complicated space-time interactions at different stages of the model's hierarchy, and the models are feasible to implement in high dimensions.

In spatial statistics, the data locations and the measurement process are commonly stochastically dependent. In Chapter 6 (*Accounting for design in the analysis of spatial data*) Brian J. Reich and Montserrat Fuentes identify and distinguish between two such situations: informative sampling and informative missingness. Informative or preferential sampling occurs when measurement locations are selected in a way that depends on the underlying process. For example, air pollution monitors may be placed in locations with high air pollution levels. On the other hand, informative missingness occurs when the measurement locations are fixed, but observations are censored in a way that is informative about the underlying process. An approach that can be used in both situations is the shared variable model which introduces random effects that are shared between the data location and the measurement processes. Conditioned on the random effects, the data location and measurement processes are assumed to be independent. These methods can be naturally implemented in a Bayesian framework using MCMC methods.

Chapter 7 (*Spatial design for knot selection in knot-based dimension reduction models*), written by Alan E. Gelfand, Sudipto Banerjee and Andrew O. Finley, investigates the challenge of fitting hierarchical models for large spatial datasets. The authors propose the use of approximation through dimension reduction models, and work in the setting of spatial random effects specified through Gaussian processes. Such specification requires development of knot locations, whence the spatial design problem emerges. More precisely, they have a post-data collection, predate analysis design problem, and need to specify knots in order to fit desired spatial models. They propose an implementation of knot design based upon an average predictive variance criterion, and show how to implement the design and model fitting in a two-step process.

Efficient data acquisition requires some prior understanding of the process to be observed, ideally in the form of a spatio-temporal model. If this is not the case, the employed design must guarantee that an appropriate model can be identified. These designs, usually called exploratory designs, typically employ minimum assumptions which make random sampling a reasonable choice. However, random sampling can be very inefficient in a spatio-temporal context for the assessment of whether spatial or temporal dependence is present or not and if present at what intensity (and form). Chapter 8 (*Exploratory designs for assessing spatial dependence*), written by Agnes Fussl, Werner G. Müller and Juan Rodríguez-Díaz, addresses these questions and possible improvements over random sampling in coping with them.

Space-time geostatistics has received increasing attention over the past decades, both on the theoretical side with the publication of extensions to the set of valid space-time covariance structures, and on the practical side with the publication of many real-world applications. In this context, the use of space-time kriging will likely increase dramatically in the near future. Chapter 9 (*Sampling design optimization for space-time kriging*), written by Gerard B.M. Heuvelink, Daniel A. Griffith, Tomislav Hengl, and Stephanie J. Melles, deals with the choice of the number and configuration of the observation points in space and time. As the authors state, the main differences are that in space-time kriging, strong anisotropies must often be taken into account because variation in time can be quite different from variation in space, and in space-time there is a much richer set of sampling designs that one may choose from (e.g., static, synchronous and rotational designs). In their chapter, the authors restrict themselves to a criterion that simultaneously minimizes the estimation error of a linear trend and the average interpolation error of the kriging residual. This boils down to minimizing the average universal kriging variance. To obtain an optimal design, a solution algorithm that minimizes the defined criterion is necessary. They work with one particular numerical optimization technique known as spatial simulated annealing.

In many applications, spatio-temporal sampling procedures allow for temporal adaptability, in the sense that the set of spatial sampling locations can be (totally or partially) modified between different observation times. Prior or inferential knowledge on structural properties, jointly with real-time sample-path observed evolution, provide the basic information sources for dynamic modeling and prediction in this context. In Chapter 10 (*Space-time adaptive sampling and data transformations*), written by José M. Angulo, María C. Bueso and Francisco J. Alonso, some related approaches, mainly based on entropy criteria, are described and illustrated. The effect of variable transformations such as averages and/or maxima over given time periods or space regions, often considered in the formulations of critical indicators, is also evaluated in terms of entropy-based information measures. Optimality criteria based on entropy and related quantities such as mutual information have been adopted for the design of spatial and spatio-temporal sampling strategies, particularly in the context of environmental applications. Here, the authors consider the sampling network real-time adaptation based on the observations, and the quantification of the information obtained by transformation of the data.

As exposure assessment technology improves, it is likely that mobile monitoring networks will be used instead of, or in addition to, permanent networks, possibly in conjunction with alternative technologies such as leaf sampling. This gives rise to the adaptive or dynamic sampling design problem. In Chapter 11 (*Adaptive sampling design for spatio-temporal prediction*) Thomas R. Fanshawe and Peter J. Diggle consider model-based design, in which the optimal design problem requires two key features to be specified: (i) a statistical or mathematical model for the process under consideration; (ii) a criterion with respect to which the design is required to be optimized. In practice, a good choice of model may not be known until data collection has begun. This highlights another benefit of adaptive designs: model fitting can take place as data become available, and future design locations chosen accordingly.

Capturing this dynamical component of the spatial structure of many environmental processes, through effective modeling strategies, can lead to efficient and cost effective designs for spatial monitoring networks. Furthermore, environmental variables related to processes are often non-Gaussian. In many cases, there also exist important environmental factors that are related nonlinearly to the process mean (or location parameter). To accommodate these types of data complexities, Scott H. Holan and Christopher K. Wikle present in Chapter 12 (*Semiparametric dynamic design of monitoring networks for non-Gaussian spatio-temporal data*) a low-rank semiparametric simulation-based design approach for data distributions from the spatio-temporal exponential family. Fundamental to their approach is that the principal random component in the conditional mean function is a spatio-temporal dynamic process modeled with a low-rank basis function expansion and that other environmental factors are potentially modeled semiparametrically using low-rank basis function representations.

Chapter 13 (*Active learning for monitoring network optimization*), written by Devis Tuia, Alexei Pozdnoukhov, Loris Foresti and Mikhail Kanevski, presents the use of active learning to design monitoring networks that are optimal for environmental modeling with machine learning algorithms. In machine learning, active learning defines a family of algorithms aimed at the selection of new data points from a pool of available or potential measurements. In the case of monitoring networks optimization, these data are the locations of new measuring stations or measurement points. A general framework of active learning is presented in this chapter. The main attention is paid to the Support Vector algorithms which recently demonstrated high efficiency in data analysis and modeling tasks and gave rise to new interesting and important approaches in active learning.

Monitoring of pollution in air or water can be improved by considering the physical laws underlying the dispersion process of such substances. Pollutants usually spread from point sources in the shape of a plume. The randomness of these fields is mainly due to the uncertainty of flow parameters or weather forecasts and to the limited information about the source characteristics. Simulations of plumes are a feasible way to study these fields for sampling design optimization. They give a detailed estimation of what sensors could measure and how good the conclusions are that we draw from such measurements. Although these phenomena are dynamic, Chapter 14 (*Stationary sampling designs based on plume simulations*), written by Kristina B. Helle and Edzer Pebesma, focuses on optimizing networks with stationary sensors, meaning that the sensors do not move during plume passage. Optimizing sampling designs with the help of plume simulations is quite common for groundwater monitoring. It is a bit less common for the monitoring of air pollution because winds change faster than ground water fluxes and therefore the dispersion parameters of the fields are less well known in most applications. Their main concern is monitoring of pollution that spreads from a few potential sources by single plumes like poisonous or radioactive gases or aerosols. The random fields of such plumes are the concentrations or dose rates in the space around the source in the time after the beginning of the release. Uncertainty in these fields is due to the incomplete knowledge of the boundary conditions of the dispersion like the wind fields and sometimes also due to uncertainty about the source. Probabilities of these conditions are very complex as is their propagation through the dispersion. However, by adopting a dispersion model we can generate realizations to test proposed sampling designs.

Acknowledgments

Werner Müller would like to acknowledge the support of the projects OEAD-WTZ FR 11/2010 and FWF/ANR I-833 N18 and his colleagues Luc Pronzato, Joao Rendas, and Milan Stehlík; and particularly Helmut Waldl, who provided the R-code for producing the kriging variance maps that are displayed on the book cover. He also wants to thank attentive students from various courses at the JKU Linz, particularly Andreas Rappold, who provided Figure 1.7. The work on this book was made easier by the supportive environment at the Department of Applied Statistics; particular gratitude goes to Andreas Quatember. Gabriele Mack provided invaluable secretarial assistance.

Jorge Mateu acknowledges the support of the projects MTM2010-14961 and P11B2008-27.

Both authors would like to thank their international colleagues for showing interest in the project, in particular Noel Cressie for fruitful discussions and providing the Foreword. We are grateful to Tomislav Hengl for providing transformations of our data between different coordinate systems. We are also grateful to the Zentralanstalt für Meteorologie und Geodynamik, Vienna and Thomas Raferzeder from the Statistics department of the Amt der Oberösterreichischen Landesregierung, Linz for providing the data.

We are thankful for the stimulating professional environment Wiley has provided conducive to this book activity. The support of the Wiley team through the Production Editor, Project Manager and Copyeditor is acknowledged with appreciation for encouragement. They have done a magnificent job. Special gratitude goes to Ilaria Meliconi, who initiated the project, Richard Davies, Heather Kay, Prachi Sinha Sahay from Wiley and Jayashree Saishankar from Laserwords.

References

Abida R, Bocquet M, Vercauteren N and Isnard O 2008 Design of a monitoring network over France in case of a radiological accidental release. *Atmospheric Environment* **42**(21), 5205–5219.

Anselin L 1988 *Spatial Econometrics: Methods and Models (Studies in Operational Regional Science)*. Springer.

Anselin L 2010 Thirty years of spatial econometrics. *Papers in Regional Science* **89**(1), 3–25.

Atkinson A, Donev A and Tobias R 2007 *Optimum Experimental Designs, with SAS (Oxford Statistical Science Series)*. Oxford University Press.

Barabesi L, Franceschi S and Marcheselli M 2012 Properties of design-based estimation under stratified spatial sampling with application to canopy coverage estimation. *Annals of Applied Statistics* **6**(1), 210–228.

Baume OP, Gebhardt A, Gebhardt C, Heuvelink GBM and Pilz J 2010 Network optimization algorithms and scenarios in the context of automatic mapping. *Computers & Geosciences* **37**, 289–294.

Baxevani A, Podgórski K and Rychlik I 2011 Dynamically evolving Gaussian spatial fields. *Extremes* **14**(2), 223–251.

Berliner LM, Lu ZQ and Snyder C 1999 Statistical design for adaptive weather observations. *Journal of the Atmospheric Sciences* **56**(15), 2536–2552.

Bevilacqua M, Gaetan C, Mateu J and Porcu E 2012 Estimating space and space-time covariance functions for large data sets: a weighted composite likelihood approach. *Journal of the American Statistical Association* **107**(497), 268–280.

Bivand RS, Pebesma EJ and Gómez-Rubio V 2008 *Applied Spatial Data Analysis with R (Use R!)*. Springer.

Bochner S 1933 A theorem on fourier-stieltjes integrals. *Mathematische Annalen* **108**, 378–410.

Borgoni R, Radaelli L, Tritto V and Zappa D 2010 Optimal reduction of a monitoring grid for SiO_2 deposition surface over a wafer for semiconductor devices. *Atti della XLV Riunione Scientifica della Società Italiana di Statistica*.

Brimkulov UN, Krug GK and Savanov VL 1980 Numerical construction of exact experimental designs when the measurements are correlated. *Zavodskaya Laboratoria* **36**, 435–442.

Brown P, Roberts G, KÅresen K and Tonellato S 2000 Blur-generated non-separable space–time models. *Journal of the Royal Statistical Society: Series B (Statistical Methodology)* **62**(4), 847–860.

Brus D and de Gruijter JJ 1997 Random sampling or geostatistical modelling? Choosing between design-based and model-based sampling strategies for soil (with discussion). *Geoderma* **80**(1–2), 1–44.

Brus D and Heuvelink G 2007 Optimization of sample patterns for universal kriging of environmental variables. *Geoderma* **138**(1–2), 86–95.

Brus DJ and de Gruijter JJ 2011 Design-based Generalized Least Squares estimation of status and trend of soil properties from monitoring data. *Geoderma* **164**(3–4), 172–180.

Brus DJ and de Gruijter JJ 2012 A hybrid design-based and model-based sampling approach to estimate the temporal trend of spatial means. *Geoderma* **173–174**, 241–248.

Caragea P and Smith R 2006 Approximate likelihoods for spatial processes. *Proceedings of the American Statistical Association Preprint*.

Carlier L, De Vliegher A, Van Cleemput O and Boeckx P 2005 Importance and functions of European grasslands. *Communications in Agricultural and Applied Biological Sciences* **70**(1), 5–15.

Caselton WF and Zidek JV 1984 Optimal monitoring network designs. *Statistics & Probability Letters* **2**(4), 223–227.

Chen L, Fuentes M and Davis J 2006 Spatial–temporal statistical modeling and prediction of environmental processes. In *Hierarchical Modelling for the Environmental Sciences* (eds Clark JS and Gelfand A). Oxford University Press, pp. 121–144.

Chilès JP and Delfiner P 1999 *Geostatistics: Modeling Spatial Uncertainty (Wiley Series in Probability and Statistics)*. Wiley-Interscience.

Corwin DL, Lesch SM, Segal E, Skaggs TH and Bradford SA 2010 Comparison of sampling strategies for characterizing spatial variability with apparent soil electrical conductivity directed soil sampling. *Journal of Environmental & Engineering Geophysics* **15**(3), 147–162.

Creelman C and Risk D 2011 Network design for soil CO_2 monitoring of the northern North American region. *Ecological Modelling* **222**(18), 3421–3428.

Cressie N 1993 *Statistics for Spatial Data (Wiley Series in Probability and Statistics)* revised. Wiley-Interscience.

Cressie N and Hawkins D 1980 Robust estimation of the variogram: I. *Mathematical Geology* **12**(2), 115–125.

Cressie N and Huang H 1999 Classes of nonseparable, spatio-temporal stationary covariance functions. *Journal of the American Statistical Association* **94**, 1330–1340.

Cressie N and Wikle CK 2011 *Statistics for Spatio-Temporal Data (Wiley Series in Probability and Statistics)*. John Wiley & Sons, Ltd.

Cressie N, Calder CA, Clark JS, Hoef JM and Wikle CK 2009 Accounting for uncertainty in ecological analysis: the strengths and limitations of hierarchical statistical modeling. *Ecological Applications* **19**(3), 553–570.

Curriero F and Lele S 1999 A composite likelihood approach to semivariogram estimation. *Journal of Agricultural, Biological, and Environmental Statistics* **4**(1), 9–28.

De Cesare L, Myers D and Posa D 2001a Estimating and modeling space–time correlation structures. *Statistics & Probability Letters* **51**(1), 9–14.

De Cesare L, Myers D and Posa D 2001b Product-sum covariance for space-time modeling: an environmental application. *Environmetrics* **12**(1), 11–23.

de Gruijter JJ and ter Braak CJF 1990 Model-free estimation from spatial samples: a reappraisal of classical sampling theory. *Mathematical Geology* **22**(4), 407–415.

de Gruijter JJ, Brus DJ, Bierkens MFP and Knotters M 2006 *Sampling for Natural Resource Monitoring*. Springer.

De Iaco S, Myers D and Posa D 2001 Space–time analysis using a general product–sum model. *Statistics & Probability Letters* **52**(1), 21–28.

De Iaco S, Myers D and Posa D 2002a Nonseparable space-time covariance models: some parametric families. *Mathematical Geology* **34**(1), 23–42.

De Iaco S, Myers D and Posa D 2002b Space–time variograms and a functional form for total air pollution measurements. *Computational Statistics & Data Analysis* **41**(2), 311–328.

De Iaco S, Myers D and Posa D 2003 The linear coregionalization model and the product–sum space–time variogram. *Mathematical Geology* **35**(1), 25–38.

Diggle P and Lophaven S 2006 Bayesian Geostatistical Design. *Scandinavian Journal of Statistics* **33**(1), 53–64.

Dobbie MJ, Henderson BL and Stevens DL 2008 Sparse sampling: spatial design for monitoring stream networks. *Statistical Survey* **2**, 113–153.

Elhorst JP 2012 Dynamic spatial panels: models, methods, and inferences. *Journal of Geographical Systems* **14**(1), 5–28.

Fedorov V 1996 Design of spatial experiments: model fitting and prediction. In *Handbook of Statistics* (eds Gosh S and Rao CR) vol. 13. Elsevier, pp. 515–553.

Fedorov VV 1971 The design of experiments in the multiresponse case. *Theory of Probability and its Applications* **16**(2), 323–332.

Fedorov VV and Müller WG 1988 Two approaches in optimization of observing networks. In *Optimal Design and Analysis of Experiments* (eds Dodge Y, Fedorov VV and Wynn HP). North Holland, pp. 239–256.

Fernández-Casal R, González-Manteiga W and Febrero-Bande M 2003 Flexible spatio-temporal stationary variogram models. *Statistics and Computing* **13**(2), 127–136.

Fewster RM 2011 Variance estimation for systematic designs in spatial surveys. *Biometrics* **67**(4), 1518–1531.

Fuentes M, Chen L, Davis J and Lackmann G 2005 Modeling and predicting complex space–time structures and patterns of coastal wind fields. *Environmetrics* **16**(5), 449–464.

Furrer R, Genton M and Nychka D 2006 Covariance tapering for interpolation of large spatial datasets. *Journal of Computational and Graphical Statistics* **15**(3), 502–523.

Furrer R, Nychka D and Sain S 2011 R-package 'fields', version 6.6.3.

Gao H, Wang J and Zhao P 1996 The updated kriging variance and optimal sample design. *Mathematical Geology* **28**(3), 295–313.

Gneiting T 2002 Nonseparable, stationary covariance functions for space-time data. *Journal of the American Statistical Association* **97**(458), 590–600.

Gregori P, Porcu E, Mateu J and Sasvári Z 2008 On potentially negative space time covariances obtained as sum of products of marginal ones. *Annals of the Institute of Statistical Mathematics* **60**(4), 865–882.

Gribik PR, Kortanek KO and Sweigart JR 1976 Designing a regional air pollution monitoring network: An appraisal of a regression experimental design approach. In *Conference on Environmental Modeling and Simulation* (ed. Ott WR). EPA, pp. 86–91.

Griffith D and Paelinck J 2007 An equation by any other name is still the same: on spatial econometrics and spatial statistics. *The Annals of Regional Science* **41**(1), 209–227.

Griffith DA and Csillag F 1993 Exploring relationships between semi-variogram and spatial autoregressive models. *Papers in Regional Science* **72**(3), 283–295.

Grimes DIF and Pardo-Igúzquiza E 2010 Geostatistical analysis of rainfall. *Geographical Analysis* **42**(2), 136–160.

Guedes LPC, Ribeiro PJ, De Stefano SM and Uribe-Opazo MA 2011 Optimization of spatial sample configurations using hybrid genetic algorithm and simulated annealing. *Chilean Journal of Statistics* **2**(2), 39–50.

Guhaniyogi R, Finley AO, Banerjee S and Gelfand AE 2011 Adaptive Gaussian predictive process models for large spatial datasets. *Environmetrics* **22**(8), 997–1007.

Hae-Ryoung S, Fuentes M and Ghosh S 2008 A comparative study of Gaussian geostatistical models and Gaussian Markov random field models. *Journal of Multivariate Analysis* **99**(8), 1681–1697.

Haining RP, Kerry R and Oliver MA 2010 Geography, spatial data analysis, and geostatistics: an overview. *Geographical Analysis* **42**(1), 7–31.

Heuvelink GBM and Egmond FM 2010 Space-time geostatistics for precision agriculture: a case study of NDVI mapping for a Dutch potato field. In *Geostatistical Applications for Precision Agriculture* (ed. Oliver MA). Springer, pp. 117–137.

Hooten MB, Wikle CK, Sheriff SL and Rushin JW 2009 Optimal spatio-temporal hybrid sampling designs for ecological monitoring. *Journal of Vegetation Science* **20**(4), 639–649.

Johnson ME, Moore LM and Ylvisaker D 1990 Minimax and maximin distance designs. *Journal of Statistical Planning and Inference* **26**(2), 131–148.

Jones R and Zhang Y 1997 Models for continuous stationary space-time processes. *Lecture Notes in Statistics* 289–298.

Kaufman C, Schervish M and Nychka D 2008 Covariance tapering for likelihood-based estimation in large spatial data sets. *Journal of the American Statistical Association* **103**(484), 1545–1555.

Kiefer J and Wolfowitz J 1960 The equivalence of two extremum problems. *Canadian Journal of Mathematics* **12**, 363–366.

Kincaid T and Olsen T 2012 R-package 'spsurvey', version 2.3.

Kolovos A, Christakos G, Hristopulos D and Serre M 2004 Methods for generating non-separable spatiotemporal covariance models with potential environmental applications. *Advances in Water Resources* **27**(8), 815–830.

Kumar N 2007 Spatial Sampling Design for a Demographic and Health Survey. *Population Research and Policy Review* **26**(5), 581–599.

Le ND and Zidek JV 2006 *Statistical Analysis of Environmental Space-Time Processes (Springer Series in Statistics)*. Springer.

Lindgren F, Rue H and Lindström J 2011 An explicit link between Gaussian fields and Gaussian Markov random fields: the stochastic partial differential equation approach. *Journal of the Royal Statistical Society: Series B (Statistical Methodology)* **73**(4), 423–498.

Lindsay B 1988 Composite likelihood methods. *Contemporary Mathematics* **80**(1), 221–239.

Ma C 2002 Spatio-temporal covariance functions generated by mixtures. *Mathematical Geology* **34**(8), 965–975.

Ma C 2003a Families of spatio-temporal stationary covariance models. *Journal of Statistical Planning and Inference* **116**(2), 489–501.

Ma C 2003b Nonstationary covariance functions that model space–time interactions. *Statistics & Probability Letters* **61**(4), 411–419.

Ma C 2003c Spatio-temporal stationary covariance models. *Journal of Multivariate Analysis* **86**(1), 97–107.

Ma C 2005a Linear combinations of space-time covariance functions and variograms. *IEEE Transactions on Signal Processing* **53**(3), 857–864.

Ma C 2005b Semiparametric spatio-temporal covariance models with the arma temporal margin. *Annals of the Institute of Statistical Mathematics* **57**(2), 221–233.

Ma C 2005c Spatio-temporal variograms and covariance models. *Advances in Applied Probability* **37**(3), 706–725.

Marchant B and Lark R 2007 Optimized sample schemes for geostatistical surveys. *Mathematical Geosciences* **39**(1), 113–134.

Marchant BP and Lark RM 2006 Adaptive sampling and reconnaissance surveys for geostatistical mapping of the soil. *European Journal of Soil Science* **57**(6), 831–845.

Mardia K and Marshall R 1984 Maximum likelihood estimation of models for residual covariance in spatial regression. *Biometrika* **71**(1), 135–146.

Martínez F, Mateu J, Montes F, Bodas-Salcedo A and López-Baeza E 2008 A comparative analysis of different spatial sampling schemes: modelling of SSRB data. *International Journal of Remote Sensing* **29**(6), 1635–1647.

Masoumi F and Kerachian R 2010 Optimal redesign of groundwater quality monitoring networks: a case study. *Environmental Monitoring and Assessment* **161**(1), 247–257.

Mateu J, Juan P and Porcu E 2007 Geostatistical analysis through spectral techniques: some words of caution. *Communications in Statistics - Simulation and Computation* **36**(5), 1035–1051.

Maurer H, Curtis A and Boerner DE 2010 Recent advances in optimized geophysical survey design. *Geophysics* **75**(5), 75A177–75A194.

McBratney A and Webster R 1981 The design of optimal sampling schemes for local estimation and mapping of regionalized variables. II Program and examples. *Computers & Geosciences* **7**(4), 335–365.

Melles SJ, Heuvelink GBM, Twenhöfel CJW, van Dijk A, Hiemstra PH, Baume O and Stöhlker U 2011 Optimizing the spatial pattern of networks for monitoring radioactive releases. *Computers & Geosciences* **37**(3), 280–288.

Müller WG 1998 *Collecting Spatial Data: Optimum Design of Experiments for Random Fields*. Springer.

Müller WG 1999 Least-squares fitting from the variogram cloud. *Statistics & Probability Letters* **43**(1), 93–98.

Müller WG 2007 *Collecting Spatial Data: Optimum Design of Experiments for Random Fields* 3rd revised and extend edn. Springer.

Müller WG and Pázman A 2003 Measures for designs in experiments with correlated errors. *Biometrika* **90**(2), 423–434.

Müller WG and Stehlík M 2010 Compound optimal spatial designs. *Environmetrics* **21**(3-4), 354–364.

Müller WG and Waldl H 2011 Discussion of Lindgren, Rue and Lindström 'An explicit link between Gaussian fields and Gaussian Markov random fields: the stochastic partial differential equation approach'. *Journal of the Royal Statistical Society: Series B (Statistical Methodology)* **73**(4), 483–485.

Müller WG and Zimmerman DL 1999 Optimal designs for variogram estimation. *Environmetrics* **10**(1), 23–37.

Müller WG, Pronzato L and Waldl H 2012 Relations between designs for prediction and estimation in random fields: an illustrative case. In *Advances and Challenges in Space-time Modelling of Natural Events* (ed. Porcu E, Montero J and Schlather M) vol. 207 of *Lecture Notes in Statistics*. Springer. pp. 125–139.

Murtojärvi M, Suominen T, Uusipaikka E and Nevalainen OS 2011 Optimising an observational water monitoring network for Archipelago Sea, South West Finland. *Computers & Geosciences* **37**(7), 844–854.

Nowak W 2010 Measures of parameter uncertainty in geostatistical estimation and geostatistical optimal design. *Mathematical Geosciences* **42**(2), 199–221.

Nowak W, de Barros FPJ and Rubin Y 2010 Bayesian geostatistical design: task-driven optimal site investigation when the geostatistical model is uncertain. *Water Resources Research* **46**(3), W03535+.

Olea RA 1984 Sampling design optimization for spatial functions. *Mathematical Geology* **16**(4), 369–392.

Oud JHL, Folmer H, Patuelli R and Nijkamp P 2012 Continuous-time modeling with spatial dependence. *Geographical Analysis* **44**(1), 29–46.

Overton WS and Stehman SV 1996 Desirable design characteristics for long-term monitoring of ecological variables. *Environmental and Ecological Statistics* **3**, 349–361.

Papamichail DM and Metaxa IG 1996 Geostatistical analysis of spatial variability of rainfall and optimal design of a rain gauge network. *Water Resources Management* **10**(2), 107–127.

Papritz A and Webster R 1995 Estimating temporal change in soil monitoring: I. Statistical theory. *European Journal of Soil Science* **46**(1), 1–12.

Porcu E and Mateu J 2007 Mixture-based modeling for space–time data. *Environmetrics* **18**(3), 285–302.

Porcu E, Gregori P and Mateu J 2006 Nonseparable stationary anisotropic space–time covariance functions. *Stochastic Environmental Research and Risk Assessment* **21**(2), 113–122.

Porcu E, Gregori P and Mateu J 2007a La descente et la montée étendues: the spatially d-anisotropic and the spatio-temporal case. *Stochastic Environmental Research and Risk Assessment* **21**(6), 683–693.

Porcu E, Mateu J and Bevilacqua M 2007b Covariance functions that are stationary or nonstationary in space and stationary in time. *Statistica Neerlandica* **61**(3), 358–382.

Porcu E, Mateu J and Christakos G 2009 Quasi-arithmetic means of covariance functions with potential applications to space-time data. *Journal of Multivariate Analysis* **100**(8), 1830–1844.

Pronzato L and Müller WG 2012 Design of computer experiments: space filling and beyond. *Statistics and Computing* **22**(3), 681–701.

Rasmussen CE and Williams CKI 2005 *Gaussian Processes for Machine Learning (Adaptive Computation and Machine Learning series)*. The MIT Press.

Reed PM and Minsker BS 2004 Striking the balance: long-term groundwater monitoring design for conflicting objectives. *Journal of Water Resources Planning and Management* **130**(2), 140–149.

Rogerson P, Delmelle E, Batta R, Akella MR, Blatt A and Wilson G 2004 Optimal sampling design for variables with varying spatial importance. *Geographical Analysis* **36**(2), 177–194.

Romary T, de Fouquet C and Malherbe L 2011 Sampling design for air quality measurement surveys: an optimization approach. *Atmospheric Environment* **45**(21), 3613–3620.

Royle J and Nychka D 1998 An algorithm for the construction of spatial coverage designs with implementation in SPLUS. *Computers & Geosciences* **24**(5), 479–488.

Rue H and Held L 2005 *Gaussian Markov Random Fields: Theory and Applications (Chapman & Hall/CRC Monographs on Statistics & Applied Probability)* Chapman and Hall/CRC.

Ruiz-Cárdenas R, Ferreira MAR and Schmidt AM 2010 Stochastic search algorithms for optimal design of monitoring networks. *Environmetrics* **21**(1), 102–112.

Schoenberg I 1938 Metric spaces and positive definite functions. *Transactions of the American Mathematical Society* **44**(3), 522–536.

Spöck G and Pilz J 2010 Spatial sampling design and covariance-robust minimax prediction based on convex design ideas. *Stochastic Environmental Research and Risk Assessment* **24**(3), 463–482.

Stehman SV 1999 Basic probability sampling designs for thematic map accuracy assessment. *International Journal of Remote Sensing* **20**(12), 2423–2441.

Stehman SV 2009 Sampling designs for accuracy assessment of land cover. *International Journal of Remote Sensing* **30**(20), 5243–5272.

Stein M 2005 Space-time covariance functions. *Journal of the American Statistical Association* **100**(469), 310–321.

Stein M, Chi Z and Welty L 2004 Approximating likelihoods for large spatial data sets. *Journal of the Royal Statistical Society: Series B (Statistical Methodology)* **66**(2), 275–296.

Stevens DL 2006 Spatial properties of design-based versus model-based approaches to environmental sampling. In *Proceedings of the 7th International Symposium on Spatial Accuracy Assessment in Natural Resources and Environmental Sciences*, pp. 119–125.

Stevens DL and Olsen AR 2004 Spatially balanced sampling of natural resources. *Journal of the American Statistical Association* **99**(465), 262–278.

Thompson SK 2002 On sampling and experiments. *Environmetrics* **13**(5–6), 429–436.

Thornley JHM and Cannell MGR 1997 Temperate grassland responses to climate change: an analysis using the Hurley Pasture Model. *Annals of Botany* **80**(2), 205–221.

Ucinski D and Chen Y 2005 Time optimal path planning of moving sensors for parameter estimation of distributed systems. In *Proceedings of the 44th IEEE Conference on Decision and Control*. IEEE, pp. 5257–5262.

Ucinski D and Patan M 2010 Sensor network design for the estimation of spatially distributed processes. *International Journal of Applied Mathematics and Computer Science* **20**(3), 459–481.

van Groenigen JW and Stein A 1998 Constrained optimization of spatial sampling using continuous simulated annealing. *Journal of Environmental Quality* **27**(5), 1078–1086.

Varin C and Vidoni P 2005 A note on composite likelihood inference and model selection. *Biometrika* **92**(3), 519–528.

Vašát R, Heuvelink GBM and Borůvka L 2009 Sampling design optimization for multivariate soil mapping. *Geoderma* **155**(3–4), 147–153.

Vecchia A 1988 Estimation and model identification for continuous spatial processes. *Journal of the Royal Statistical Society: Series B (Statistical Methodological)* **50**, 297–312.

Vellinga T, van den Pol-van Dasselaar A and Kuikman PJ 2004 The impact of grassland ploughing on CO_2 and N_2O emissions in the Netherlands. *Nutrient Cycling in Agroecosystems* **70**(1), 33–45.

Ver Hoef JM 2012 Practical considerations for experimental designs of spatially autocorrelated data using computer intensive methods. *Statistical Methodology* **9**(1–2), 172–184.

Whittle P 1954 On stationary processes in the plane. *Biometrika* **41**, 434–449.

Wiens DP 2005 Robustness in spatial studies II: minimax design. *Environmetrics* **16**(2), 205–217.

Wikle CK and Royle JA 1999 Space-time dynamic design of environmental monitoring networks. *Journal of Agricultural, Biological, and Environmental Statistics* **4**(4), 489–507.

Wikle CK and Royle JA 2005 Dynamic design of ecological monitoring networks for non-Gaussian spatio-temporal data. *Environmetrics* **16**(5), 507–522.

Wu L and Bocquet M 2011 Optimal redistribution of the background ozone monitoring stations over France. *Atmospheric Environment* **45**(3), 772–783.

Wynn HP 1970 The sequential generation of D-optimum experimental designs. *The Annals of Mathematical Statistics* **41**(5), 1655–1664.

Yang Y, Feng Z, Liu D and Zhang J 2006 The Spatial-temporal change of grassland in Qinghai-Tibet Plateau. In *2006 IEEE International Symposium on Geoscience and Remote Sensing*. IEEE, pp. 3099–3102.

Zhu Z and Stein ML 2006 Spatial sampling design for prediction with estimated parameters. *Journal of Agricultural, Biological, and Environmental Statistics* **11**(1), 24–44.

Zidek JV and Zimmerman DL 2010 *Monitoring Network Design*. CRC Press, pp. 131–148.

Zimmerman D 1989 Computationally efficient restricted maximum likelihood estimation of generalized covariance functions. *Mathematical Geology* **21**(7), 655–672.

Zimmerman DL 2006 Optimal network design for spatial prediction, covariance parameter estimation, and empirical prediction. *Environmetrics* **17**(6), 635–652.

Zimmermann B, Zimmermann A, Lark RM and Elsenbeer H 2010 Sampling procedures for throughfall monitoring: a simulation study. *Water Resources Research* **46**(1), W01503+.

2

Model-based frequentist design for univariate and multivariate geostatistics

Dale L. Zimmerman[1] **and Jie Li**[2]

[1]*Department of Statistics and Actuarial Science, University of Iowa, USA*

[2]*Department of Statistics, Virginia Tech University, USA*

2.1 Introduction

It is widely recognized that the quality of inferences made from geostatistical data are affected substantially by the spatial configuration of the network of sites where measurements are taken. Naturally, then, an extensive literature exists on spatial network design; for relatively recent reviews see, e.g., Müller (2007) and Zidek and Zimmerman (2010). This literature encompasses a variety of design objectives and statistical perspectives. In this chapter, which extends portions of Zimmerman (2006) and Li (2009), we consider spatial network design for four design objectives from a frequentist, parametric model-based perspective. The design objectives considered pertain to optimal estimation of model parameters and optimal spatial prediction, both for the univariate case (one spatially varying quantity of interest) and multivariate case (two or more such quantities). Virtually no exact optimal design theory exists for such problems, so these design approaches are illustrated using contrived 'toy' examples, chosen

Spatio-temporal design: Advances in efficient data acquisition, First Edition.
Edited by Jorge Mateu and Werner G. Müller.

to be sufficiently small that optimal or near-optimal designs may be obtained by complete enumeration. We also indicate how the features exemplified by these designs extend to more realistic situations.

2.2 Design for univariate geostatistics

2.2.1 Data-model framework

Consider the following commonly used data-model framework for univariate geostatistics. Suppose that a continuous, spatially varying quantity, Z, is to be observed at a predetermined number, n, of point sites in a two-dimensional region of interest, A. The observations on Z are regarded statistically as a spatially incomplete sample of one realization of a random field $\{Z(\mathbf{s}) : \mathbf{s} \in A\}$. Assume that the mean function of the random field exists and is linear in its parameters, i.e., $E[Z(\mathbf{s})] = [\mathbf{x}(\mathbf{s})]'\boldsymbol{\beta}$ where $\mathbf{x}(\mathbf{s})$ is a vector of known functions of the spatial coordinates or other covariates observed at $\mathbf{s}$ and $\boldsymbol{\beta}$ is a vector of unknown parameters. Also assume that the covariance function, $\text{cov}[Z(\mathbf{s}), Z(\mathbf{u})] = C(\mathbf{s}, \mathbf{u}; \boldsymbol{\theta})$, exists, where C is a known positive definite function of pairs of location vectors and $\boldsymbol{\theta}$ is a vector of known or unknown parameters. If the random field is second-order stationary, then we may write $C(\mathbf{s}, \mathbf{u}; \boldsymbol{\theta}) = C(\mathbf{s} - \mathbf{u}; \boldsymbol{\theta})$; if the random field is isotropic as well, then we may write

$$C(\mathbf{s} - \mathbf{u}; \boldsymbol{\theta}) = C(\|\mathbf{s} - \mathbf{u}\|; \boldsymbol{\theta}) = \begin{cases} \sigma^2 + \tau^2 & \text{if } \mathbf{s} = \mathbf{u} \\ \sigma^2\rho(\|\mathbf{s} - \mathbf{u}\|; \boldsymbol{\theta}^*) & \text{otherwise} \end{cases}$$

where $\boldsymbol{\theta} = (\sigma^2, \tau^2, \boldsymbol{\theta}^{*\prime})'$, $\sigma^2 > 0$ is the partial sill, $\tau^2 \geq 0$ is the nugget effect, ρ is a known positive definite correlation function, and $\boldsymbol{\theta}^*$ is a vector of correlation parameters.

Let $\mathcal{S}$ be the set of candidate design points, i.e., the set of points where it is feasible to take observations on Z. Although in principle $\mathcal{S}$ could coincide with A, administrative, political, economic, or other practical considerations will usually restrict $\mathcal{S}$ to be a finite subset of points in A. The problem of this section is to choose the 'design' D, i.e. the set $D = \{\mathbf{s}_1, \ldots, \mathbf{s}_n\}$, where each $\mathbf{s}_i \in \mathcal{S}$, to optimize, or come close to optimizing, a chosen criterion. Here all criteria are expressed in smaller-is-better form for the sake of uniformity, so designs will be chosen to minimize the chosen criterion. The choice of criterion depends primarily on the inferential objectives of the investigator and the model assumptions under which the objectives are to be accomplished.

2.2.2 Design criteria

2.2.2.1 Estimation of mean parameters assuming known covariance parameters

Suppose that the covariance parameter $\boldsymbol{\theta}$ is known, and we wish to estimate $\boldsymbol{\beta}$ well, in some sense. The best (in the sense of minimizing the estimator's variance) linear unbiased estimator (BLUE) of $\boldsymbol{\beta}$, based on observations taken at points of the design $D = \{\mathbf{s}_1, \ldots, \mathbf{s}_n\}$, is

$$\hat{\boldsymbol{\beta}}(\mathbf{z}, \boldsymbol{\theta}, D) = (\mathbf{X}'\boldsymbol{\Sigma}^{-1}\mathbf{X})^{-1}\mathbf{X}'\boldsymbol{\Sigma}^{-1}\mathbf{z}. \tag{2.1}$$

Here $\mathbf{X}$ is the matrix whose ith row is $[\mathbf{x}(\mathbf{s}_i)]'$; $\boldsymbol{\Sigma}$ is the matrix whose ijth element is $C(\mathbf{s}_i, \mathbf{s}_j; \boldsymbol{\theta})$; and $\mathbf{z}$ is the data vector with ith element $Z(\mathbf{s}_i)$. Each of $\mathbf{X}$, $\boldsymbol{\Sigma}$, and $\mathbf{z}$ depend on the design D, but this is not explicitly indicated by the notation. The assumed invertibility of $\mathbf{X}'\boldsymbol{\Sigma}^{-1}\mathbf{X}$ in (2.1) requires that $\mathbf{X}$ have full column rank, which may somewhat restrict the set of candidate designs. The covariance matrix of $\hat{\boldsymbol{\beta}}$ is given by $\mathbf{V}(\boldsymbol{\theta}, D) = \text{var}_{\boldsymbol{\theta}}(\hat{\boldsymbol{\beta}}(\mathbf{z}, \boldsymbol{\theta}, D)) = (\mathbf{X}'\boldsymbol{\Sigma}^{-1}\mathbf{X})^{-1}$. As $\mathbf{V}(\boldsymbol{\theta}, D)$ is matrix-valued, typically a scalar-valued function of it is used as a design criterion. Here we feature its determinant; thus a design is said to be locally MPE-optimal (where MPE stands for mean parameter estimation) if it minimizes $|(\mathbf{X}'\boldsymbol{\Sigma}^{-1}\mathbf{X})^{-1}|$, or equivalently, if it minimizes

$$\gamma_{MPE}(D; \boldsymbol{\theta}) = 1/|\mathbf{X}'\boldsymbol{\Sigma}^{-1}\mathbf{X}|$$

with respect to D. Here and throughout, 'locally' refers to the fact that the optimal design depends on $\boldsymbol{\theta}$, so that a design that is optimal for some value of $\boldsymbol{\theta}$, say $\boldsymbol{\theta}_0$, is likely to also be optimal for $\boldsymbol{\theta}$ in sufficiently close proximity to $\boldsymbol{\theta}_0$, but may be suboptimal for less proximate values of $\boldsymbol{\theta}$.

2.2.2.2 Prediction assuming known covariance parameters

Let $Z(\mathbf{s}_0)$ denote the realized but unobserved value of Z at an arbitrary point $\mathbf{s}_0 \in A$. If the covariance parameter vector $\boldsymbol{\theta}$ is known, the best (in the sense of minimizing the prediction error variance) linear unbiased predictor (BLUP) of $Z(\mathbf{s}_0)$, based on observations taken at points of the design $D = \{\mathbf{s}_1, \ldots, \mathbf{s}_n\}$, is given by

$$p_1(\mathbf{z}, \mathbf{s}_0, \boldsymbol{\theta}, D) = [\mathbf{c} + \mathbf{X}(\mathbf{X}'\boldsymbol{\Sigma}^{-1}\mathbf{X})^{-1}(\mathbf{x}_0 - \mathbf{X}'\boldsymbol{\Sigma}^{-1}\mathbf{c})]'\boldsymbol{\Sigma}^{-1}\mathbf{z}.$$

Here, $\mathbf{c}$ is the vector whose ith element is $C(\mathbf{s}_i, \mathbf{s}_0; \boldsymbol{\theta})$, and $\mathbf{x}_0 = \mathbf{x}(\mathbf{s}_0)$. The minimized prediction error variance associated with $p_1(\mathbf{z}, \mathbf{s}_0, \boldsymbol{\theta}, D)$ is given by

$$\begin{aligned} v_1(\mathbf{s}_0, \boldsymbol{\theta}, D) &= \text{var}_{\boldsymbol{\theta}}(p_1(\mathbf{z}, \mathbf{s}_0, \boldsymbol{\theta}, D) - Z(\mathbf{s}_0)) \\ &= C(\mathbf{s}_0, \mathbf{s}_0; \boldsymbol{\theta}) - \mathbf{c}'\boldsymbol{\Sigma}^{-1}\mathbf{c} + \\ &\quad (\mathbf{x}_0 - \mathbf{X}'\boldsymbol{\Sigma}^{-1}\mathbf{c})'(\mathbf{X}'\boldsymbol{\Sigma}^{-1}\mathbf{X})^{-1}(\mathbf{x}_0 - \mathbf{X}'\boldsymbol{\Sigma}^{-1}\mathbf{c}). \end{aligned} \tag{2.2}$$

Observe that $v_1(\mathbf{s}_0, \boldsymbol{\theta}, D)$ depends on neither the mean function parameters $\boldsymbol{\beta}$ nor the observed data $\mathbf{z}$; it does, however, depend on the prediction site $\mathbf{s}_0$, the covariance parameter $\boldsymbol{\theta}$, and the design D. Thus, for a given prediction site and covariance parameter, the prediction performance of the design can be optimized by minimizing this quantity with respect to D. It is a well known and completely unsurprising fact that if $\mathbf{s}_0 \in \mathcal{S}$, then the unique design of size $n = 1$

that minimizes $v_1(\mathbf{s}_0, \boldsymbol{\theta}, D)$ is simply $D = \{\mathbf{s}_0\}$ (and that the minimized value of $v_1(\mathbf{s}_0, \boldsymbol{\theta}, D)$ is zero). More generally, if $\mathbf{s}_{0j} \in \mathcal{S}$ for $j = 1, \ldots, m$, then the unique design of size m that minimizes $m^{-1} \sum_{j=1}^{m} v_1(\mathbf{s}_{0j}, \boldsymbol{\theta}, D)$ is $D = \{\mathbf{s}_{01}, \ldots, \mathbf{s}_{0m}\}$; this design also minimizes $\max_j v_1(\mathbf{s}_{0j}, \boldsymbol{\theta}, D)$.

Usually, however, it is much more relevant to summarize the suitability of a design for prediction over points whose number exceeds the predetermined sample size n–perhaps even over all points in A. Let $\mathcal{P} \subset A$ denote the set of points where BLUPs are desired. A design is said to be locally APEV-optimal if it minimizes the average prediction error variance

$$\gamma_{APEV}(D; \boldsymbol{\theta}) = \begin{cases} \frac{1}{|\mathcal{P}|} \int_{P} v_1(\mathbf{s}, \boldsymbol{\theta}, D)\, ds & \text{if } \mathcal{P} \text{ is a region} \\ \frac{1}{|\mathcal{P}|} \sum_{\mathbf{s} \in \mathcal{P}} v_1(\mathbf{s}, \boldsymbol{\theta}, D) & \text{if } \mathcal{P} \text{ is a finite point set,} \end{cases}$$

and a design is said to be locally MPEV-optimal if it minimizes the maximum prediction error variance,

$$\gamma_{MPEV}(D; \boldsymbol{\theta}) = \max_{\mathbf{s} \in \mathcal{P}} v_1(\mathbf{s}, \boldsymbol{\theta}, D).$$

Here $|\mathcal{P}|$ is either the volume of $\mathcal{P}$ (if $\mathcal{P} = A$ or $\mathcal{P}$ is a subregion of A) or the number of points in $\mathcal{P}$ (if $\mathcal{P}$ is finite).

2.2.2.3 Estimation of covariance parameters

Now suppose that the covariance function is known only up to the parameter vector $\boldsymbol{\theta}$. Either because we are interested in $\boldsymbol{\theta}$ for its own sake, or because mean parameter estimation or spatial prediction requires us to estimate $\boldsymbol{\theta}$, we may wish to determine a design that is optimal, in some sense, for estimating $\boldsymbol{\theta}$. Of course, such a design may depend on the particular estimator, $\hat{\boldsymbol{\theta}} = \hat{\boldsymbol{\theta}}(\mathbf{z}, D)$, and on the particular measure of that estimator's quality that are to be used, among other quantities. One reasonable approach is to maximize the determinant of the information matrix associated with either the maximum likelihood (ML) estimator or residual maximum likelihood (REML) estimator of $\boldsymbol{\theta}$; here we consider the REML estimator only. The rationale for such a criterion is that the inverse of the information matrix is, under regularity conditions, the asymptotic covariance matrix of the corresponding estimator of $\boldsymbol{\theta}$; thus for sufficiently large samples the determinant of the inverse information matrix should be a reasonable approximation to the determinant of the intractable mean squared error matrix

$$\mathbf{M}(\boldsymbol{\theta}, D) = E_{\boldsymbol{\theta}}[(\hat{\boldsymbol{\theta}} - \boldsymbol{\theta})(\hat{\boldsymbol{\theta}} - \boldsymbol{\theta})'].$$

Let $\mathbf{I}(\boldsymbol{\theta}, D)$ represent the information matrix associated with the REML estimator of $\boldsymbol{\theta}$, based on design D. A design that minimizes

$$\gamma_{CPE}(D; \boldsymbol{\theta}) = 1/|\mathbf{I}(\boldsymbol{\theta}, D)|$$

with respect to D is said to be a locally CPE-optimal design (where CPE stands for covariance parameter estimation).

Tractable expressions for the elements of $\mathbf{I}(\boldsymbol{\theta}, D)$ are available when the random field is Gaussian. For this case, the ijth element of $\mathbf{I}(\boldsymbol{\theta}, D)$ is given by

$$\frac{1}{2}\mathrm{tr}\left(\mathbf{P}\frac{\partial \boldsymbol{\Sigma}}{\partial \theta_i}\mathbf{P}\frac{\partial \boldsymbol{\Sigma}}{\partial \theta_j}\right),$$

where $\mathbf{P} = \boldsymbol{\Sigma}^{-1} - \boldsymbol{\Sigma}^{-1}\mathbf{X}(\mathbf{X}'\boldsymbol{\Sigma}^{-1}\mathbf{X})^{-1}\mathbf{X}'\boldsymbol{\Sigma}^{-1}$.

2.2.2.4 Prediction using estimated covariance parameters

The design criterion for prediction discussed previously does not fully address the prediction problem of greatest practical interest (and unfortunately of much greater difficulty), which is to predict the unobserved value of $Z(\mathbf{s}_0)$ when the spatial dependence parameters are unknown. The standard predictor in this situation, known as the E-BLUP (empirical BLUP), is given by an expression identical to the BLUP when $\boldsymbol{\theta}$ is known but with the covariance function evaluated at $\hat{\boldsymbol{\theta}} = \hat{\boldsymbol{\theta}}(\mathbf{z})$, an estimate of $\boldsymbol{\theta}$, rather than at the hitherto-assumed-known $\boldsymbol{\theta}$. That is, with $p_1(\mathbf{z}, \mathbf{s}_0, \boldsymbol{\theta}, D)$ defined, as before, as the known-covariance-parameter BLUP at $\mathbf{s}_0$ for design D, then the E-BLUP at $\mathbf{s}_0$ is given by $p_2(\mathbf{z}, \mathbf{s}_0, D) = p_1(\mathbf{z}, \mathbf{s}_0, \hat{\boldsymbol{\theta}}, D)$.

A natural design criterion in this situation would be the mean squared prediction error of the E-BLUP,

$$m_2(\mathbf{s}_0, \boldsymbol{\theta}, D) \equiv E_{\boldsymbol{\theta}}\left[(p_2(\mathbf{z}, \mathbf{s}_0, D) - Z(\mathbf{s}_0))^2\right].$$

However, this is usually not practicable as it stands, for an exact general expression for the E-BLUP's mean squared prediction error is unknown and simulation-based approaches to approximate it at more than a few sites are computationally prohibitive at present. Some aspects of the E-BLUP's sampling distribution *are* known; in particular, the E-BLUP is known to be unbiased, and thus its mean squared prediction error and its prediction error variance coincide, under quite weak conditions. For this reason, as well as for simplicity, in practice $m_2(\mathbf{s}_0, \boldsymbol{\theta}, D)$ is usually estimated by substituting $\hat{\boldsymbol{\theta}}$ for $\boldsymbol{\theta}$ in (2.2), the expression for the BLUP's prediction error variance. Harville and Jeske (1992) and Zimmerman and Cressie (1992) considered the performance of this so-called 'plug-in' estimator and also proposed the following approximation:

$$m_2(\mathbf{s}_0, \boldsymbol{\theta}, D) \doteq \widetilde{v}_2(\mathbf{s}_0, \boldsymbol{\theta}, D) \equiv v_1(\mathbf{s}_0, \boldsymbol{\theta}, D) + \mathrm{tr}\left[\mathbf{A}(\mathbf{s}_0, \boldsymbol{\theta}, D)\mathbf{H}(\boldsymbol{\theta}, D)\right] \quad (2.3)$$

where $\mathbf{A}(\mathbf{s}_0, \boldsymbol{\theta}, D) = \mathrm{var}_{\boldsymbol{\theta}}[\partial p_1(\mathbf{z}, \mathbf{s}_0, \boldsymbol{\theta}, D)/\partial\boldsymbol{\theta}]$, $\mathbf{H}(\boldsymbol{\theta}, D)$ is a matrix that either equals or approximates the MSE matrix $\mathbf{M}(\boldsymbol{\theta}, D)$ of the covariance parameters, and $\mathrm{tr}(\cdot)$ denotes the trace of a square matrix. If $\boldsymbol{\theta}$ is estimated by REML, then a natural choice for $\mathbf{H}(\boldsymbol{\theta}, D)$ is $\mathbf{I}^{-1}(\boldsymbol{\theta}, D)$. Zimmerman and Cressie (1992) and Abt (1999) found $\widetilde{v}_2(\mathbf{s}_0, \boldsymbol{\theta}, D)$ to be much closer than $v_1(\mathbf{s}_0, \boldsymbol{\theta}, D)$ to $m_2(\mathbf{s}_0, \boldsymbol{\theta}, D)$ in many, though not all, situations they considered.

Accordingly, as criteria for optimal design for spatial prediction with estimated covariance parameters, we consider either the average or maximum value of the E-BLUP's approximate prediction error variance over all sites in $\mathcal{P}$,

$$\gamma_{AEPEV}(D;\boldsymbol{\theta}) = \frac{1}{|\mathcal{P}|}\int_{P}\{v_1(\mathbf{s},\boldsymbol{\theta},D) + \mathrm{tr}[\mathbf{A}(\mathbf{s},\boldsymbol{\theta},D)\mathbf{I}^{-1}(\boldsymbol{\theta},D)]\,ds\}$$

$$\text{or}\quad \frac{1}{|\mathcal{P}|}\sum_{\mathbf{s}\in\mathcal{P}}\{v_1(\mathbf{s},\boldsymbol{\theta},D) + \mathrm{tr}[\mathbf{A}(\mathbf{s},\boldsymbol{\theta},D)\mathbf{I}^{-1}(\boldsymbol{\theta},D)]\},$$

$$\gamma_{MEPEV}(D;\boldsymbol{\theta}) = \max_{\mathbf{s}\in\mathcal{P}}\{v_1(\mathbf{s},\boldsymbol{\theta},D) + \mathrm{tr}[\mathbf{A}(\mathbf{s},\boldsymbol{\theta},D)\mathbf{I}^{-1}(\boldsymbol{\theta},D)]\}.$$

Designs that minimize these criteria for a given $\boldsymbol{\theta}$ are called locally AEPEV-optimal and locally MEPEV-optimal designs (where AEPEV and MEPEV stand for average and maximum empirical prediction error variance). Observe that these criteria combine a measure of the quality of a design for prediction using known covariance parameters, i.e. $v_1(\mathbf{s},\boldsymbol{\theta},D)$, with a measure of the quality of a design for covariance parameter estimation, $\mathrm{tr}[\mathbf{A}(\mathbf{s},\boldsymbol{\theta},D)\mathbf{I}^{-1}(\boldsymbol{\theta},D)]$. It may therefore be expected that locally AEPEV/MEPEV-optimal designs may lie somewhere between the 'extremes' of locally APEV/MPEV-optimal and CPE-optimal designs, having some features of both.

2.2.3 Algorithms

In many real situations some data have already been collected without formal consideration of design criteria, and the spatial design problem is to select additional sites where observations should be taken (network expansion) or to select which existing sites to delete (network contraction) for the future. An example of the latter, using the Austrian climate data, is presented in Section 2.4. For relatively small expansion and contraction problems, as well as very small 'build-from-scratch' problems, it may be feasible to find the optimal design by enumerating all possible designs. In problems too large to be solved by complete enumeration, any of several algorithms, including exchange algorithms and simulated annealing , may be useful; see Ko *et al.* (1995), van Groenigen and Stein (1998), and Lark (2002) for details.

2.2.4 Toy example

Now we provide an example, with the goal of describing noteworthy features of, and differences between, optimal or near-optimal designs corresponding to the criteria just presented. In this example, $\mathcal{S}$ is a 5×5 square grid of points with unit spacing and $n = 5$; these are sufficiently small that the optimal design(s) with respect to a given criterion can be determined by complete enumeration of all possible designs. There are 53130 five-point subsets of $\mathcal{S}$, but due to the symmetry of $\mathcal{S}$ with respect to rotations of 90°, 180°, and 270° and with respect to reflections, many of these are equivalent to each other. In fact, the designs can be partitioned into equivalence classes, and in the figures we shall always display just one member of a class. It is assumed initially that the observations corresponding to any design arise from a second-order stationary process on A with unknown constant mean and an isotropic nuggetless exponential covariance function given by $C(\|\mathbf{s}-\mathbf{u}\|;\sigma^2,\rho)=\sigma^2\rho^{\|\mathbf{s}-\mathbf{u}\|}$; thus ρ represents the correlation between variables at adjacent points in the grid. For designs for which prediction is the objective, we take $\mathcal{P}=\mathcal{S}$.

Figure 2.1 displays locally optimal designs, when $\sigma^2=1$ and $\rho\in\{0.25, 0.50, 0.75\}$, corresponding to the criteria $\gamma_{MPE}(D;\boldsymbol{\theta})$ for mean parameter estimation, $\gamma_{MPEV}(D;\boldsymbol{\theta})$ for prediction with known covariance parameters, $\gamma_{CPE}(D;\boldsymbol{\theta})$ for covariance parameter estimation, and $\gamma_{MEPEV}(D;\boldsymbol{\theta})$ for prediction with estimated covariance parameters. The locally optimal designs for mean parameter estimation are quite similar across all correlation strengths. These designs might be characterized as 'regular,' i.e. evenly and widely dispersed, which renders observations as uncorrelated as possible. The locally optimal designs for prediction likewise are regular, but accomplish something that the designs for mean parameter estimation do not: all but four prediction locations are within one grid spacing unit of an observation, and the four exceptional locations are only $\sqrt{2}$ units from an observation. This is a reasonable feature, as proximity of a point to at least one observation is a crucial determinant of the magnitude of the prediction error variance at that point. In stark contrast, the locally optimal designs for covariance parameter estimation are distinctly nonregular, occurring instead in two widely separated clusters. Thus, it appears that the presence of some small lags in the design is beneficial for covariance parameter estimation but not for mean parameter estimation or prediction. Finally, the locally optimal designs for empirical prediction exhibit considerable variety and in some cases defy simple characterizations. None of them are accurately characterized as regular, nor do they consist of just two widely separated clusters. However, locally optimal empirical prediction designs corresponding to additional values of ρ (not shown) suggest that the magnitude of ρ affects the clustering and dispersion of the design points: when ρ is small the optimal design is more clustered than dispersed, and as ρ increases the optimal design is more dispersed than clustered. Regardless of the magnitude of ρ, the designs exhibit at least a few very small intersite distances together with good overall spatial coverage.

Table 2.1 provides relative efficiency gains for each of the locally optimal designs in Figure 2.1, computed in two ways: (1) as 1.0 minus the ratio of the minimized value of the criterion to its average value over all five-point designs, times 100%; (2) as 1.0 minus the ratio of the minimized value of the criterion to its value for the regular five-point design displayed in the upper left panel of Figure 2.1. The gains in efficiency relative to the average are sufficiently large to suggest that a design that optimizes a criterion is generally preferable to a randomly or arbitrarily chosen design. Gains relative to the regular design are inconsequential when the objective is estimation of a (constant) mean or prediction with known covariance function, but can be huge when the objective is covariance parameter estimation or empirical prediction, especially if the spatial correlation is weak.

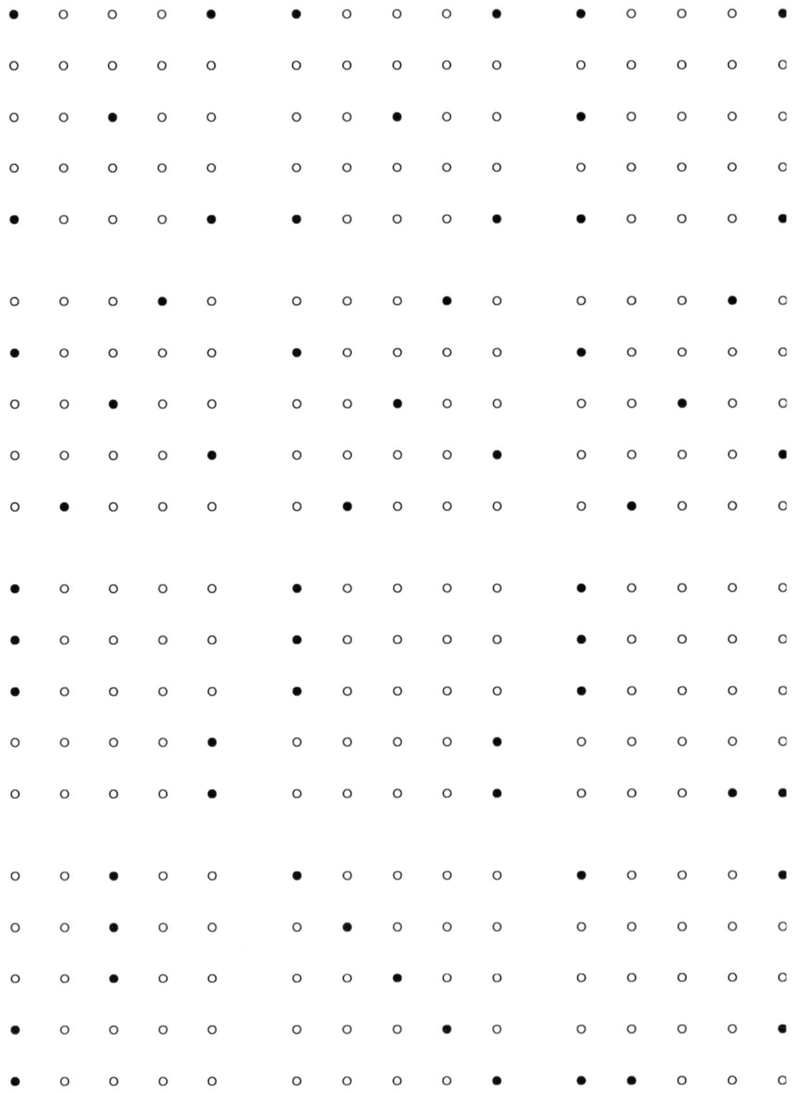

Figure 2.1 *Locally optimal designs for univariate mean estimation (first row of designs), prediction with known covariance function (second row), covariance function estimation (third row), and empirical prediction (fourth row). Columns correspond to values of the correlation strength parameter,* ρ *(*$\rho = 0.25, 0.50, 0.75$ *for columns 1, 2, and 3, respectively). Candidate design points within each* 5×5 *grid are represented by open circles, and points actually included in the design are represented by closed circles.*

Table 2.1 Minimized values of four univariate design criteria, for spatial processes with constant mean, constant variance, and exponential correlation function.

Criterion	ρ	Minimum	Efficiency gain (%) relative to	
			Random design	Regular design
Mean estimation	0.25	0.2077	17	0.0
	0.50	0.2652	26	0.0
	0.75	0.4545	20	0.5
Prediction	0.25	1.1032	10	2.1
	0.50	0.8504	26	4.1
	0.75	0.4524	43	4.4
Covariance estimation	0.25	0.2427	60	98.2
	0.50	0.2215	49	82.3
	0.75	0.2053	44	39.2
Empirical prediction	0.25	1.3884	35	92.6
	0.50	1.1696	25	41.1
	0.75	0.5964	51	0.4

Strength of spatial correlation is indexed by ρ.

Results of larger, more realistic examples are qualitatively similar to the results presented here; see, for example, Zimmerman and Homer (1991), Müller and Zimmerman (1999), Zhu and Stein (2005, 2006) and Zimmerman (2006). It is worth noting, however, that optimal and near-optimal designs can be quite different than those presented here if the mean is not spatially constant (Zimmerman 2006), or if the nugget effect of the covariance function is nonzero (Zhu and Stein 2005, 2006; Zimmerman 2006).

The overall conclusion that can be made from the toy example presented here and larger, more realistic examples presented in the literature just cited is that the design objectives of prediction with known covariance parameters, and covariance parameter estimation, are largely antithetical: good designs for the former objective have good coverage of the study area and are quite regular, while those for the latter objective consist of multiple, regularly dispersed clusters. Good designs for empirical prediction combine features of both, having good overall coverage but including a few small clusters.

2.3 Design for multivariate geostatistics

2.3.1 Data-model framework

Consider a data-model framework for multivariate geostatistics that is a natural extension of the univariate framework described in Section 2.2.1. Suppose that there are m spatially varying quantities to be observed, and the kth of these is to be observed at n_k point sites $\{\mathbf{s}_1^k, \ldots, \mathbf{s}_{n_k}^k\}$; the overall design is then denoted by $D = \{\mathbf{s}_1^k, \ldots, \mathbf{s}_{n_k}^k : k = 1, \ldots, m\}$. Thus the sites at which Z_k is to be observed need not coincide, in number or location, with the sites at which Z_j is to be observed; if such coincidence is imposed, however, the design is said to be *collocated*. Let $\mathbf{z}_k \equiv [Z_k(\mathbf{s}_1^k), \ldots, Z_k(\mathbf{s}_{n_k}^k)]'$ denote the data vector for the kth variable, and put $\mathbf{z} = (\mathbf{z}_1', \ldots, \mathbf{z}_m')'$. Also, assume that $E[Z_k(\mathbf{s})] = [\mathbf{x}_k(\mathbf{s})]'\boldsymbol{\beta}_k$, where $\mathbf{x}_k(\mathbf{s})$ is a vector of observed explanatory variables and $\boldsymbol{\beta}_k$ is a p_k-dimensional vector of unknown parameters. Denote the covariance function of the kth variable by $\text{cov}[Z_k(\mathbf{s}), Z_k(\mathbf{u})] = C_{kk}(\mathbf{s}, \mathbf{u}; \boldsymbol{\theta}_{kk})$, and the cross-covariance function of the kth and jth variables by $\text{cov}[Z_k(\mathbf{s}), Z_j(\mathbf{u})] = C_{kj}(\mathbf{s}, \mathbf{u}; \boldsymbol{\theta}_{kj})$. Finally, put $\boldsymbol{\beta} = (\boldsymbol{\beta}_1', \ldots, \boldsymbol{\beta}_m')'$ and $\boldsymbol{\theta} = (\boldsymbol{\theta}_{11}', \boldsymbol{\theta}_{12}', \ldots, \boldsymbol{\theta}_{1m}', \boldsymbol{\theta}_{21}', \ldots, \boldsymbol{\theta}_{kk}')'$.

Some additional notation is required in what follows. Let

$$\mathbf{X} = \begin{pmatrix} \mathbf{X}_1 & \mathbf{0} & \ldots & \mathbf{0} \\ \mathbf{0} & \mathbf{X}_2 & \ldots & \mathbf{0} \\ \vdots & \vdots & \ddots & \vdots \\ \mathbf{0} & \mathbf{0} & \ldots & \mathbf{X}_m \end{pmatrix}$$

and

$$\boldsymbol{\Sigma} = \begin{pmatrix} \boldsymbol{\Sigma}_{11} & \boldsymbol{\Sigma}_{12} & \ldots & \boldsymbol{\Sigma}_{1m} \\ \boldsymbol{\Sigma}_{21} & \boldsymbol{\Sigma}_{22} & \ldots & \boldsymbol{\Sigma}_{2m} \\ \vdots & \vdots & & \vdots \\ \boldsymbol{\Sigma}_{m1} & \boldsymbol{\Sigma}_{m2} & \ldots & \boldsymbol{\Sigma}_{mm} \end{pmatrix},$$

where $\mathbf{X}_k$ is the $n_k \times p_k$ matrix whose lth row is $[\mathbf{x}_k(\mathbf{s}_l^k)]'$ $(l = 1, \ldots, n_k;\ k = 1, \ldots, m)$ and $\boldsymbol{\Sigma}_{kj}$ is the matrix whose ilth element is $C_{kj}(\mathbf{s}_i, \mathbf{s}_l; \boldsymbol{\theta}_{kj})$. For prediction of the vector of variables at a site $\mathbf{s}_0$, i.e. $\mathbf{Z}(\mathbf{s}_0)$, let $\boldsymbol{\Sigma}_0$ be the matrix with kjth element $C_{kj}(\mathbf{s}_0, \mathbf{s}_0; \boldsymbol{\theta})$, and define

$$\mathbf{X}_0 = \begin{pmatrix} \mathbf{x}_1(\mathbf{s}_0) & \mathbf{0} & \cdots & \mathbf{0} \\ \mathbf{0} & \mathbf{x}_2(\mathbf{s}_0) & \cdots & \mathbf{0} \\ \vdots & \vdots & & \vdots \\ \mathbf{0} & \mathbf{0} & \cdots & \mathbf{x}_m(\mathbf{s}_0) \end{pmatrix}$$

and

$$\mathbf{C} = \begin{pmatrix} C_{11}(\mathbf{s}_1^1, \mathbf{s}_0) & C_{12}(\mathbf{s}_1^1, \mathbf{s}_0) & \cdots & C_{1m}(\mathbf{s}_1^1, \mathbf{s}_0) \\ \vdots & \vdots & & \vdots \\ C_{11}(\mathbf{s}_{n_1}^1, \mathbf{s}_0) & C_{12}(\mathbf{s}_{n_1}^1, \mathbf{s}_0) & \cdots & C_{1m}(\mathbf{s}_{n_1}^1, \mathbf{s}_0) \\ C_{21}(\mathbf{s}_1^2, \mathbf{s}_0) & C_{22}(\mathbf{s}_1^2, \mathbf{s}_0) & \cdots & C_{2m}(\mathbf{s}_1^2, \mathbf{s}_0) \\ \vdots & \vdots & & \vdots \\ C_{21}(\mathbf{s}_{n_2}^2, \mathbf{s}_0) & C_{22}(\mathbf{s}_{n_2}^2, \mathbf{s}_0) & \cdots & C_{2m}(\mathbf{s}_{n_2}^2, \mathbf{s}_0) \\ \vdots & \vdots & & \vdots \\ C_{m1}(\mathbf{s}_1^m, \mathbf{s}_0) & C_{m2}(\mathbf{s}_1^m, \mathbf{s}_0) & \cdots & C_{mm}(\mathbf{s}_1^m, \mathbf{s}_0) \\ \vdots & \vdots & & \vdots \\ C_{m1}(\mathbf{s}_{n_m}^m, \mathbf{s}_0) & C_{m2}(\mathbf{s}_{n_m}^m, \mathbf{s}_0) & \cdots & C_{mm}(\mathbf{s}_{n_m}^m, \mathbf{s}_0) \end{pmatrix}.$$

2.3.2 Design criteria

Straightforward extensions of the previously defined design criteria for univariate geostatistics to the multivariate setting are as follows (here the notation is as defined in Section 2.3.1).

2.3.2.1 Estimation of mean parameters assuming known covariance and cross-covariance parameters

$$\gamma_{MPE}(D; \boldsymbol{\theta}) = 1/|\mathbf{X}'\boldsymbol{\Sigma}^{-1}\mathbf{X}|.$$

2.3.2.2 Prediction assuming known covariance and cross-covariance parameters

$$\gamma_{AEPEV}(D; \boldsymbol{\theta}) = \begin{cases} \frac{1}{|\mathcal{P}|} \int_P |\mathbf{V}_1(\mathbf{s}, \boldsymbol{\theta}, D)|\, ds & \text{if } \mathcal{P} \text{ is a region} \\ \frac{1}{|\mathcal{P}|} \sum_{\mathbf{s} \in \mathcal{P}} |\mathbf{V}_1(\mathbf{s}, \boldsymbol{\theta}, D)| & \text{if } \mathcal{P} \text{ is a finite point set,} \end{cases}$$

$$\gamma_{MEPEV}(D; \boldsymbol{\theta}) = \max_{\mathbf{s} \in \mathcal{P}} |\mathbf{V}_1(\mathbf{s}, \boldsymbol{\theta}, D)|,$$

where $\mathbf{V}_1(\mathbf{s}, \boldsymbol{\theta}, D) = \boldsymbol{\Sigma}_0 - \mathbf{C}'\boldsymbol{\Sigma}^{-1}\mathbf{C} + (\mathbf{X}_0 - \mathbf{X}'\boldsymbol{\Sigma}^{-1}\mathbf{C})'(\mathbf{X}'\boldsymbol{\Sigma}^{-1}\mathbf{X})^{-1}(\mathbf{X}_0 - \mathbf{X}'\boldsymbol{\Sigma}^{-1}\mathbf{C})$. Note that $\mathbf{V}_1(\mathbf{s}, \boldsymbol{\theta}, D)$ is the covariance matrix of prediction errors associated with the vector $\mathbf{p}_1(\mathbf{z}, \mathbf{s}, \boldsymbol{\theta}, D) = [\mathbf{C} + \mathbf{X}(\mathbf{X}'\boldsymbol{\Sigma}^{-1}\mathbf{X})^{-1}(\mathbf{X}_0 - \mathbf{X}'\boldsymbol{\Sigma}^{-1}\mathbf{C})]'\boldsymbol{\Sigma}^{-1}\mathbf{z}$ of BLUPs of $Z(\mathbf{s}_0)$.

2.3.2.3 Estimation of covariance and cross-covariance parameters

$$\gamma_{CPE}(D; \boldsymbol{\theta}) = 1/|\mathbf{I}(\boldsymbol{\theta}, D)|.$$

2.3.2.4 Prediction using estimated covariance and cross-covariance parameters

$$\gamma_{AEPEV}(D; \boldsymbol{\theta}) = \begin{cases} \frac{1}{|\mathcal{P}|} \int_P |\widetilde{\mathbf{V}}_2(\mathbf{s}, \boldsymbol{\theta}, D)|\, ds & \text{if } \mathcal{P} \text{ is a region} \\ \frac{1}{|\mathcal{P}|} \sum_{\mathbf{s} \in \mathcal{P}} |\widetilde{\mathbf{V}}_2(\mathbf{s}, \boldsymbol{\theta}, D)| & \text{if } \mathcal{P} \text{ is a finite point set,} \end{cases}$$

$$\gamma_{MEPEV}(D; \boldsymbol{\theta}) = \max_{\mathbf{s} \in \mathcal{P}} |\widetilde{\mathbf{V}}_2(\mathbf{s}, \boldsymbol{\theta}, D)|,$$

where

$$\widetilde{\mathbf{V}}_2(\mathbf{s}, \boldsymbol{\theta}, D) = \mathbf{V}_1(\mathbf{s}, \boldsymbol{\theta}, D) + \begin{pmatrix} \mathrm{tr}[\mathbf{A}_{11}(\mathbf{s}, \boldsymbol{\theta}, D)\mathbf{I}^{-1}(\boldsymbol{\theta}, D)] & \cdots & \mathrm{tr}[\mathbf{A}_{1m}(\mathbf{s}, \boldsymbol{\theta}, D)\mathbf{I}^{-1}(\boldsymbol{\theta}, D)] \\ \vdots & & \vdots \\ \mathrm{tr}[\mathbf{A}_{m1}(\mathbf{s}, \boldsymbol{\theta}, D)\mathbf{I}^{-1}(\boldsymbol{\theta}, D)] & \cdots & \mathrm{tr}[\mathbf{A}_{mm}(\mathbf{s}, \boldsymbol{\theta}, D)\mathbf{I}^{-1}(\boldsymbol{\theta}, D)] \end{pmatrix},$$

$$\mathbf{A}_{kj}(\mathbf{s}, \boldsymbol{\theta}, D) = \mathrm{cov}\left(\frac{\partial p_{1k}(\mathbf{z}, \mathbf{s}, \boldsymbol{\theta}, D)}{\partial \boldsymbol{\theta}}, \frac{\partial p_{1j}(\mathbf{z}, \mathbf{s}, \boldsymbol{\theta}, D)}{\partial \boldsymbol{\theta}}\right),$$

and $p_{1k}(\mathbf{z}, \mathbf{s}, \boldsymbol{\theta}, D)$ is the kth element of $\mathbf{p}_1(\mathbf{z}, \mathbf{s}, \boldsymbol{\theta}, D)$. Observe that $\widetilde{\mathbf{V}}_2(\mathbf{s}, \boldsymbol{\theta}, D)$ is a multivariate extension of the approximate covariance matrix of E-BLUP prediction errors given by (2.3).

2.3.3 Toy example

To indicate the characteristics of multivariate designs obtained using the various criteria, we extend the toy example of Section 2.2.4. Consider the same spatial configuration of candidate design points, but now suppose there are $m = 2$ variables. It is assumed that the variables follow a second-order stationary bivariate process on A with spatially constant (but possibly different for the two variables) means, covariance functions $C_{11}(\|\mathbf{s} - \mathbf{u}\|; \sigma_1^2, \rho) = \sigma_1^2 \rho^{\|\mathbf{s}-\mathbf{u}\|}$ and $C_{22}(\|\mathbf{s} - \mathbf{u}\|; \sigma_2^2, \rho) = \sigma_2^2 \rho^{\|\mathbf{s}-\mathbf{u}\|}$, and cross-covariance function $C_{12}(\|\mathbf{s} - \mathbf{u}\|; \sigma_1^2, \sigma_2^2, \rho_c, \rho) = \sigma_1 \sigma_2 \rho_c \rho^{\|\mathbf{s}-\mathbf{u}\|}$ where $0 < \rho_c < 1$. This choice of cross-covariance function yields a spatially separable process, simplifying computation. Several values were considered for each parameter. Unfortunately, enumeration of all 10-point multivariate designs is computationally prohibitive, so instead we find optimal designs within two restricted subclasses that are small enough to be enumerated. The first such subclass is the set of collocated designs, and the second consists of the locally optimal five-point design for the first variable with respect to a given univariate objective, which was obtained by enumeration in Section 2.2.4, supplemented by the five-point design for the second variable (allowing for, but not imposing, collocation) that minimizes the corresponding multivariate criterion. We call optimal design within the second subclass the *supplementary* approach. Of course, neither approach necessarily yields the locally optimal 10-point bivariate design. However, they are not unrealistic: collocated designs are sometimes required for reasons of cost or practicality, and the supplementary approach occurs in practice when a sampling network for one of the two variables already exists and the design problem is to determine where to take observations on the second variable.

Locally optimal collocated and supplementary designs for three of the four objectives are displayed in Figure 2.2 for one set of parameter values: $\sigma_1 = \sigma_2 = 1$, $\rho = 0.50$, $\rho_c = 0.50$. The locally optimal collocated and supplementary bivariate designs for mean parameter estimation are completely collocated copies of the corresponding locally optimal univariate design, so we do not display them. The optimal collocated and supplementary designs for bivariate prediction with known covariance and cross-covariance parameters are not identical, but like the design for the analogous univariate problem, are widely dispersed. In fact, the design points of the optimal collocated design coincide with those of the analogous optimal univariate design. The optimal supplementary bivariate design exhibits little collocation; however, this is not universal and when ρ is small relative to ρ_c the optimal supplementary design is highly collocated. The optimal collocated

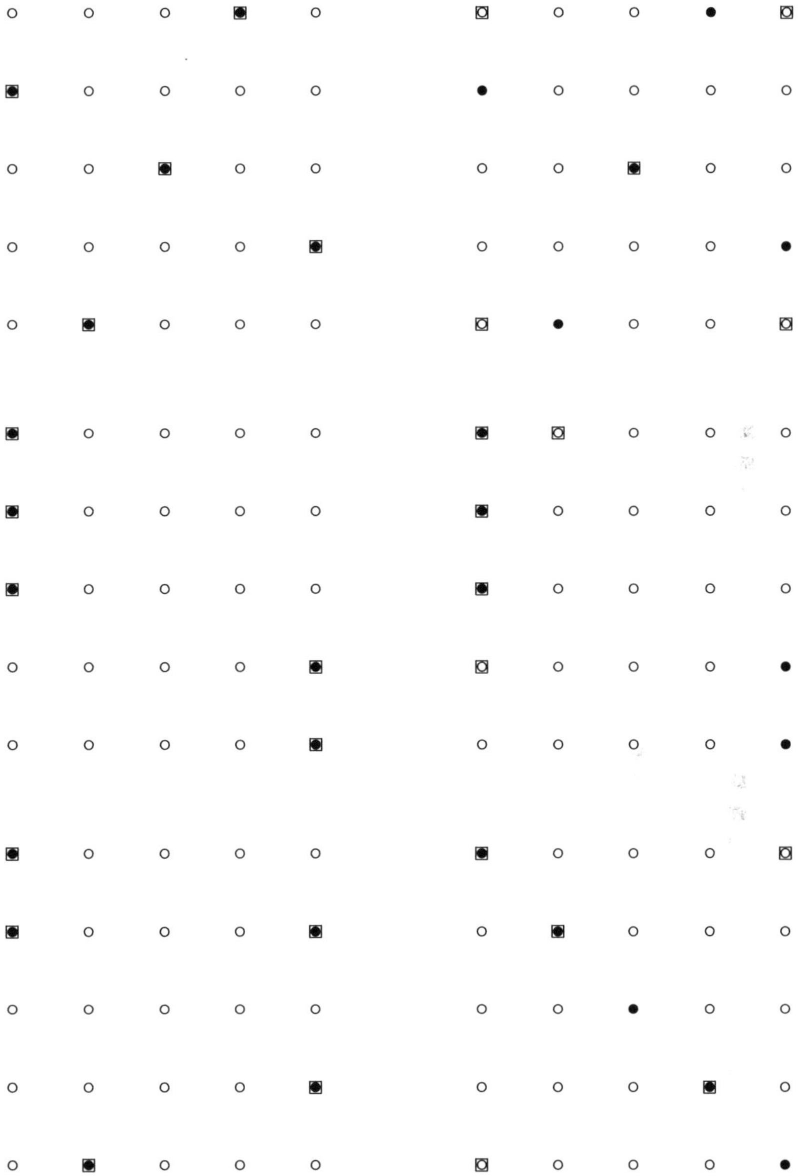

Figure 2.2 Locally optimal collocated (left column) and supplementary (right column) designs for multivariate prediction with known covariance function (top panel), covariance and cross-covariance function estimation (middle panel), and empirical multivariate prediction (bottom panel). Here $\sigma_1 = \sigma_2 = 1.0$, $\rho = 0.50$, and $\rho_c = 0.50$. Candidate design points within each 5×5 grid are represented by open circles, points included in the design for the first variable are represented by closed circles, and points included in the design for the second variable are represented by open squares.

design for estimation of covariance and cross-covariance parameters coincides with the analogous optimal univariate design. Its supplementary counterpart exhibits even stronger clustering than the analog univariate design, with extensive collocation. Finally, the optimal collocated and supplementary designs for bivariate empirical prediction are as enigmatic as their univariate analogs, exhibiting some clustering, some dispersion, and (in the case of the supplementary design) some, but not complete, collocation. However, in contrast to optimal designs with respect to the other criteria, the optimal collocated design with respect to this criterion does not coincide with the analogous optimal univariate design.

2.4 Application: Austrian precipitation data network

As a final example, we consider a hypothetical yet more realistic problem involving the optimal contraction of the upper Austrian precipitation data network. In particular, we consider the network as it existed in 1998, and we suppose that it was necessary (perhaps as a result of budget cuts) to eliminate two sites from the network for the next year. The variable of interest is the annual rainfall total (in cm) in 1998, which was measured at 35 sites. A Gaussian spatial process model with constant mean and exponential covariance function was fitted to these data by the method of REML. REML estimates of

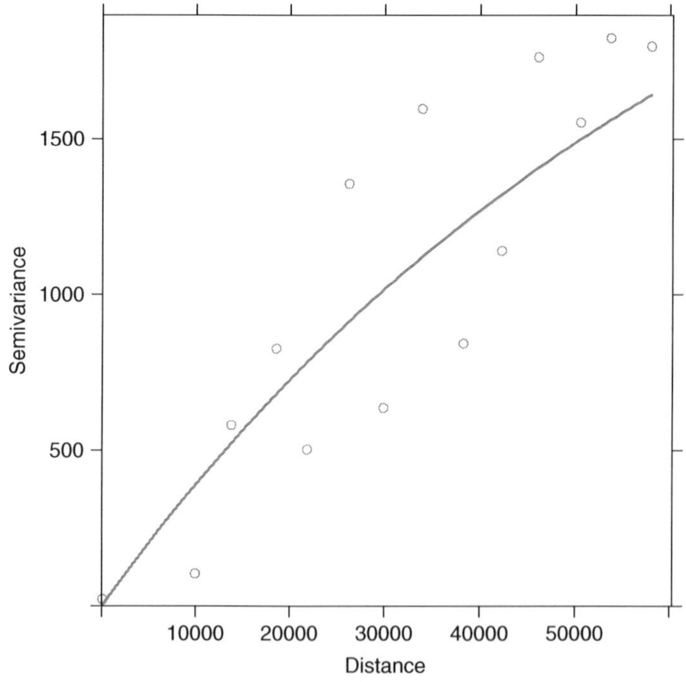

Figure 2.3 Sample semivariogram (open circles) and REML estimated exponential semivariogram (smooth curve) for the 1998 Austrian precipitation data.

the sill and range parameters were 2873 cm^2 and 68751, respectively. Figure 2.3 is a plot of the sample semivariogram and REML estimated exponential semivariogram. The corresponding estimated covariance function for 1998 was then regarded as the true covariance function for 1999, and the 33-point design (chosen by deleting two sites from the existing 35-point network) that resulted in the smallest increase in the prediction criterion $\gamma_{MPEV}(D; \boldsymbol{\theta})$ was determined by complete enumeration. This was repeated using the covariance parameter estimation criterion $\gamma_{CPE}(D; \boldsymbol{\theta})$ and the empirical prediction criterion $\gamma_{MEPEV}(D; \boldsymbol{\theta})$. For the two prediction criteria, the set of prediction sites $\mathcal{P}$ was taken to be a 100 × 100 grid of points overlaid upon the study area.

Locally optimal designs corresponding to each criterion are displayed in Figure 2.4. It is difficult to extract general conclusions from these results. One

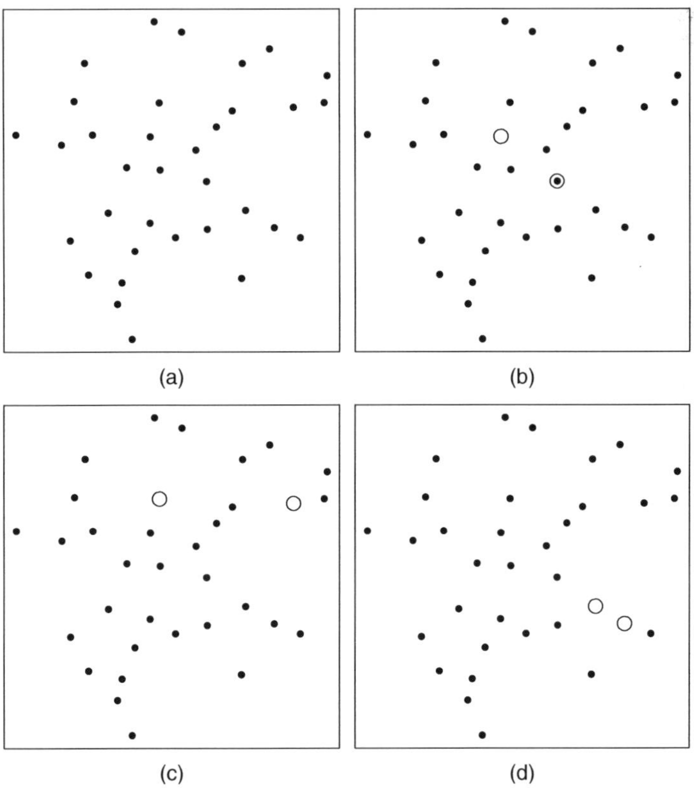

Figure 2.4 Locally optimal contracted designs (by two sites) for the Austrian rainfall network. Closed circles indicate existing observation sites; open circles indicate sites to be removed. (a) is the original network; remaining plots display the designs that minimize the increase in the prediction criterion $\gamma_{MPEV}(D; \boldsymbol{\theta})$ (b), the covariance parameter estimation criterion $\gamma_{CPE}(D; \boldsymbol{\theta})$ (c), and the empirical prediction criterion $\gamma_{MEPEV}(D; \boldsymbol{\theta})$ (d).

unsurprising feature pertains to two sites, Kremsmuenster 1 and Kremsmuenster 2, that are nearly coincident–so much so that they are indistinguishable in the figure. The locally optimal design for prediction with known covariance parameters removes one of these sites, but the locally optimal designs for the other two criteria retain both of them. Having both present yields a very small lag, which is very valuable for covariance parameter estimation and empirical prediction, especially since no other sites are separated by lags nearly as small. It is worth noting the extent by which each design criterion increases upon optimal contraction by two sites. This increase is less than one one-thousandth of a percent for the first prediction criterion, roughly 8% for the covariance parameter estimation criterion, and less than one one-hundreth of a percent for the empirical prediction criterion.

2.5 Conclusions

The statistical perspective taken in this chapter is one that is frequentist and (parametric) model-based. Although this perspective leads to reasonable criteria and sensible designs, by no means is it the only perspective that could be (or has been) taken to geostatistical design. One alternative design paradigm is geometry-based; no model is assumed and designs are chosen for having a regular geometry (e.g., a triangular or square lattice) or good spatial coverage (so-called space-filling designs) (see e.g., Yfantis *et al.* 1987, or Royle and Nychka 1998). This approach typically yields designs that are good for prediction with known covariance parameters but not so good for other objectives. Another paradigm, based on classical probabilistic sampling theory, adopts a model for the sampling process but not for the spatial field of attribute variable(s), which is regarded as fixed; a good summary of this approach (sometimes called the design-based approach) may be found in Dobbie *et al.* (2008). Even within the model-based approach, several perspectives and objective criteria are possible. For example, a Bayesian model-based approach is taken by Diggle and Lophaven (2006), yielding the same general conclusion as our frequentist approach: for the goal of efficient spatial prediction using estimated covariance parameters, good designs should include very small intersite distances as well as good regional coverage. As for alternative objectives, many authors have considered design to maximize the reduction in entropy, a measure of the uncertainty associated with the observations; see Le and Zidek (2006) for a comprehensive summary. Entropy represents one attempt to summarize all the uncertainty in a given situation into a single numerical objective criterion, and as such it extends naturally to multivariate problems. Another such attempt is to optimize weighted averages of several design criteria, which is known as compound optimal design . This last approach has been used recently to combine criteria for mean and covariance parameter estimation in univariate geostatistical design (Müller and Stehlik 2010). It has potential to also be used to combine criteria in multivariate problems, possibly differentially weighting the variables as well as the criteria.

References

Abt M 1999 Estimating the prediction mean squared error in Gaussian stochastic processes with exponential correlation structure. *Scandinavian Journal of Statistics* **26**, 563–578.

Diggle P and Lophaven S 2006 Bayesian geostatistical design. *Scandinavian Journal of Statistics* **33**, 53–64.

Dobbie MJ, Henderson BL and Stevens DL 2008 Sparse sampling: spatial design for monitoring stream networks. *Statistics Surveys* **2**, 113–153.

Harville DA and Jeske DR 1992 Mean squared error of estimation or prediction under a general linear model. *Journal of the American Statistical Association* **87**, 724–731.

Ko CW, Lee J and Queyranne M 1995 An exact algorithm for maximum entropy sampling. *Operations Research* **43**, 684–691.

Lark R 2002 Optimized spatial sampling of soil for estimation of the variogram by maximum likelihood. *Geoderma* **105**, 49–80.

Le NC and Zidek JV 2006 *Statistical Analysis of Environmental Space-Time Processes*. Springer.

Li J 2009 Spatial Multivariate Design in the Plane and on Stream Networks. PhD thesis, University of Iowa, Iowa City, IA.

Müller WG 2007 *Collecting Spatial Data: Optimum Design of Experiments for Random Fields*, 3rd edn. Springer.

Müller WG and Stehlik M 2010 Compound optimal spatial designs. *Environmetrics* **21**, 354–364.

Müller WG and Zimmerman DL 1999 Optimal designs for variogram estimation. *Environmetrics* **10**, 23–37.

Royle JA and Nychka D 1998 An algorithm for the construction of spatial coverage designs with implementation in Splus. *Computers and Geosciences* **24**, 2081–2096.

van Groenigen JW and Stein A 1998 Constrained optimization of spatial sampling using continuous simulated annealing. *Journal of Environmental Quality* **43**, 684–691.

Yfantis EA, Flatman GT and Behar JV 1987 Efficiency of kriging estimation for square, triangular, and hexagonal grids. *Mathematical Geology* **19**, 183–205.

Zhu Z and Stein ML 2005 Spatial sampling design for parameter estimation of the covariance function. *Journal of Statistical Planning and Inference* **134**, 583–603.

Zhu Z and Stein ML 2006 Spatial sampling design for prediction with estimated parameters. *Journal of Agricultural, Biological, and Environmental Statistics* **11**, 24–44.

Zidek JV and Zimmerman DL 2010 Monitoring network design. In *Handbook of Spatial Statistics* (eds. Gelfand AE, Diggle PJ, Fuentes M and Guttorp P), pp. 131–148. CRC Press.

Zimmerman, D. L. (2006). Optimal network design for spatial prediction, covariance parameter estimation, and empirical prediction. *Environmetrics* **17**, 635–652.

Zimmerman, D. L. and Cressie, N. (1992). Mean squared prediction error in the spatial linear model with estimated covariance parameters. *Annals of the Institute for Statistical Mathematics* **44**, 27–43.

Zimmerman, D. L. and Homer, K. E. (1991). A network design criterion for estimating selected attributes of the semivariogram. *Environmetrics* **2**, 425–441.

3

Model-based criteria heuristics for second-phase spatial sampling

Eric M. Delmelle
Geography and Earth Sciences, University of North Carolina at Charlotte, USA

3.1 Introduction

In sampling, available time and budget usually determines the number of samples that can be collected, and careful attention must be paid to design an appropriate network of observations. It is generally recognized that a sampling configuration which will minimize the variance associated with the estimation is more desirable. For one-dimensional problems ($\mathbb{R}$), Cochran (1963) has suggested that a *stratified random* design will always be more efficient than a *random* design to provide an unbiased estimate of the sampling variance. Cochran's initial work was later extended to two-dimensional sampling designs ($\mathbb{R}^2$) by Quenouille (1949) and Das (1950) who used a linear autocorrelation model. With nonlinear autocorrelation function however, *systematic* sampling is the most efficient technique, followed by *stratified random* sampling and *random* sampling (Zubrzycki

Spatio-temporal design: Advances in efficient data acquisition, First Edition.
Edited by Jorge Mateu and Werner G. Müller.

1958). Those results were later confirmed in a series of articles by Berry and Baker (1968), Ripley (1981) and Iachan (1985). In spatial sampling, measurements at specific areas must be acquired instead of trying to obtain information at every possible location (see, e.g., Cochran 1963; Dalton *et al.* 1975; Ripley 1981; Arbia 1989; Haining 1990; Hedayat and Sinha 1991; Cressie 1991; Stehman and Overton 1996; Müller 1998; Thompson 2002). Although a detailed inventory documents the variation of the variable of interest quite well, this process is rather time-consuming and constrained by the available budget. On the other hand, sparse sampling is less costly (and faster of course), but at the price that the spatial variation of the variable may not be captured properly (Berry and Baker 1968). This chapter is concerned with two-dimensional (and second-phase) spatial sampling designs to acquire (and complement, respectively) information on the spatial variation of a variable, for instance in the form of a map, or as a summary measure, which highlights the scales of variation.

In spatial sampling, the location of the samples is critical and may be influenced by the spatial structure of the variable: for phenomena with little variation, samples can be spaced more evenly without the risk of not detecting variations at smaller ranges. Unfortunately, this variation must be estimated, and an objective is to design optimal sampling patterns that will capture a maximum amount of information. If we undersample in some areas, the spatial variation of the variable of interest is not captured. In spatially autocorrelated fields, oversampling may result in redundant data (Griffith 2005).

Once samples of the primary variable have been collected, it is possible to augment the initial set by collecting additional measurements at other locations, a method known as second-phase sampling (Cressie 1991; Müller 1998; van Groenigen and Stein 1998; de Gruitjer *et al.* 2006). The inherent objective of a second-phase sampling strategy is to improve spatial prediction with added information. Following a first sampling phase, spatial covariance structure is summarized through a variogam function, and the kriging variance computed at each location. Generally, additional observations are gathered away from existing points, that is where the kriging variance is large. However, when the process under study is not stationary, sampling efforts should be directed in those strategic locations exhibiting strong spatial variation (Rogerson *et al.* 2004; Delmelle and Goovaerts 2009). Whether the function accounts primarily for the minimization of the kriging variance, or increases sampling efforts in areas of strong nonstationarity, the optimization problem is nonlinear, and calls for robust heuristic methods, for instance simulated annealing.

This chapter is structured as follows. First, geometric and geostatistical designs are presented in Sections 3.2 and 3.3. Secondly, strategies to support a second-phase sampling design are discussed. Thirdly, the framework to implement a simulated annealing procedure is presented in Section 3.4, applied to a second-phase sampling design. This arises due to the nonlinearity of the problem and that a complete enumeration of all possible solutions is not feasible in a timely manner. This is illustrated in Section 3.5 with an application to the Austrian dataset.

3.2 Geometric and geostatistical designs

There exists different sampling schemes for the purpose of two-dimensional sampling with no prior information available (these are generally termed random, systematic or stratified approach). Assuming that a limited number of samples m are allocated in a study area denoted $\mathfrak{D}$, the spatial variable Z is then measured on m supports, $\{z(\mathbf{s}_i)|i = 1, 2, \ldots m\}$. In a *simple random sampling* design, m sample points in $\mathfrak{D}$ are selected randomly (King 1969; Ripley 1981), and the selection of a sample set should not influence the selection of any other one. Practically, the coordinates of the sample the pair $\{x_i, y_i\}$ are randomly drawn on the interval [(minX, maxX),(minY, maxY)]. For nonsquare boundaries (or nonrectangular region), a point can still be drawn at random in the minimum bounding rectangle, followed by a `inside` algorithm to determine whether the point will fall within the study region. In those situations, there is still a risk to experience edge effects. In a *systematic sampling* design, the population of interest is divided into m intervals of equal size (the same applies for nonsquared areas). The first element is randomly or purposively chosen within the first interval $\frac{L}{\sqrt{m}}$ (as long as that interval is within the boundary of the study region), starting at the origin, while the location of the remaining $m - 1$ elements are aligned regularly by the size of the interval.

$$x_i = x_1 + (i-1)\frac{L}{\sqrt{m}} \quad y_i = y_1 + (j-1)\frac{L}{\sqrt{m}} \quad \forall i, j = 1, \ldots, \sqrt{m}. \tag{3.1}$$

The most common regular geometric configurations are the equilateral triangular grid, the rectangular (square) grid, and the hexagonal grid (Cressie 1991). The advantage of a systematic design lies in a good coverage of the observations. This design presents two inconveniences, however:

(i) the distribution of distances between points of $\mathfrak{D}$ is not represented fairly since many pairs of points are separated by the same distance,

(ii) there is a danger that the spatial process shows evidence of recurring periodicities that will remain uncaptured.

Samples from a systematic design may coincide in frequency with a regular pattern in the landscape (Overton and Stehman 1993; Griffith and Amrhein 1997). A *systematic random method* approach, which combines both systematic and random procedures (King 1969; Dalton *et al.* 1975) can prevent the latter. *Stratified random* partitions the population (or $\mathfrak{D}$) into nonoverlapping strata, and for each stratum, a specific set of samples is collected, for instance a greater number of samples may be collected in a geographic area due to greater population.

3.2.1 Efficiency of spatial sampling designs

Different criteria have been proposed to evaluate the merits of sampling designs (Müller 1998). An example is an estimator for the global mean that estimates the accuracy of the global mean estimates. Careful attention must be paid to the fact that some designs may be optimal for one criterion, yet not very efficient for other criteria. Another example is the minimization of the prediction error of the true surface (as discussed later). A design that leads to an accurate estimation of the global mean $z_{\mathfrak{D}}$:

$$z_{\mathfrak{D}} = \frac{1}{[\mathfrak{D}]} \int_{\mathfrak{D}} z(\mathbf{s}) ds \tag{3.2}$$

In this respect, it is desirable to select a configuration that minimizes the prediction error of $z_{\mathfrak{D}}$ for a given estimator, for instance the arithmetic mean:

$$\overline{z} = \frac{1}{m} \sum_{i=1}^{m} z(\mathbf{s}_i) \tag{3.3}$$

Efficiency is calculated for all possible realizations of the variable z by $var_{\xi}[z^*_{\mathfrak{D}} - z_{\mathfrak{D}}]$. This is generally implemented using σ_k^2, the geostatistical kriging variance σ^2, defined in the next section. In terms of the sampling variance, *stratified random sampling* is at least always equally or more accurate than *random sampling*; its relative efficiency is a monotone increasing function of sample size.

3.2.2 Sampling spatial variables in a geostatistical context

Spatial information from closeby measurements is generally not accounted for in classical sampling theory. In Geostatistics, the variable z is modeled as a random process that can take a series of outcome values, according to some probability distribution (Goovaerts 1997). Central to geostatistics is kriging, an interpolation method which predicts the value of z at unsampled locations (usually on a set G of grid points $\{\mathbf{s}_g | g = 1, 2 \ldots G\}$), keeping the kriging variance (also called prediction error) to a minimum. Kriging is based on a variogram, which summarizes the variance of values separated by a particular distance lag (h):

$$\widehat{\gamma}(h) = \frac{1}{2d(h)} \sum_{|\mathbf{s}_i - \mathbf{s}_j| = h} \left(z(\mathbf{s}_i) - z(\mathbf{s}_j) \right)^2, \tag{3.4}$$

where $d(h)$ is the number of pairs of points for a given lag value, and $z(\mathbf{s}_i)$ is the measured attribute value at location $\mathbf{s}_i$, a the nugget effect and σ^2 the sill, where $\widehat{\gamma}(h)$ levels out (Cressie 1991). Once the lag distance exceeds the range r, there is no spatial dependence between sample sites. The interpolated, kriged value at a location $\mathbf{s}$ in $\mathfrak{D}$ is a weighted mean of surrounding values; each value is weighted according to the variogram model:

$$\widehat{z}(\mathbf{s}) = \sum_{i=1}^{I} w_i(\mathbf{s}) z(\mathbf{s}_i), \tag{3.5}$$

with I the set of neighboring points used to estimate the interpolated value at location $\mathbf{s}$, and $w_i(\mathbf{s})$ is the weight associated with each neighboring point. It is critical to have a wide range of distances in order to estimate the variogram accurately, that is a good coverage of samples is desirable (Russo 1984; van Groenigen *et al.* 1999). But this configuration should be supplemented by clustered samples to guarantee that a few pairs of points are separated by very small distances, critical to estimate the nugget effect. Webster and Oliver (1993) have indicated a minimum of $m = 150$ samples over the study area is necessary to estimate the variogram, but this is also influenced by the spatial variation over $\mathfrak{D}$ (a phenomenon with less variation needs less samples). The strength of the variogram is also partly dependent on the number of pairs of points available within each distance class. The Warrick-Myers (WM) criterion attempts at reproducing an a priori defined ideal distribution of pairs of points for estimating the covariogram:

$$J_{w/m}(S) = \mathbf{a}\sum_{i=1}^{K} \mathbf{w}_i \left(\xi_i^* - \xi_i\right)^2 + \mathbf{b}\sum_{i=1}^{K} \sigma(m_i), \tag{3.6}$$

$$\sum_{i=1}^{K} \xi_i^* = \sum_{i=1}^{K} \xi_i^* = \frac{m \cdot (m-1)}{2}, \tag{3.7}$$

where i denotes a given lag class of the covariogram, K represents the total number of classes, the parameters $\mathbf{a}$, $\mathbf{b}$, and $\mathbf{w}_i$ are user-defined weights. The term ξ_i^* is a prespecified number of point-pairs for the ith class, ξ_i is the actual number of distances within that class, and $\sigma(m_i)$ is the standard deviation from the median of the distance lag class (Warrick and Myers 1987; Delmelle 2009). Equation (3.7) expresses the total number of possible distance pairs, given the number of samples. Another similar criterion suggested in the literature is the *Minimization of the Mean of the Shortest Distances*, requiring sampling points to be spread evenly over the study region (van Groenigen *et al.* 1999).

3.2.3 Sampling designs minimizing the kriging variance

The kriging variance quantifies the prediction uncertainty at a particular location in space. This uncertainty is minimal at existing sampling points and increases with distance to the nearest samples. One common criterion is to design a sampling configuration, which minimizes the kriging variance, with a known, a priori (or estimated) variogram structure (van Groenigen *et al.* 1999). Equation (3.8) formulates the kriging variance at a location $\mathbf{s}$, where $\mathbf{C}_M^{-1}$ is the inverse of the covariance matrix $\mathbf{C}_M$, based on the covariogram function (Bailey and Gatrell 1995). M denotes the set of initial samples and has cardinality m. The term $\mathbf{c}$ is a column vector and $\mathbf{c}^T$ the corresponding row vector, as given in Equation (3.10).

$$\sigma_k^2(\mathbf{s}) = \sigma^2 - \mathbf{c}^T(\mathbf{s}) \cdot \mathbf{C}_M^{-1} \cdot \mathbf{c}(\mathbf{s}) \tag{3.8}$$

$$\mathbf{C}_M = \begin{bmatrix} \sigma^2 & C_{1,2} & \cdots & C_{1,m} \\ C_{2,1} & \sigma^2 & \cdots & C_{2,m} \\ \vdots & \vdots & \ddots & \vdots \\ C_{m,1} & C_{m,2} & \cdots & \sigma^2 \end{bmatrix} \tag{3.9}$$

$$\mathbf{c} = \begin{bmatrix} \sigma^2 \\ C_{21} \\ \vdots \\ C_{m1} \end{bmatrix}, \quad \mathbf{c}^T = \begin{bmatrix} \sigma^2 & C_{12} & \cdots & C_{1m} \end{bmatrix} \tag{3.10}$$

Computationally, the study area $\mathfrak{D}$ is discretized and the kriging variance summed over all grid points $\mathbf{s}_g$. Alternatively, it can be computed at each sample candidate location. The (average) kriging variance becomes (Delmelle 2009):

$$AKV = \frac{1}{G} \sum_{g \in G} \sigma_k^2(\mathbf{s}_g) \tag{3.11}$$

The only requirement to calculate the kriging variance is to have an initial (co)variogram and the locations of the m initial sample points. It then depends solely on the spatial dependence and configuration of the observations (Cressie 1991). In a *first stage*, initial measurements of the variable are collected to calibrate the variogram. Designs minimizing the kriging variance tend to spread samples evenly in the study region. An example optimizing initial samples for the minimization of the kriging variance is given in Section 3.5.

3.3 Augmented designs: Second-phase sampling

Second-phase sampling occurs when there is a need to go out in the field to gather additional information about the variable of interest. One important aspect is to capitalize on the spatial covariance information obtained during a first sampling phase, for instance under the form of a variogram. Based on this covariance structure, the kriging variance is computed at each grid node. Different sampling objectives are then possible: (a) maximizing the change in kriging variance (that is, locating those samples away from existing ones) and, (b) increasing sampling efforts in locally varying areas.

3.3.1 Additional sampling schemes to maximize change in the kriging variance

Our first objective $J[S]$ is to select a set of n additional points to our exisiting set of m samples, which will maximize the change in kriging variance by as much as possible. This process can be thought as a simulation of what the change in kriging variance is expected to be, without having to collect additional points, assuming the variogram structure would remain constant (Burgess *et al.* 1981; Cressie 1991). Specifically:

$$\underbrace{\text{MAXIMIZE}}_{\{\mathbf{s}_{m+1},\ldots,\mathbf{s}_{m+n}\}} \quad J[S] = \frac{1}{\lfloor G \rfloor} \sum_{g \epsilon G} \left(\sigma_k^{\text{old}}(\mathbf{s}_g)\right)^2 - \left(\sigma_k^{\text{new}}(\mathbf{s}_g)\right)^2, \tag{3.12}$$

with S the sampling scheme. Objective (3.12) aims to collect new samples to reduce the kriging variance or uncertainty by as much as possible. Equation (3.13) formulates the change in kriging variance $\Delta\sigma_k^2$ over all grid points $\mathbf{s}_g$, with the addition of a new sampling set N of size n. The change $\Delta\sigma_k^2$ is the difference between the kriging variance of the initial set σ_k^{old} with the kriging variance σ_k^{new} of the augmented set:

$$\Delta\sigma_k^2 = \frac{1}{G}\left[\sum_{g \epsilon G} \sigma_k^{\text{old}}(\mathbf{s}_g) - \sum_{g \epsilon G} \sigma_k^{\text{new}}(\mathbf{s}_g)\right] \tag{3.13}$$

$$\sigma_k^{\text{old}}(\mathbf{s}_g) = \sigma^2 - \underbrace{\mathbf{c}(\mathbf{s}_g)}_{[1,m]} \times \underbrace{\mathbf{C}^{-1}}_{[m]} \times \underbrace{\mathbf{c}^T(\mathbf{s}_g)}_{[m,1]} \tag{3.14}$$

$$\sigma_k^{\text{new}}(\mathbf{s}_g) = \sigma^2 - \underbrace{\mathbf{c}(\mathbf{s}_g)}_{[1,m+n]} \times \underbrace{\mathbf{C}^{-1}}_{[m+n]} \times \underbrace{\mathbf{c}^T(\mathbf{s}_g)}_{[m+n,1]} \tag{3.15}$$

The objective function [Equation (3.16)] is to find the optimal set S^* containing $m+n$ points that will maximize this change in kriging variance (Christakos and Olea 1992; van Groenigen *et al.* 1999; Rogerson *et al.* 2004), where S is a specific sampling scheme.

$$\underbrace{\text{MAXIMIZE}}_{\{\mathbf{s}_{m+1},\ldots,\mathbf{s}_{m+n}\}} \quad J(S) = \frac{1}{G} \sum_{g \epsilon G} \Delta\sigma_k^2(\mathbf{s}_g; S) \tag{3.16}$$

The set of new points is selected from the set of candidate location P. With a total of $\binom{P}{n}$ possible sampling combinations, it is too time-consuming to find the optimal set using combinatorics and requires heuristic methods as defined in Section 3.4. The initial sampling problem, which minimizes the kriging variance (or maximizes the change in kriging variance) is a simplification of Equation (3.16) with the set $n = 0$.

3.3.2 A weighted kriging variance approach

Many authors have advocated the use of the kriging variance as a measure of uncertainty. It can be misused as a measure of reliability of the kriging estimate, as noted by several authors (Armstrong 1994; Deutsch and Journel 1997). The rationale for this criticism is that the kriging variance is merely a function of the sample pattern, sample density, the number of samples and their covariance structure (Delmelle and Goovaerts 2009). The kriging variance assumes that the errors are independent of each other, a situation referred to as homoscedasticity. This means that the process is stationary, an assumption violated in practice, since the response surface will exhibot many changes. Stationarity entails that the variation of the primary variable between two points remains similar at different locations in space, as long as their separating distance remains unchanged. Mathematically, if we measure the difference in absolute value between two points $\mathbf{s}_i$ and $\mathbf{s}_j$ separated by a distance $d(\mathbf{s}_i, \mathbf{s}_j)$, and if we measure that difference again at two other points $\mathbf{s}_k$ and $\mathbf{s}_l$ separated by a similar distance $d(\mathbf{s}_k, \mathbf{s}_l)$, the results should be the same. In other words, $|y(\mathbf{s}_i) - y(\mathbf{s}_j)| \approx |y(\mathbf{s}_k) - y(\mathbf{s}_l)|$ for $d(\mathbf{s}_i, \mathbf{s}_j) = d(\mathbf{s}_k, \mathbf{s}_l)$. The letter $\mathbf{s}$ denotes a row vector containing the $\{x, y\}$ coordinate of the point. Figure 3.1 illustrates the nonstationarity problem in one dimension, where ten hypothetical temperature values have been randomly simulated. Figure 3.1(b) depicts the interpolated temperature values calculated using an exponential fitting model. From this graph it is possible to determine the values at locations $\mathbf{s}_1$ and $\mathbf{s}_2$. As Goovaerts (1997) points out, the variation in the close neighborhood of location $\mathbf{s}_1$ is much greater than $\mathbf{s}_2$, because it is surrounded by a very low and a very high value. However, the kriging variance is similar at both $\mathbf{s}_1$ and $\mathbf{s}_2$, since their neighbors are at an equal distance. If a sample point is added to the initial set of 10 points, it should preferably be point $\mathbf{s}_1$, as it exhibits a greater amount of spatial variation around its location than $\mathbf{s}_2$. As we can see, one issue pertains to choosing a good indicator to quantify the spatial uncertainty at a data point. The kriging variance may not be the sole indicator, and can be combined with other sampling criteria. Rogerson *et al.* (2004), Brus and Heuvelink (2007) and recently Delmelle and Goovaerts (2009) have proposed different criterion alternatives. For instance, in Delmelle and Goovaerts (2009), the kriging variance is weighted where the weights reflect the spatial variation. Their application to an exhaustive dataset led to better reconstruction of the image than if using the kriging variance as the sole criterion of locating additional samples.

There are two different approaches to formulate the second-phase sampling problem. Either the problem is defined as a single-weighted objective with no constraints, where the weights reflect the sampling objectives, or as a single objective where the weights become constraints. The single-weighted approach was suggested by Cressie (1991) and has been applied by van Groenigen *et* al. (1999) and Rogerson *et al.* (2004) to weight the kriging variance by a suitable weighting function $w(\cdot)$. The importance of a location to be sampled is represented by a weight $w(\mathbf{s})$, that is location-specific. The objective is to find the optimal sampling scheme S^* containing $m + n$ points that will maximize this change in weighted kriging variance:

$$\underbrace{\text{Maximize}}_{\{\mathbf{s}_{m+1},\ldots,\mathbf{s}_{m+n}\}} \quad J(S) = \frac{1}{G}\sum_{g \epsilon G} w(\mathbf{s}_g) \cdot \Delta\sigma_k^2(\mathbf{s}_g; S) \tag{3.17}$$

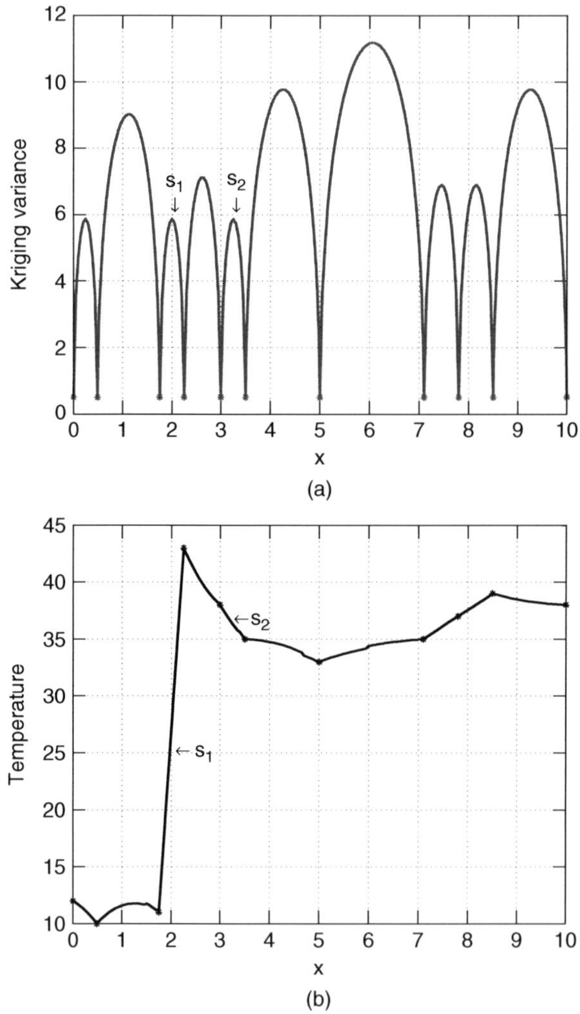

Figure 3.1 Example of nonstationarity. Points $\mathbf{s}_1$ *and* $\mathbf{s}_2$ *have the same kriging variance. However, the local variance at* $\mathbf{s}_1$ *is much greater than at* $\mathbf{s}_2$ *because there is much greater temperature variation among its neighboring points. After Goovaerts (1997).*

3.3.2.1 Computational implementation

The inverse of the covariance matrix $\mathbf{C}_{M \cup N}$ must be calculated to compute the kriging variance. The initial matrix $\mathbf{C}_M$ [Equation (3.9)] has been augmented by n rows and n columns. However, it is possible to calculate $\mathbf{C}_{M \cup N}^{-1}$ without having to invert the entire matrix. Consider in Equation (3.18) the augmented matrix $\mathbf{C}_{M\cup N}$

$$\mathbf{C}_{M\cup N} = \begin{bmatrix} \mathbf{C}_M & \mathbf{B} \\ \mathbf{B}^T & \mathbf{C}_N \end{bmatrix} \tag{3.18}$$

$$= \begin{bmatrix} \sigma^2 & \cdots & C_{1,m} & C_{1,m+1} & \cdots & C_{1,m+n} \\ \vdots & \ddots & \vdots & \cdots & \cdots & \cdots \\ C_{m,1} & \cdots & \sigma^2 & C_{m,m+1} & \cdots & C_{m,m+n} \\ C_{m+1,1} & \cdots & C_{m+1,m} & \sigma^2 & \cdots & C_{m+1,m+n} \\ \vdots & \vdots & \vdots & \vdots & \ddots & \vdots \\ C_{m+n,1} & \cdots & C_{m+n,m} & C_{m+n,m+1} & \cdots & \sigma^2 \end{bmatrix}$$

where $\mathbf{B}$ is the covariance matrix between initial and new added points, $\mathbf{B}^T$ is the transpose of $\mathbf{B}$ ad $\mathbf{C}_N$ is the covariance matrix among the new, added samples. A new matrix of size $n \times n$ is introduced in Equation (3.19) for simplification purposes and called the $\mathbf{P}$-matrix.

$$\mathbf{P} = \left[\mathbf{C}_N - (\mathbf{B}^T \cdot \mathbf{C}_M^{-1} \cdot \mathbf{B})\right] \tag{3.19}$$

Keeping in mind this matrix simplification notification, the inverse of the augmented matrix $\mathbf{C}_{M\cup N}$ can be formulated as (Horn and Johnson 1985):

$$\mathbf{C}_{M\cup N}^{-1} = \begin{bmatrix} \mathbf{C}_M^{-1} & \mathbf{0} \\ \mathbf{0}^T & 0 \end{bmatrix} + \begin{bmatrix} -\mathbf{C}_M^{-1} \cdot \mathbf{B} \\ I \end{bmatrix} \cdot \mathbf{P}^{-1} \cdot \begin{bmatrix} -\mathbf{C}_M^{-1} \cdot \mathbf{B} \\ I \end{bmatrix}^T \tag{3.20}$$

where I is an identity matrix of size $n \times n$, and the $\mathbf{0}$-matrix filled with zero elements. $\mathbf{C}_{M\cup N}^{-1}$ is a function of $\mathbf{C}_M^{-1}$ that remains constant, regardless of which new samples and how many of those are added. Only the matrices $\mathbf{P}$ and $\mathbf{B}$ vary when new points are added.

3.4 A simulated annealing approach

Whether for initial sampling or second-phase sampling, the objective is to find the optimal location of samples which will minimize a defined objective function. That set of optimal samples is selected from a set of candidate locations P, which is relatively large in practice, forbidding a total enumeration for the optimal set (Michalewicz and Fogel 2000). A heuristic H guides the search for an optimal sample set S^* (or near optimal S^+) $\subset P$. The set S^* is optimal to the objective function J, as defined in Equation (3.16). The efficiency of a heuristic depends on its capacity to give as often as possible a solution S^+ close to S^* (Grötschel and Lovàsz 1995). A *naïve optimization* heuristic selecting n points at random would not return a very good value for J, and is therefore not efficient. A *simultaneous* addition selects a set of samples one time. The major concern lies in the selection of those points: for an initial sampling set of, say $m = 50$ with a candidate set $p = 445$, there is a combinatorial explosion $\binom{p}{n} = \binom{445}{50}$, while in additional sampling, with an initial set $m = 50$ and an additional set of $n = 20$, the number of combinations becomes $\binom{p}{n} = \binom{445-50}{20}$. With such a high number of possible combinations, it is recommended to use intelligent search techniques, for instance simulated annealing, which is discussed next.

Simulated annealing has been used as an optimization procedure in initial and second-phase sampling; van Groenigen and Stein (1998), Lark (2002) and Delmelle (2005) further discuss the implementation of simulated annealing in second-phase spatial sampling. Simulated annealing is a method by which a metal cools and freezes into a minimum energy crystalline structure. The algorithm was originally proposed as a means of finding the equilibrium configuration of a collection of atoms at a given temperature. Kirkpatrick *et al.* (1983) made the connection between the cooling technique and the mathematical minimization problem. The major advantage of SA simulated annealing is its ability to avoid becoming trapped at local maxima. The algorithm employs a random search that accepts changes improving the objective function, but also nonimproving moves. The latter is accepted with probability P_T where T stands for the current temperature. T cools down as the algorithm progresses, and so does the probability of acceptance.

The optimization starts with a randomly selected sampling scheme $S(i) = M \cup N$ where M is the initial set, and N the set of additional samples ($N \subset P \setminus M$). The set N can be selected at random or is determined from a greedy heuristic, for instance using samples corresponding to the peaks of the surface area (Delmelle and Goovaerts 2009). The objective function $J[S(i)]$ is evaluated and called the incumbent solution. $S(i) = \{\mathbf{s}_1 + \ldots + \mathbf{s}_m + \mathbf{s}^j_{m+i} + \ldots + \mathbf{s}^j_{m+n}\}$ is created as the union of the initial sample set M, augmented by a set of new sample points $\mathbf{s}^j_{m+i}$. From the first iterations, $S(i)$ becomes $S(j)$ and will serve as the initial sampling scheme for simulated annealing. Since the space of solution has not been explored yet, the value of $J[S(i)]$ is kept in memory as the best solution $S(i) = S^+ = S^\diamond$ found so far. $S(i)$ becomes $S(j)$. One sample $i \ni M$ is swapped for a point $i \in M$, and at that time $S(j)$ becomes $S(j+1)$. Following a sequence of random perturbations $S(j+1)$ of $S(j)$ that have a probability $P_T\{S(j) \to S(j+1)\}$ of being accepted, where $\mathbf{s}^j_{m+i}$ is swapped in favor of $\mathbf{s}^{j+1}_{m+i}$:

$$P_T\{S(j) \to S(j+1)\} = 1 \qquad \text{if } J[S(j+1)] \le J[S(j)] \tag{3.21}$$

$$P_T\{S(j) \to S(j+1)\} = \frac{1}{1+e^{\left(\frac{\Delta J}{T}\right)}} \qquad \text{if } J[S(j+1)] > J[S(i)] \tag{3.22}$$

The sampling scheme $S(j)$ becomes $S(j+1)$ when $J[S(j+1)] \ge J[S(j)]$, or when $J[S(j+1)] \le J[S(j)]$ a test must be conducted as follows:

$exp\left[-\frac{(candObj-incObj)}{\xi \times T}\right] > rand()$, with ξ a parameter reflecting changes in the objective function J. If $S(j+1)$ is accepted, it becomes the incumbent solution $S^{\diamond} = S(j+1)$, and serves as a starting point for a next scheme $S(j+2)$. The process continues in a similar fashion, until a certain level of iterations T_{fin} has been reached. Note that the temperature and step size κ decrease as the algorithm progresses. As the system cools down, the probability of accepting nonimproving moves decreases with temperature decrease. The algorithm stops when the temperature T has reached its cut-off value T_{fin}. To find the set of optimal points $\mathbf{s}^*$ to be added–or nearly optimal $\mathbf{s}^+$, a high starting temperature T and a cooling factor T_{dec} close to 1 are necessary. In these conditions, simulated annealing can escape from a local maximum, but a slow cooling schedule will increase the running time of the algorithm, and as such, a trade-off must be determined between optimality gap and running time. As a rule of thumb, it is advisable to have a large step size δ in the beginning of the algorithm, allowing wide jumps across $\mathfrak{D}$, when there might be different maxima across the map. Simulated annealing is therefore sensitive to the choice of the cooling factor, that governs the search procedure. Additionally, it is suggested to conduct the simulated annealing in two stages, where the second one consists of concentrating the search around the best solution found so far during the first stage (Delmelle 2005), that is during the second stage a smaller step size κ is used. Note that the simulated annealing algorithm does not always converge. A pseudocode for the algorithm is given below:

```
  def kv = function computing kriging variance
set incSet;incObj←kv(incSet)
set T, K, T_fin
While T> T_fin
  candSet←neighbor(incSet,K)
  candObj←kv(candSet)
  If candObj < bestObj
  incObj←candObj, incSet←candSet
  bestObj←candObj, bestSet←candSet
  ElseIf exp[-(candObj-incObj)/(ξ×T)] >rand()
  incObj←candObj, incSet←candSet
  EndIf
  T=T×T_dec, K=K×K_dec
End While
```

3.5 Illustration

In this section, we apply simultaneous simulated annealing to the Austrian data set to gain insight into the problem structure and behavior. The goal is to find the location of (initial and additional) sample points which will keep the kriging variance to a minimum. All reported computational results were obtained using Matlab v. 2010.

3.5.1 Initial sampling designs

Different sampling objectives are given in the literature when little (or no) a priori spatial information is known for the variable of interest. Assuming some spatial covariance information (range r, nugget a and sill σ^2), the sampling design can be optimized to minimize the kriging variance. Information on the covariance structure from a previous study can have a strong influence on the sampling design configuration, with large range values spreading observations throughout the study region, while short ranges bring these closer to one another. In the Austrian dataset, the centroid of each statistical unit is used as a candidate location in the optimization, that is $p = 445$. In this example, we use the following parameters: $r = 50000\,\text{m}$, $\sigma^2 = 200$ and a nugget $a = 1$. We optimize the location of initial samples ($m = 30$ and $m = 50$, respectively) for the minimization of the kriging variance using randomly selected points as the starting set (see Figure 3.2 for an illustration) The kriging variance σ_k^2 is measured

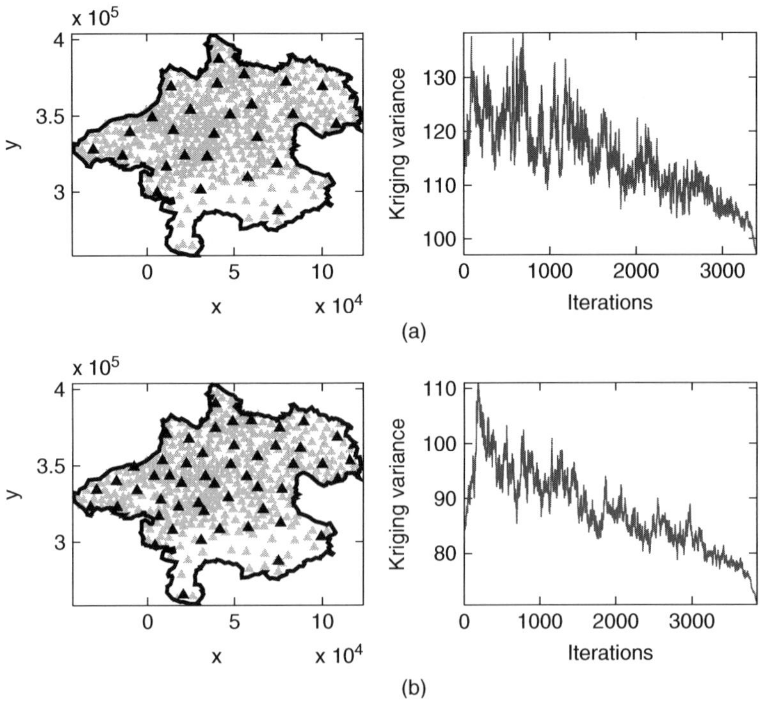

Figure 3.2 Two initial sampling designs with m = 30 *samples (a) and* m = 50 *(b) from* p = 445 *candidate locations, minimizing the kriging variance. Light grey triangles represent potential locations while black triangles are selected samples. The graphs on the right illustrate the convergence of the optimization process with simulated annealing.*

and displayed on each statistical unit support (**s**). The kriging variance is measured on the points which can also serve as candidate location to be included in the sampling set. Since the statistical supports are not exactly distributed in a grid pattern in the region (especially in the outer edge), careful attention must be paid when evaluating the kriging variance over the study region. The average kriging variance [Equation (3.11)] is lower when more measurements are used in the initial sampling phase. Due to a large range, the optimization algorithm tends to spread the points as far as possible from one another, guaranteeing a certain level of coverage. The graphs on the right illustrate the simulated annealing algorithm, which accepts several nonimproving moves in the beginning of the optimization, but eventually converge to a minimum kriging variance.

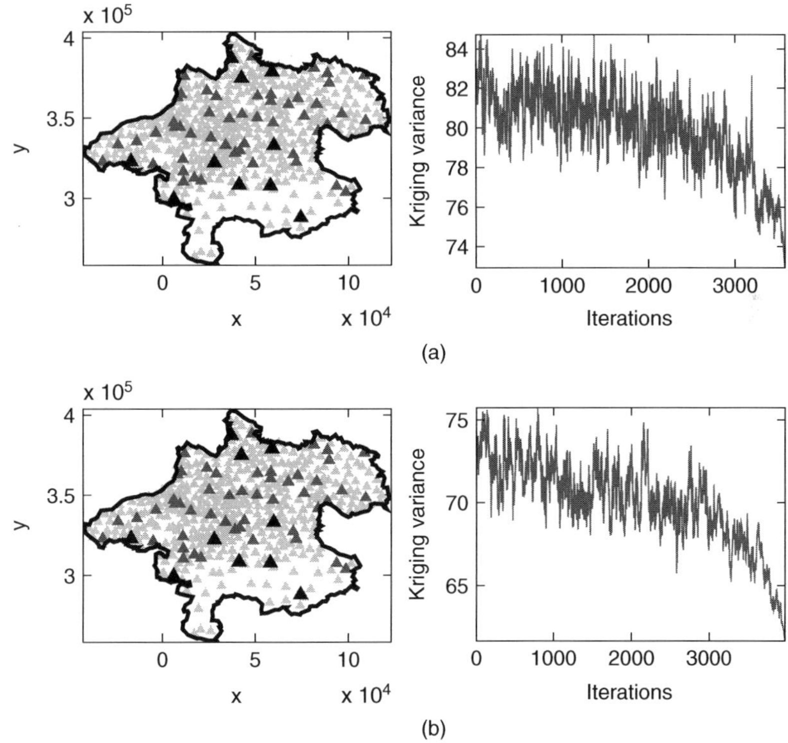

Figure 3.3 An example of two augmented sampling designs; n = 10 *samples (a) and* n = 25 *(b) from* p = 445 − 50 *candidate locations, minimizing the kriging variance. Light gray triangles represent potential locations, darker gray triangles initial samples and black triangles second-phase samples. The graphs on the right illustrate the optimization process with simulated annealing.*

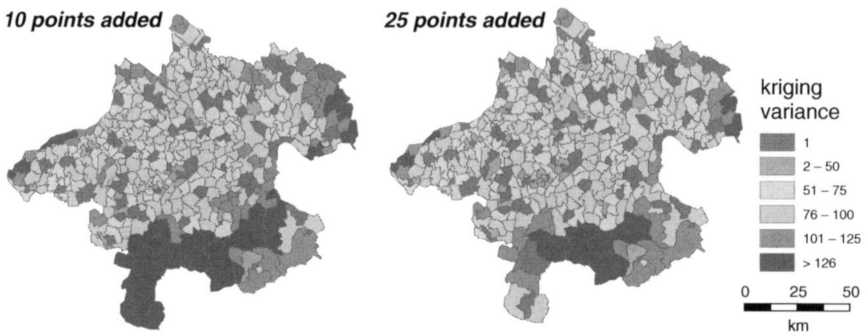

Figure 3.4 The kriging variance after the addition of new samples to the initial set (use in conjunction with Figure 3.3). Darker areas denote regions of high uncertainty since no points were located in those regions. (Please see plate section for color version of the figure.)

3.5.2 Augmented designs

We illustrate the example of augmented designs for the minimization of the kriging variance. Additional samples are usually taken away from existing ones (van Groenigen and Stein 1998; Delmelle 2005), that is where the kriging variance is maximum. The kriging variance at unsampled locations can be determined with multiple design scenarios. An augmented design is said to be optimal to Equation (3.16) when the configuration maximizes the change in kriging variance. Figure 3.3 illustrates the addition of $n = 10$ and $n = 25$ points to an initial randomly selected sample set $m = 50$. The large, black triangles are the locations of additional samples. In both cases, the initial sampling set (smaller, dark, gray triangles) is the same, but the locations of new samples change. The algorithm tends to fill in these areas with additional samples away from initial samples. When the number of additional samples increases, the kriging variance drops significantly. For visualization purposes, the kriging variance at each statistical support is assigned to the polygon it belongs to, and a choropleth map in Figure 3.4 indicates that the kriging variance remains high in the outer edges of the region.

3.6 Discussion

In this chapter, we have illustrated a framework for initial and second-phase spatial sampling problems based on the change in kriging variance. Different objectives exist to account for heterogeneity of the spatial variable, for instance by weighting the kriging variance by locally varying priorities. These weights can reflect local non-homogeneity. Another approach consists of using locally varying variograms (Haas 1990). Due to the nonlinearity of the objective, it is recommended to use a heuristic technique to find a suitable set. Results of our application to the Austrian dataset illustrates the suitability of simulated annealing to optimize an initial and augmented sampling set.

References

Arbia G. (1989). *Spatial Data Configuration in Statistical Analysis, Regional Economics and Related Problems*. Kluwer.

Armstrong M. (1994). Is research in mining geostats as dead as a dodo? *In*: Dimitrakopoulos R. (Ed.) *Geostatistics for the Next Century*. Kluwer: 303–312.

Bailey T. C. and A. C. Gatrell (1995). *Interactive Spatial Data Analysis*. Longman, 413p.

Berry B. J. L. and A. M. Baker (1968). Geographic sampling. *In:* Berry B. J. L. and Marble D. F. (Eds) *Spatial Analysis: a Reader in Statistical Geography*. Prentice Hall: 91–100.

Brus D. and G. B. M. Heuvelink (2007). Optimization of sample patterns for universal kriging of environmental variables. *Geoderma*, vol. **138**: 86–95

Burgess T. M., R. Webster and A. B. McBratney (1981). Optimal interpolation and isarithmic mapping of soil properties: IV. Sampling strategy. *Journal of Soil Science*, vol. **32**: 643–659.

Cochran W. G. (1963). *Sampling Techniques*. Second edition. John Wiley & Sons, Ltd, 413p.

Christakos G. and R. A. Olea (1992). Sampling design for spatially distributed hydrogeologic and environmental processes. *Advanced Water Resources*, vol. **15**: 219–237.

Cressie N. (1991). *Statistics for Spatial Data*. John Wiley and Sons, Ltd, 900p.

Dalton R., J. Garlick, R. Minshull and A. Robinson (1975). *Sampling Techniques in Geography*. Goerges Philip and Son Limited, 95p.

Das A. C. (1950). Two-dimensional systematic sampling and the associated stratified and random sampling. *Sankhyā*, vol. **10**: 95–108.

de Gruijter, J., D. J. Brus, M. F. P. Bierkens and M. Knotters (2006). *Sampling for Natural Resource Monitoring*. Springer, 332p.

Delmelle E. (2005). *Optimizing Second-Phase Spatial Sampling Using Auxiliary Information*. PhD dissertation, SUNY at Buffalo.

Delmelle E. (2009). Spatial sampling. *In*: Fotheringham S. and Rogerson P. (Eds.) *Handbook of Spatial Analysis*. Sage Publications: 183–206.

Delmelle E. and P. Goovaerts (2009). Second-phase sampling designs for non-stationary spatial variables. *Geoderma*, vol. **153**: 205–216

Deutsch C. V. and A. G. Journel (1997) *Gslib: Geostatistical Software Library and User's Guide*. Second edition Oxford University Press, 369p.

Goovaerts P. (1997). *Geostatistics for Natural Resources Evaluation*. Oxford University Press, 483p.

Griffith D. (2005). Effective geographic sample size in the presence of spatial autocorrelation. *Geographical Analysis*, vol. **95**: 740–760.

Griffith D. and C. Amrhein (1997). *Multivariate Statistical Analysis for Geographers*, Prentice Hall, 275p.

Grötschel M. and L. Lovàsz (1995). Combinatorial optimization. *In:* Graham R. L., Grötschel M. and Lovàsz L. (Eds.) *Handbook of Combinatorics*, vol. 2. Elsevier: 1541–1579.

Haas T. C. (1990). Kriging and automated variogram modeling within a moving window. *Atmospheric Environment*, vol. **24**: 1759–1769.

Haining R. P. (1990). Sampling spatial populations. *In:* Haining R. P. (Ed.) *Spatial Data Analysis in the Social and Environmental Sciences*. Cambridge University Press: 171–196.

Hedayat A. S. and B. K. Sinha (1991). *Design and Inference in Finite Population Sampling*. John Wiley & Sons, Ltd, 377p.

Iachan R. (1985). Plane sampling. *Statistics and Probability Letters*, vol. **3**: 151–159.

Horn R. A. and C. R. Johnson (1985). *Matrix Analysis*, Cambridge University Press.

King L. J. (1969). *Statistical Analysis in Geography*. Prentice-Hall, Chapter 4.

Kirkpatrick S., C. D. Gelatt and M. P. Vecchi (1983). Optimization by simmulated annealing. *Science*, vol. **220**: 671–680.

Lark R. M. (2002). Optimized spatial sampling of soil for estimation of the variogram by maximum likelihood. *Geoderma*, vol. **105**: 49–80.

Michalewicz Z. and D. Fogel (2000). *How to Solve It: Modern Heuristics*. Springer. 467p.

Müller, W. (1998). *Collecting Spatial Data: Optimal Design of Experiments for Random Fields*. Physica-Verlag.

Overton W. S. and S. V. Stehman (1993). Properties of designs for sampling continuous spatial resources from a triangular grid. *Communications in Statistics - Theory and Methods*, vol. **21**: 2641–2660.

Quenouille M. H. (1949). Problems in plane sampling. *Annals of Mathematical Statistics*, vol. **20**: 355–375.

Ripley B. D. (1981). *Spatial Statistics*. John Wiley & Sons, Ltd, 252p.

Rogerson P. A., E. M. Delmelle, R. Batta, M. R. Akella, A. Blatt and G. Wilson (2004). Optimal sampling design for variables with varying spatial importance. *Geographical Analysis*, vol. **36**: 177–194.

Russo D. (1984). Design of an optimal sampling network for estimating the variogram. *Soil Science Society of America Journal*, vol. **48**: 708–716.

Stehman S. V. and S. W. Overton (1996). Spatial sampling. *In*: Arlinghaus, S. (Ed.) *Practical Handbook of Spatial Statistics*. CRC Press: 31–64.

Thompson S. K. (2002). *Sampling*. Second edition. John Wiley & Sons, Ltd, 367p.

van Groenigen, J. W. and A. Stein (1998). Constrained optimization of spatial sampling using continuous simulated annealing. *Journal of Environmental Quality*, vol. **27**: 1078–1086.

van Groenigen J. W., W. Siderius and A. Stein (1999). Constrained optimisation of soil sampling for minimisation of the kriging variance. *Geoderma*, vol. **87**: 239–259.

Warrick A. W. and D. E. Myers (1987). Optimization of sampling locations for variogram calculations. *Water Resources Research*, vol. **23**: 496–500.

Webster R. and M. A. Oliver (1993). How large a sample is needed to estimate the regional variogram adequately. *In:* Soares A. (Ed.) *Geostatistics Tróia '92*. Kluwer: 155–166.

Zubrzycki S. (1958). Remarks on random, stratified and systematic sampling in a plane. *Colloquium Mathematicum*, vol. **6**: 251–264.

4

Spatial sampling design by means of spectral approximations to the error process

Gunter Spöck and Jürgen Pilz
Department of Statistics, University of Klagenfurt, Austria

4.1 Introduction

Geostatistics is the science of continuous stochastic processes or so-called random fields that are defined over some region in two- or three-dimensional geographic space $\mathbf{X}$ or in space-time. A random field $\{Y(x) : x \in \mathbf{X}\}$ is a set of random variables or so-called regionalized variables $Y(x)$ that are attached to every location $x \in \mathbf{X}$. The probability law of the random field is uniquely determined by its so-called projective family, the set of all finite dimensional distributions of any finite set of $Y(x)$ obeying symmetry and consistency with respect to marginalization. Very often, only the first- and second-order characteristics, the trend or mean function $m(x) = \mathrm{E}(Y(x))$ and the covariance function $C(x, y) = \mathrm{Cov}(Y(x), Y(y))$ of the random field are considered. Usually, a linear trend function $m(x) = \mathbf{f}(x)^T \beta$, where $\mathbf{f}(x)$ is a fixed vector-valued function and β is a

Spatio-temporal design: Advances in efficient data acquisition, First Edition.
Edited by Jorge Mateu and Werner G. Müller.

regression parameter vector to be estimated, is sufficient for modelling purposes. For $x \in \Re^2$, $\mathbf{f}(x)$ could be for example a vector of polynomials in the coordinates $x = (x_1, x_2)$. The covariance function $C(.,.)$ must be positive semidefinite, meaning that it must provide a positive variance for any linear combination of $Y(x_1), Y(x_2), \ldots, Y(x_n)$. Most often an additional assumption of second-order stationarity must be met, meaning that $C(x, x+h) = C(h)$ is dependent only on the lag h and not on the locations x and $x+h$ themselves. The next stronger assumption is the assumption of isotropy, meaning that $C(x, x+h) = C(||h||_2)$, only depends on the Euclidean length of h. Both these assumptions of second-order stationarity and of isotropy assure that the covariance function can be estimated from only a single realization of the random field, i.e. from a finite collection of data $y(x_1), y(x_2), \ldots, y(x_n)$. The task of geostatistics is to produce a prediction map of all $y(x_0), x_0 \in \mathbf{X}$, based on the available data $y(x_1), y(x_2), \ldots, y(x_n)$ and to report on the accuracy of these predictions. The best known methodology for this task of interpolation is kriging, also known as best linear unbiased prediction. The so-called universal kriging predictor depends on the data $\mathbf{y} = (y(x_1), y(x_2), \ldots, y(x_n))^T$, the covariance matrix $\mathbf{K} = Cov(\mathbf{y})$ and the covariance vector $\mathbf{c}_0 = (Cov(Y(x_0), Y(x_1)), \ldots, Cov(Y(x_0), Y(x_n)))^T$. It can be written

$$\hat{Y}(x_0) = \mathbf{f}(x_0)^T \hat{\beta} + \mathbf{c}_0^T \mathbf{K}^{-1}(\mathbf{y} - \mathbf{F}\hat{\beta}), \tag{4.1}$$

where

$$\hat{\beta} = (\mathbf{F}^T \mathbf{K}^{-1} \mathbf{F})^{-1} \mathbf{F}^T \mathbf{K}^{-1} \mathbf{y} \tag{4.2}$$

is the generalized least squares estimate of the regression parameter vector β and $\mathbf{F}$ is the design matrix corresponding to $\mathbf{f}(.)$ and the locations $x_1, x_2, \ldots, x_n$. The mean squared error (MSEP) of this unbiased predictor is given by

$$\begin{aligned} \sigma^2(x_0) &= \mathbf{E}(Y(x_0) - \hat{Y}(x_0))^2 \\ &= C(\mathbf{0}) - \mathbf{c}_0^T \mathbf{K}^{-1} \mathbf{c}_0 + \mathbf{g}(x_0)^T (\mathbf{F}^T \mathbf{K}^{-1} \mathbf{F})^{-1} \mathbf{g}(x_0), \end{aligned} \tag{4.3}$$

where

$$\mathbf{g}(x_0) = \mathbf{f}(x_0) - \mathbf{F}^T \mathbf{K}^{-1} \mathbf{c}_0.$$

Obviously, the mean squared error $\sigma^2(x_0)$ is dependent on the arrangement of the sampling locations $x_1, x_2, \ldots, x_n$ via the covariance matrix $\mathbf{K}$, the covariance vector $\mathbf{c}_0$ and the design matrix $\mathbf{F}$.

As with any statistical prediction method, our aim is to make best use of the available data, possibly under certain budget constraints or constraints on the number of locations that can be sampled. Thus, the goal is to obtain as good predictions as possible for the complete area of investigation $\mathbf{X}$. One possibility to formalize this goal is to try to select the sampling locations $x_1, x_2, \ldots, x_n$ in such a way that the sum of all kriging MSEPs

$$\int_{\mathbf{X}} \sigma^2(x_0)dx_0 \tag{4.4}$$

becomes a minimum over the area of investigation **X**. Notably, this is a very complicated optimization problem and becomes still more complicated by the fact that the covariance matrix **K** enters the kriging MSEP in its inverse form $\mathbf{K}^{-1}$. Besides other criteria which also measure the accuracy of kriging predictions, this design problem has been tackled in a number of recent research papers. The aim of the next section is to discuss some of these papers.

We first consider the integrated kriging variance averaged over the area of investigation and then, as a combined design criterion, the averaged expected lengths of predictive intervals are considered. Finally, it is shown how these design criteria can be generalized to spatial variables having skewed distributions. In this context a criterion for spatial sampling design with Box–Cox transformed spatial variables is proposed.

Mathematically speaking, we will approximate the investigated spatial random field by means of a large regression model consisting of cosine-sine Bessel surface harmonics with random amplitudes (Sections 4.3 and 4.4). This approximating regression model is a direct result of the polar spectral representation theorem for isotropic random fields (Section 4.3). We show that kriging prediction in the original random field model is equivalent to trend prediction in this approximating Bayesian linear regression model (Section 4.4). Thus, standard convex experimental design theory for Bayesian linear regression may be used to find spatial sampling designs. Based on this theory we propose two algorithms for the calculation of exact designs (Section 4.7.3):

- An algorithm for removing redundant design locations.
- An algorithm for adding sampling locations to an existing network.

In Section 4.5 we generalize our approach to the case of an uncertain covariance function estimated by restricted maximum likelihood (REML). The result is that the actual predictive intervals are wider than those obtained when disregarding this uncertainty. The Smith and Zhu design criterion adjusts to this uncertainty by considering the average of the expected lengths of predictive intervals as a design criterion. Section 4.6 contains a further generalization; here we assume the data distribution to be skewed and to be transformable to a Gaussian distribution by means of a Box–Cox transformation. Once again the average of the expected lengths of predictive intervals is considered as a design criterion. In Section 4.7 we describe a MATLAB and Octave toolbox which can perform all the tasks of spatial sampling design and spatial interpolation, and which is freely available from the first author. Finally, the proposed methodologies are illustrated by a true data example of rainfall measurements in Upper Austria (Section 4.8).

4.2 A brief review on spatial sampling design

The importance of (optimal) spatial sampling design considerations for environmental applications and soil science has been demonstrated in several papers and monographs (Brus and de Gruijter 1997; Cox Jr 1999; van Groenigen *et al.* 1999; Lark 2002; US-EPA 2002; Caeiro *et al.* 2003; Hengl *et al.* 2003; Diggle and Lophaven 2006; Brus and Heuvelink 2007; Dobbie *et al.* 2008; Delmelle and Goovaerts 2009). The papers on spatial sampling design may be divided into several categories, some of which are overlapping. First of all we may differentiate between design criteria for spatial prediction and for estimation of the covariance function, and criteria combining both objectives. Contributions falling into the category of criteria for prediction are provided by Fedorov and Flanagan (1997), Müller and Pazman (1998, 1999), Pazman and Müller (2001), Müller (2005) and Brus and Heuvelink (2007). Criteria for the estimation of the covariance function are considered by Zimmerman and Homer (1991), Müller and Zimmerman (1999) and Lark (2002). Combined criteria can be found in Zhu and Stein (2006), who consider the minimization of the average expected length of predictive intervals. Further papers falling into this category of combined criteria are Bayesian approaches specifying prior distributions over covariance functions such as those by Brown *et al.* (1994), Müller *et al.* (2004), Diggle and Lophaven (2006) and Fuentes *et al.* (2007). Actually, Brown *et al.* (1994) and Fuentes *et al.* (2007) consider the covariance function to be nonstationary and deal with an entropy-based design criterion according to which the determinant of the covariance matrix between locations to be added to the design must be maximized. Both make use of simulated annealing algorithms to find optimal designs obeying their criteria.

On the algorithmic side, we can distinguish beween stochastic search algorithms like simulated annealing (Aarts and Korst 1989), or evolutionary genetic algorithms and deterministic algorithms for optimizing the investigated design criteria. With the exception of Fedorov and Flanagan (1997), Müller and Pazman (1998, 1999), Pazman and Müller (2001), Müller (2005), Spöck and Pilz (2010) almost all algorithms for spatial sampling design optimization use stochastic search algorithms for finding optimal configurations of sampling locations $x_1, x_2, \ldots, x_n$. The term spatial simulated annealing (SSA) finds its first manifestation in the work of van Groenigen *et al.* (1999). Trujillo-Ventura and Ellis (1991) consider multiobjective sampling design optimization.

Free software for spatial sampling design optimization is quite rare. To the authors' knowledge there are only six sampling design toolboxes freely available, including our own one: Gramacy (2007) has implemented sampling design for treed Gaussian random fields in the R-package tgp. In treed Gaussian random fields the area $\mathbf{X}$ of investigation is partitioned by means of classification trees into rectangular sub-areas with sides parallel to the coordinate axes. This software is especially useful for the design of computer simulation experiments, where parameters guiding the computer simulation output are identified as spatial coordinates. Another software especially useful for computer simulation experiments and sequential design is the DAKOTA package (http://dakota.sandia.gov). Further papers falling into the category of computer simulation experiments are those by Mitchell and Morris (1992), Morris *et al.* (1993), Schonlau (1997), Lim *et al.* (2002), Kleijnen and van Beers (2004), Chen *et al.* (2006) and Gramacy and Lee (2010). Software freely available upon request for research purposes and monitoring network optimization is provided by Le and Zidek (2006). This software implements the entropy-based design criterion mentioned above. Gebhardt (2003) implements a branch and bound algorithm for designing with the criterion (4.4). Baume *et al.* (2011) compare different greedy algorithms for spatial design. This small list of software for spatial sampling design may not be complete; hopefully more software can be obtained from the different authors of research papers upon request. Since for the practioner there is a strong need for spatial sampling design, and almost no software is freely available, this was an impetus for us to program a toolbox for spatial sampling design. The spatDesign Matlab toolbox of the first author is freely available online at: http://wwwu.uni-klu.ac.at/guspoeck/spatDesignMatlab.zip.

An Octave version not implementing as many design criteria as the Matlab toolbox is available at: http://wwwu.uni-klu.ac.at/guspoeck/spatDesignOctave.zip.

4.3 The spatial mixed linear model

This section presents our approach to spatial sampling design. No stochastic search algorithms like simulated annealing to optimize the design criterion are needed in this approach because we make use of the mathematical structure of the investigated design criteria.

We consider a mean square continuous (m.s.c.) and isotropic random field $\{Y(x) : x \in \mathbf{X} \subseteq \Re^2\}$ such that

$$Y(x) = \mathbf{f}(x)^T \beta + \varepsilon(x), \quad \mathrm{E}\varepsilon(x) = 0, \tag{4.5}$$

where $\mathbf{f}(x)$ is a known vector of regression functions, $\beta \in R^r$ a vector of unknown regression parameters and denote

$$\mathrm{Cov}(Y(x), Y(y)) = C(||x - y||); \quad x, y \in \mathbf{X}.$$

Then, according to Yaglom (1987), the covariance function can be represented in the form

$$C(t) = \int_0^\infty J_0(t\omega)dG(\omega), \ t \geq 0,$$

where $J_0(.)$ is the Bessel function of the first kind and order 0, $t = ||x - y||$ is the Euclidean distance between x and y, and $G(.)$ is the so-called (polar) spectral distribution function associated with $C(.)$. As such $G(.)$ is positive, monotonically increasing and bounded from above. On the other hand, knowing

$C(.)$, the spectral distribution can be obtained from the inversion formula

$$\frac{G(\omega^+) + G(\omega^-)}{2} = \int_0^\infty J_1(t\omega)\omega C(t)dt,$$

where $G(\omega^+)$ and $G(\omega^-)$ denote the right- and left-hand side limits at ω and J_1 denotes the Bessel function of first kind and order 1. Approximating $G(.)$ by means of a step function with positive jumps $a_i^2 = G(\omega_{i+1}) - G(\omega_i)$ at preselected points $\omega_i, i = 0, 1, \ldots, n-1$, and changing to polar coordinates $(t, \varphi) = (radius, angle)$, the polar spectral representation theorem for m.s.c. isotropic random fields tells us that the error process may be approximated as

$$\varepsilon(t, \varphi) \approx \tag{4.6}$$

$$\textstyle\sum_{m=0}^{\infty} \{\cos(m\varphi) \displaystyle\sum_{i=1}^{n} J_m(\omega_i t) U_{m,i}\} + \sum_{m=1}^{\infty} \{\sin(m\varphi) \sum_{i=1}^{n} J_m(\omega_i t) V_{m,i}\},$$

where all the random variables $U_{m,i}$ and $V_{m,i}$ are uncorrelated, have mean zero, and their variances are $\text{var}(U_{m,i}) = \text{var}(V_{m,i}) = d_m a_i^2$; and $d_m = 1$ for $m = 0$ and $d_m = 2$ for $m \geq 1$. By truncating the above series at a sufficiently large $m = M$, we get an approximation of our random field in the form of a mixed linear model

$$Y(x) \approx \mathbf{f}(x)^T \beta + \mathbf{g}(x)^T \alpha + \varepsilon_0(x) \tag{4.7}$$

From the above it becomes clear that the components of the additional regression vector $\mathbf{g}(\cdot)$ are made up of the following radial basis functions (cosine-sine-Bessel-harmonics)

$$g_{m,i}(t, \varphi) = cos(m\varphi) J_m(\omega_i t); \tag{4.8}$$

$$m = 0, \ldots, M; i = 1, \ldots, n$$

$$g_{m,i}(t, \varphi) = sin((m - M)\varphi) J_{m-M}(\omega_i t);$$

$$m = M + 1, \ldots, 2M; i = 1, \ldots, n.$$

The idea to approximate the spatial random field by means of a large regression model with random coefficients was first proposed by Fedorov (1996), who approximated the random field by the so-called Karhunen–Loeve–Eigen expansion, which is in general much more complicated to calculate than our approach via the polar spectral approximation.

4.4 Classical Bayesian experimental design problem

Starting from our spatial mixed linear model (4.7) we may gain further flexibility with a Bayesian approach incorporating prior knowledge on the trend. We assume that the regression parameter vector β is random with

$$\mathrm{E}(\beta) = \mu \in R^r, \quad \mathrm{Cov}(\beta) = \mathbf{\Phi}.$$

This is exactly in the spirit of Omre (1987) who introduced Bayesian kriging this way. He used physical process knowledge to arrive at 'qualified guesses' for the first- and second-order moments, μ and $\mathbf{\Phi}$. On the other hand, the state of prior ignorance or noninformativity can be modelled by setting $\mu = \mathbf{0}$ and letting $\mathbf{\Phi}^{-1}$ tend to the matrix of zeroes, thus passing the 'Bayesian bridge' to universal kriging (Omre and Halvorsen 1989).

Now, combining (4.7) with Bayesian prior knowledge, we arrive at the *Bayesian spatial linear model* (BSLM)

$$Y(x) = \mathbf{h}(x)^T \gamma + \varepsilon_0(x), \tag{4.9}$$

where

$$\mathbf{h}(x) = \begin{pmatrix} \mathbf{f}(x) \\ \mathbf{g}(x) \end{pmatrix}, \gamma = \begin{pmatrix} \beta \\ \alpha \end{pmatrix},$$

$$\mathrm{E}\gamma = \begin{pmatrix} \mu \\ 0 \end{pmatrix} =: \gamma_0, \mathrm{Cov}(\gamma) = \begin{pmatrix} \mathbf{\Phi} & 0 \\ 0 & \mathbf{A} \end{pmatrix} =: \mathbf{\Gamma}.$$

Here $\varepsilon_0(x)$ is white-noise with variance σ_0^2 and $\mathbf{A}$ denotes the covariance matrix of α, resulting from the polar spectral approximation of the random field. Spöck and Pilz (2010) demonstrate that Bayesian linear trend estimation in the above BSLM actually approximates Bayesian linear kriging in the original model abitrarily closely. The same is true for the total mean squared error (TMSEP) of the trend prediction and the TMSEP of Bayesian kriging.

Thus taking the TMSEP of the trend prediction in the approximating model as a substitute for the Bayes kriging TMSEP we arrive at the following classical experimental design problem for so-called I-optimality:

$$\int_{\mathbf{X}} \mathbf{h}(x)^T (\mathbf{H}^T(d_n)\mathbf{H}(d_n) + \sigma_0^2 \mathbf{\Gamma}^{-1})^{-1} \mathbf{h}(x) dx \to \min_{d_n}. \tag{4.10}$$

Here $d_n = \{x_1, x_2, \ldots, x_n\}$ collects either the design points to be added to the monitoring network or in the case of reducing the network the design points remaining in the monitoring network. $\mathbf{H}(d_n)$ expresses the dependence of the design matrix $\mathbf{H} = (\mathbf{h}(x_i)^T)_{i=1,2,\ldots,n}$ on the design points in the set d_n.

At this point we advise the reader not familiar with Bayesian experimental design theory to read the Appendix of Spöck and Pilz (2010). The key point in this theory is that the above so-called concrete design problem, which does not have a convenient mathematical structure, may be extended to a so-called continuous design problem that has the nice feature to be a convex optimization problem. Thus, the whole apparatus of convex optimization theory is available to approximately solve the above design problem for I-optimality. In particular, directional derivatives may be calculated and optimal continuous designs may be found by steepest descent algorithms. Continuous designs are just probability measures ξ on $\mathbf{X}$ and may be rounded to exact designs d_n. Defining the so-called continuous Bayesian information matrix

$$\mathbf{M}_B(\xi) = \int_{\mathbf{X}} \mathbf{h}(x)\mathbf{h}(x)^T \xi(dx) + \frac{\sigma_0^2}{n}\mathbf{\Gamma}^{-1} \tag{4.11}$$

and

$$\mathbf{U} = \int_{\mathbf{X}} \mathbf{h}(x)\mathbf{h}(x)^T dx, \tag{4.12}$$

it may be shown that the set of all such information matrices is convex and compact and that the extended design functional

$$\Psi(\mathbf{M}_B(\xi)) = \mathrm{tr}(\mathbf{U}\mathbf{M}_B(\xi)^{-1}) \tag{4.13}$$

is convex and continuous in $\mathbf{M}_B(\xi)$. The above design functional $\Psi(.)$ thus attains its minimum at a design $\xi^* \in \Xi$, where Ξ is the set of all probability measures defined on the compact design region $\mathbf{X}$ (Pilz 1991). The minimization of $\Psi(\cdot)$ is the continuous analogue to our original design problem (4.10). The closeness of exact designs d_n to the optimal continuous design ξ^* may be judged by means of a well-known efficiency formula, see Appendix of Spöck and Pilz (2010).

4.5 The Smith and Zhu design criterion

In real world applications the isotropic covariance function $C_\theta(t)$ is always uncertain and must be estimated. The kriging predictor used is then based on this estimated covariance function $C_{\hat{\theta}}(t)$. Thus, the kriging predictor is always a plug-in predictor and the reported (plug-in) kriging variance underestimates the true variance of this plug-in predictor.

Smith and Zhu (2004) consider spatial sampling design by means of minimizing the average of the expected lengths of $1-\alpha$ predictive intervals:

$$\int_{\mathbf{X}} \mathrm{E}(\text{length of predictive interval at } x_0) dx_0. \tag{4.14}$$

Their predictors of the $\alpha/2$ and $1-\alpha/2$ quantiles of the predictive distributions are selected in such a way that the corresponding predictive intervals have coverage probability bias 0. The predictors of the mentioned quantiles are essentially the plug-in kriging predictors based on REML estimation of the covariance function plus/minus a scaled plug-in kriging standard error term that is corrected to take account of REML estimation. Based on Laplace approximation they show that this design criterion, up to the order $O(n^{-2})$, where n is the number of data, is equivalent to:

$$\int_{\mathbf{X}} [\sigma_\theta^2(x_0) + \mathrm{tr}\left(\kappa_\theta^{-1}\left\{\frac{\partial \lambda_\theta(x_0)}{\partial \theta^T}\right\}^T \mathbf{K}_\theta \frac{\partial \lambda_\theta(x_0)}{\partial \theta^T}\right) \tag{4.15}$$

$$+ z_{1-\alpha/2}^2 \left\{\frac{\partial \sigma_\theta(x_0)}{\partial \theta}\right\}^T \kappa_\theta^{-1} \frac{\partial \sigma_\theta(x_0)}{\partial \theta}]dx_0 \to \underset{d_n=\{x_1,\ldots,x_n\}}{Min}$$

Here

$$(K_\theta)_{ij} = \mathrm{tr}\left(\mathbf{W}_\theta \frac{\partial \mathbf{K}_\theta}{\partial \theta_i} \mathbf{W}_\theta \frac{\partial \mathbf{K}_\theta}{\partial \theta_j}\right) \tag{4.16}$$

is the Fisher information matrix for REML,

$$\mathbf{W}_\theta = \mathbf{K}_\theta^{-1} - \mathbf{K}_\theta^{-1}\mathbf{F}(\mathbf{F}^T\mathbf{K}_\theta^{-1}\mathbf{F})^{-1}\mathbf{F}^T\mathbf{K}_\theta^{-1},$$

$z_{1-\alpha/2}$ is the $1-\alpha/2$-quantile of the standard normal distribution, $\sigma_\theta^2(x_0)$ is the universal kriging variance at x_0 and $\lambda_\theta(x_0)$ is the universal kriging weights vector for prediction at x_0. This design criterion takes account of both prediction accuracy and covariance uncertainty.

Sections 4.3 and 4.4 have demonstrated that by using the BSLM (4.9) as approximation to the true isotropic random field the design criterion of I-optimality can be completely expressed in terms of the (concrete) Bayesian information matrix

$$\mathbf{M}_B = \mathbf{H}^T\mathbf{H} + \sigma_0^2\mathbf{\Gamma}^{-1}.$$

Going from this information matrix to its continuous version $\mathbf{M}_B(\xi)$ according to (4.11), the extended design functional $\Psi(\mathbf{M}_B(\xi)) = \mathrm{tr}(\mathbf{U}\mathbf{M}_B(\xi)^{-1})$ becomes continuous and convex on the compact and convex set of all such information matrices $\mathbf{M}_B(\xi)$. This was the reason why classical convex experimental design algorithms could be used to find optimal spatial sampling designs minimizing the criterion (4.13).

In Spöck *et al.* (2012) it is shown that the Smith and Zhu design criterion has also some favourable properties, so that classical convex experimental design theory can be applied to this design criterion, too:

- Expression (4.15) can be expressed completely in terms of the Bayesian information matrix $\mathbf{M}_B$.
- The design functional is continuous on the convex and compact set of all $\mathbf{M}_B(\xi)$ and has some advantageous properties according to which classical experimental design algorithms may be used in order to find spatial sampling designs.

Assuming the BSLM (4.9) the covariance function is actually parametrized through the diagonal matrix $\mathbf{A}$ and the nugget variance σ_0^2. Since the Smith and Zhu design criterion assumes the covariance parameters to be estimated by REML we actually estimate this diagonal matrix $\mathbf{A}$ and σ_0^2 by this methodology. The a priori covariance matrix $\mathbf{\Phi} = \text{cov}(\beta)$ must be given almost infinite diagonal values because the Smith and Zhu (2004) approach assumes the trend parameter vector β to be estimated by generalized least squares and $\mathbf{\Phi} \to \infty$ bridges the gap from Bayesian linear to generalized least squares trend estimation. The a priori mean $\mu = \text{E}(\beta)$ can be set to $\mathbf{0}$ then.

According to the polar spectral representation (4.6) several values in the diagonal matrix $\mathbf{A}$ are identical:

$$\mathbf{A} = \text{diag}(\{d_m a_i^2\}_{m=0,\ldots,M;i=1,\ldots,n;k=1,2}), \tag{4.17}$$

where the definitions of d_m and a_i^2 and the indexing derive from the polar spectral representation (4.6). For REML estimation of $\mathbf{A}$ we have two possibilities:

- We can leave the a_i's unspecified: this approach is almost nonparametric because a lot of a_i's and corresponding frequencies w_i are needed to get the isotropic random field properly approximated, and corresponds to a semiparametric estimation of the spectral distribution function via a step function.
- We can specify a parametric model for the a_i^2's: the polar spectral density function for an isotropic random field over $\Re^2$ possessing for example an exponential covariance function $B(h) = C\exp(-\frac{3h}{\alpha})$ is given by

$$g(w) = \frac{C\frac{3}{\alpha}w}{((\frac{3}{\alpha})^2 + w^2)^{3/2}}. \tag{4.18}$$

  The polar spectral density function is defined just as the first derivative of the polar spectral distribution function $G(w)$. A possible parametrization for the a_i^2's then is

$$a_i^2 = \frac{g(w_i) + g(w_{i-1})}{2}(w_i - w_{i-1}), i = 1, 2, \ldots, n, \tag{4.19}$$

  where $0 = w_0 < w_1 < \ldots, w_n$ are fixed frequencies.

For the optimization of the Smith and Zhu design criterion we make use of the same exchange design algorithms as described in Section 4.7.3. We only have to replace $\Psi(\mathbf{M}_B(\xi))$ by the Smith an Zhu design functional given in Spöck *et al.* (2012).

4.6 Spatial sampling design for trans-Gaussian kriging

In trans-Gaussian kriging the originally positive valued data $Z(x_i)$, $i = 1, 2, \ldots, n$ are transformed to Gaussianity by means of the Box–Cox transformation

$$g_\lambda(z) = \begin{cases} \frac{z^\lambda - 1}{\lambda} & : \lambda \neq 0 \\ \log(z) & : \lambda = 0 \end{cases}.$$

Let $\mathbf{Z} = (Z(x_1), Z(x_2), \ldots, Z(x_n))^T$ be the vector of original data and

$$\mathbf{Y} = (g_\lambda(Z(x_1)), g_\lambda(Z(x_2)), \ldots, g_\lambda(Z(x_n)))^T \tag{4.20}$$

be the vector of transformed data. The predictive density for trans-Gaussian kriging at a location x_0 then may be written:

$$\varphi(g_\lambda(z); \hat{Y}_{OK}(x_0), \sigma^2_{OK}(x_0)) * z^{\lambda-1}, \tag{4.21}$$

where $\varphi(.; \hat{Y}_{OK}(x_0), \sigma^2_{OK}(x_0))$ is the Gaussian density with mean equal to the ordinary kriging predictor $\hat{Y}_{OK}(x_0)$ at x_0 and based on the transformed variables $\mathbf{Y}$, and variance equal to the ordinary kriging variance $\sigma^2_{OK}(x_0)$, $z^{\lambda-1}$ is the Jacobian of the Box–Cox transformation.

For spatial sampling design we can consider again the average expected length of $(1-\alpha)$-predictive intervals. In order to make the expected lengths of predictive intervals also dependent on REML estimation of the covariance function, we can consider instead of the Gaussian density $\varphi(.; \hat{Y}_{OK}(x_0), \sigma^2_{OK}(x_0))$ that unique Gaussian density φ whose 0.025- and 0.975-quantiles are given by the Smith and Zhu (2004) 95% predictive interval

$$\begin{aligned} &\hat{Y}_{OK}(x_0) \\ &\pm 1.96\ \sigma_\theta(x_0) \left\{ 1 + \frac{1}{2\sigma^2_\theta(x_0)} \left[\mathrm{tr}\left(\kappa_\theta^{-1} \left\{ \frac{\partial \lambda_\theta(x_0)}{\partial \theta^T} \right\}^T \mathbf{K}_\theta \frac{\partial \lambda_\theta(x_0)}{\partial \theta^T} \right) \right.\right. \\ &\left.\left. + 1.96^2 \left\{ \frac{\partial \sigma_\theta(x_0)}{\partial \theta} \right\}^T \kappa_\theta^{-1} \frac{\partial \sigma_\theta(x_0)}{\partial \theta} \right] \right\}. \end{aligned} \tag{4.22}$$

Last but not least, to get expected predictive intervals we must replace in the statistic $t(\mathbf{Y}) = \hat{Y}_{OK}(x_0)$ every variable $Y(x_i)$ for which we do not have data by its ordinary kriging predictor based on the available data. Furthermore, we note that in the above approach we have not taken into account the fact that the transformation parameter λ itself is estimated too, i.e. by maximum likelihood, and then is plugged into the ordinary kriging predictor.

4.7 The spatDesign toolbox

The spatial sampling design and geostatistics toolbox spatDesign has been developed since 2003. It can be run in both MATLAB and Octave and can be downloaded from:

- http://wwwu.uni-klu.ac.at/guspoeck/spatDesignMatlab.zip
- http://wwwu.uni-klu.ac.at/guspoeck/spatDesignOctave.zip

The toolbox underlies the GNU Public Licence Version 3 or higher and thus is freely available. In MATLAB (www.mathworks.com) the toolbox is fully functional but assumes that also the MATLAB Optimization and Statistics toolboxes are installed. In Octave the Smith and Zhu criterion is not implemented. The spatial sampling design functions corresponding to the Smith and Zhu design criterion need (on a standard PC) a lot of computing time. For this reason this part of the toolbox has been parallelized to work with NVIDIA GPUs and the freely available MATLAB parallelization package GPUmat (www.gp-you.org). If there is a CUDA (www.nvidia.com) compatible graphics card installed then this will be automatically detected and the parallelized algorithms for the Smith and Zhu design criterion will be used. Unfortunately, almost no effort has been made so far to parallelize Octave, so the Smith and Zhu part of the Octave toolbox will not work unless there is ample time to wait for the results in the unparallelized version.

Wolfgang Nowak from the Institute of Hydraulic Engineering (IWS), University of Stuttgart, has added his FFT-Kriging Toolbox to the software package (Fritz *et al.* 2009). This is really fast MATLAB and Octave code for Bayesian linear kriging with external drift and works in both two- and three-dimensional Euclidean space. The original kriging code of the spatDesign toolbox is not so fast as this code and implements only two-dimensional interpolation.

The spatDesign Toolbox provides code for all three areas of geostatistics:

- covariance estimation and variography
- spatial interpolation and kriging
- spatial sampling design and planning of monitoring networks.

However, the largest emphasis of the toolbox is on spatial sampling design and the planning of monitoring networks.

4.7.1 Covariance estimation and variography software

For covariance estimation and variography the empirical semivariogram estimator of Matheron (1962)

$$\hat{\gamma}(h) = \frac{1}{2|N(h)|} \sum_{(i,j)\in N(h)} (Y(x_i) - Y(x_j))^2, \tag{4.23}$$

where

$$N(h) = \begin{cases} \{(i, j) : x_i - x_j \approx h\} : \textit{stationarity} \\ \{(i, j) : ||x_i - x_j||_2 \approx h\} : \textit{isotropy} \end{cases} \tag{4.24}$$

is implemented in both forms, the isotropic one and the general stationary one. The corresponding MATLAB and Octave functions are EMPVARIOGRAM.m and EMPVARIOGRAMANISO.m.

Having calculated the empirical semivariogram a theoretical semivariogram model can be fitted to the empirical semivariogram by means of weighted least squares. As the theoretical semivariogram model a nested model or convex combination of an exponential and a Gaussian semivariogram model with different ranges and nugget is implemented. The corresponding MATLAB and Octave functions are WEIGHTEDLEASTSQUARES.m for the isotropic case and WEIGHTEDLEASTSQUARESANISO.m for the geometrically anisotropic case.

The toolbox also implements transformed-Gaussian kriging. For this reason a maximum likelihood estimation procedure has been implemented that estimates both the Box–Cox transformation parameter λ and the eventually geometrically anisotropic covariance function by means of a profile likelihood approach. This function is quite slow since the likelihood function is iteratively maximized either in λ, the covariance parameters or the anisotropy parameters. The function performing these tasks is called ESTIMATE_TRANSFO_COV_ML.m.

In the functions for covariance estimation mentioned before the trend of the random field is assumed to be constant. For the case that the trend is of the form

$$m(x) = \mathbf{f}(x)^T \beta \tag{4.25}$$

it must be eliminated before the covariance function or the semivariogram can be estimated. For this case the function CALCULATE_RESIDUALS_TREND_COVARIANCE.m has been implemented. Starting with a least squares estimate of β and the corresponding residuals, empirical semivariograms of residuals, corresponding weighted least squares fits to the empirical semivariogram, generalized least squares estimates of β, and again residuals and semivariograms, are calculated iteratively, until convergence.

4.7.2 Spatial interpolation and kriging software

As simplest interpolation routine, Voronoi interpolation is implemented in the function VORONOIPOLYGONALINTERPOLATIONONGRID.m. Ungauged locations are given the same value as the closest datum location. In addition to Voronoi interpolation Bayesian linear kriging with external drift and transformed-Gaussian kriging based on the Box–Cox transformation are also implemented.

The corresponding functions are named KRIGELINEARBAYESONGRID.m and TRANSGAUSSIANKRIGINGONGRID.m. The output of the first function consists of the Bayesian kriging predictions and corresponding TMSEPs on a grid. The output of the second function are the skew predictive distributions from transformed-Gaussian kriging on a grid. From the complete predictive distributions several statistics may be derived: the function VISUALIZEPOSTQUANTILE.m allows to calculate and visualize several quantiles of the predictive distributions as well as the means, medians and modes of these distributions. The function VISUALIZEPROBGREATER.m allows to calculate maps that give the probabilities that certain thresholds are exceeded. Finally, the functions CROSSVALIDATION.m and VISUALIZECROSSVALIDATION.m calculate crossvalidation statistics like mean absolute errors, percentages of actual data below the quantiles of the predictive distribution and percentages of actual data above threshold versus expected percentages of data above threshold.

4.7.3 Spatial sampling design software

spatDesign implements three design criteria for Bayesian linear kriging, where the first two are criteria for prediction only, with the covariance function assumed to be certain. The third criterion is the Smith and Zhu (2004) criterion taking also account of the fact that the covariance function is estimated. The Smith and Zhu design criterion has also been implemented for transformed-Gaussian kriging. The actual version of the toolbox is spatDesign V.2.2.0.

The implemented criteria for prediction only are:

- I-optimality:

$$\Psi(\mathbf{M}_B(d_n)) = \mathrm{tr}(\mathbf{U}\mathbf{M}_B(d_n)^{-1}) \rightarrow \min_{d_n} \tag{4.26}$$

$$\mathbf{U} = \sum_{i,j=1}^{m} \mathbf{h}(x_{i,j})\mathbf{h}(x_{i,j})^T; \tag{4.27}$$

  where the integral in (4.12) has been replaced by the sum over a fine grid of locations $x_{i,j} \in \mathbf{X}$.

- D-optimality:

$$\Psi(\mathbf{M}_B(d_n)) = |(\mathbf{H}^T(d_n)\mathbf{H}(d_n) + \sigma_0^2\mathbf{\Gamma}^{-1})^{-1}| \rightarrow \min_{d_n} \tag{4.28}$$

The Smith and Zhu design criterion can be expressed (according to Sections 4.5 and 4.6) also in the form

$$\Psi(\mathbf{M}_B(d_n)) \rightarrow \min_{d_n}, \tag{4.29}$$

The interested reader is referred to Spöck *et al.* (2012).

The basic algorithm for calculating spatial sampling designs is an exchange algorithm from experimental design theory going back to Fedorov (1972). Contrary to the construction of optimal discrete designs, here we cannot prove convergence of the exact designs to the functional value $\Psi(d^*)$ of an optimal exact design d^*; we can only guarantee stepwise improvement of a given exact starting design, i.e. the sequence of functional values $\Psi(d_{n,s})$ decreases monotonically with increasing iteration index s. The algorithm is an exchange algorithm improving n-point designs and starting from an initial design.

4.7.3.1 Exchange algorithm

Step 1. Use some initial design $d_{n,1} = \{x_{1,1}, \dots, x_{n,1}\} \in \mathbf{X}^n$ of size n.

Step 2. Beginning with $s = 1$ form the design $d_{n+1,s} = d_{n,s} + (x_{n+1,s})$ by adding the point

$$x_{n+1,s} = \arg \min_{x \in \mathbf{X}} \Psi(\mathbf{M}_B(d_{n,s} + (x)))$$

to $d_{n,s}$.

Then form $d_{n,s}^j = d_{n+1,s} - (x_{j,s})$, $j = 1, 2, \dots, n+1$ and delete that point $x_{j^*,s}$ from $d_{n+1,s}$ for which

$$\Psi(\mathbf{M}_B(d_{n,s}^{j^*})) = \min_{j \in \{1,\dots,n+1\}} \Psi(\mathbf{M}_B(d_{n,s}^j))).$$

Step 3. Repeat Step 2 until the point to be deleted is equivalent to the point to be added.

For the design functional (4.13) Step 2 is determined as follows:

$$x_{n+1,s} = \arg \max_{x \in \mathbf{X}} \frac{\mathbf{h}(x)^T \mathbf{M}_B(d_{n,s})^{-1} \mathbf{U} \mathbf{M}_B(d_{n,s})^{-1} \mathbf{h}(x)}{n + \mathbf{h}(x)^T \mathbf{M}_B(d_{n,s})^{-1} \mathbf{h}(x)}$$

$$j^* = \arg \min_{1 \le j \le n+1} \frac{\mathbf{h}(x_{j,s})^T \mathbf{Q}_B(d_{n+1,s}) \mathbf{h}(x_{j,s})}{n + 1 - \mathbf{h}(x_{j,s})^T \mathbf{M}_B(d_{n+1,s})^{-1} \mathbf{h}(x_{j,s})},$$

where

$$\mathbf{Q}_B(d_{n+1,s}) = \mathbf{M}_B(d_{n+1,s})^{-1} \mathbf{U} \mathbf{M}_B(d_{n+1,s})^{-1}.$$

For the Smith and Zhu design criterion no such simplification exists and the complete design functional $\Psi(.)$ must be recalculated in every step.

4.7.3.2 Generation of an initial design

The initial design is a one-point design which minimizes the design functional among all designs of size $n = 1$. Note that such a design exists since the Bayesian information matrix is positive definite even for designs of size $n = 1$.

Step 1. Choose $x_1 \in \mathbf{X}$ such that $x_1 = \arg\min_{x \in \mathbf{X}} \Psi(\mathbf{M}_B((x)))$, and set $d_1 = (x_1)$.

Step 2. Beginning with $i = 1$, find x_{i+1} such that $x_{i+1} = \arg\min_{x \in \mathbf{X}} \Psi(\mathbf{M}_B(d_i + (x)))$ and form $d_{i+1} = d_i + (x_{i+1})$.
Continue with i replaced by $i + 1$ until $i + 1 = n$.

Step 3. If $i + 1 = n$ then stop and take $d_{n,1} = \{x_1, \dots, x_n\}$ as an initial design.

4.7.3.3 Combination of the above two algorithms

It is a good idea to combine the initial design algorithm and the exchange algorithm in the following way:

Step 1. Start with the initial design algorithm and find a design with one first design point.

Step 2. Having found a design with $m \geq 1$ design points apply the exchange algorithm to this design to improve it.

Step 3. Add to the design from Step 2 one further design point by means of the initial design algorithm to get $m + 1$ design points.

Step 4. Go back to Step 2 and iterate Step 2 and Step 3 until you have found n desired design points.

4.7.3.4 Reduction of experimental designs

Often it is desired to reduce a given experimental design $d = \{x_1, x_2, \ldots, x_n\}$ to one including only $m < n$ design points from d:

Step 1. Delete that design point x_{j*} from d for which $x_{j*} = \arg\min_{x_j \in d} \Psi(\mathbf{M}_B(d - (x_j)))$, and set $d := d - (x_{j*})$.

Step 2. Iterate Step 1 until the design d contains only m design points.

Also this algorithm may be combined with an improvement step similar to the exchange algorithm. In the exchange algorithm merely the calculation of $x_{n+1,s}$ has to be replaced by

$$x_{n+1,s} = \arg \min_{x \in d - d_{n,s}} \Psi(\mathbf{M}_B(d_{n,s} + (x))),$$

where d is the inital design that has to be reduced. This improved algorithm has the advantage that design points once deleted can reenter the design in the exchange step.

4.7.3.5 Inverse of the information matrix

The calculation of exact designs requires in every step the calculation of the inverses of the information matrices $\mathbf{M}_B(d_{n,s})$ or $\mathbf{M}_B(d_{n+1,s})$. In the next sections we will see that these information matrices can have a quite high dimension of about 3000×3000. So, how can one invert such large matrices in affordable time?

There is computationally no need to make explicit use of numerical matrix inversion algorithms, when one considers the update formulae (13.26) and (13.28) in Pilz (1991):

$$\mathbf{M}_B(d_{n,s} + (x))^{-1} = \tfrac{n+1}{n} \left\{ \mathbf{M}_B(d_{n,s})^{-1} - \frac{\mathbf{M}_B(d_{n,s})^{-1}\mathbf{h}(x)\mathbf{h}(x)^T\mathbf{M}_B(d_{n,s})^{-1}}{n + \mathbf{h}(x)^T\mathbf{M}_B(d_{n,s})^{-1}\mathbf{h}(x)} \right\},$$

$$\mathbf{M}_B(d_{n,s}^j)^{-1} = \frac{n}{n+1}\left\{\mathbf{M}_B(d_{n+1,s})^{-1} + \frac{\mathbf{M}_B(d_{n+1,s})^{-1}\mathbf{h}(x_{j,s})\mathbf{h}(x_{j,s})^T\mathbf{M}_B(d_{n+1,s})^{-1}}{n+1-\mathbf{h}(x_{j,s})^T\mathbf{M}_B(d_{n+1,s})^{-1}\mathbf{h}(x_{j,s})}\right\}$$

Obviously, only matrix and vector multiplications are needed in these update formulae.

4.7.3.6 Computation

The I-optimality criterion has the additional advantage that the above optimization algorithms can be computationally further simplified. Even more important, the averaging over the design area $\mathbf{X}$ corresponding to Equation (4.12) has to be done only once. The Smith and Zhu design criterion and the design criterion for trans-Gaussian kriging no longer have this advantage and the averaging over the design area $\mathbf{X}$ corresponding to Equations (4.15) and (4.22) must be done whenever $\Psi(\mathbf{M}_B(d))$ is calculated. Both Equations (4.12) and (4.15) have been simplified by means of replacing the integrals by discrete sums over a fine grid of locations $x_{i,j} \in \mathbf{X}$. Also, the calculation of the minima in the above algorithms has been simplified by means of searching for the minimum over a fine grid of locations. When the minimum is found over the discrete grid we further iterate with a line search algorithm with this minimum as a starting value to reach the actual global minimum. However, the averaging for calculating the integral in the Smith and Zhu design criterion and the design criterion for trans-Gaussian kriging is computationally too intensive to be performed on a standard PC.

Recently, CUDA technology (www.nvidia.com) has been developed for NVIDIA graphic cards. This allows to make use of the parallel performance of these graphic cards and to put intensive floating point operations to these GPUs. Actually, we are now doing the averaging operations correponding to expressions (4.15) and (4.22) in parallel on a multiprocessor NVIDIA GTX 580 GPU. To this, we have installed GPUmat (www.gp-you.org), a free software for MATLAB (www.mathworks.com) that automatically can thread operations to the NVIDIA GPU. The usage of this software is quite easy. We just have to specify MATLAB objects that we want to calculate on the GPU as GPUdouble or GPUsingle. The gain in performance is tremendous: processing on the NVIDIA GTX 580 is between 100 and 200 times faster than on a standard Intel i7 8 Core 3.06 Ghz CPU.

4.7.3.7 Basic sampling design functions

The basic spatial sampling design functions are:

- OPTIMALLY_DELETE_N_LOCATIONS_FROM_POOLDELETE.m
- OPTIMALLY_ADD_N_LOCATIONS_FROM_POOLCOMPLETE.m
- OPTIMALLY_ADD_N_LOCATIONS_FROM_POOLADD.m

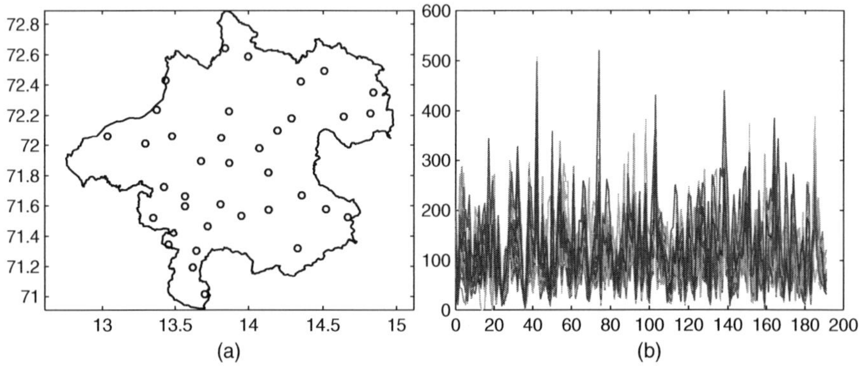

Figure 4.1 *(a) The 36 sampling locations of the Upper Austria rainfall dataset. (b) The 36 time series of average monthly rainfall at each station.*

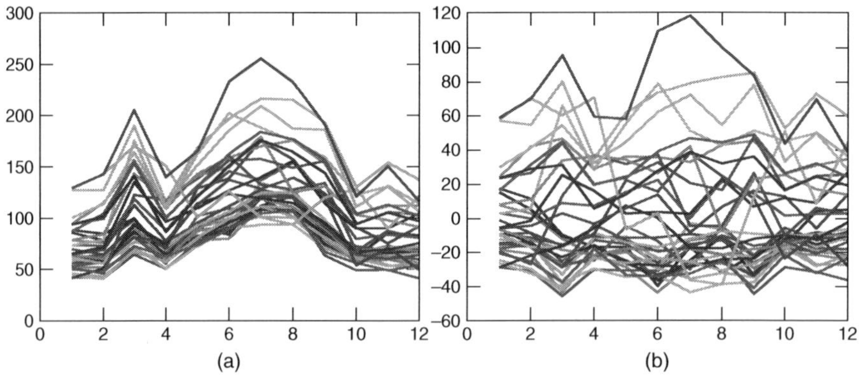

Figure 4.2 *(a) The average monthly rainfall over the years at each of the 36 stations, for each of the 12 months. (b) The residual rainfall at each of the 36 stations, for each of the 12 months. (Please see plate section for color version of the figure.)*

- trendrainmeanmonth=mean(rainmeanmonth,1);
- resrainmeanmonth=rainmeanmonth-ones(36,1)*trendrainmeanmonth;
  figure
  plot(resrainmeanmonth')
  title('monthly mean of rain residuals')

Let us now calculate from the rain residuals for each of the 12 months the empirical semivariogram and the semivariogram standardized by the variance of each month (Figure 4.3):

- for i=1:12
  empvarioresrainmeanmonth{i}= empvariogram(xyelevation(:,1),
  xyelevation(:,2),resrainmeanmonth(:,i),0:0.25:4);
  end

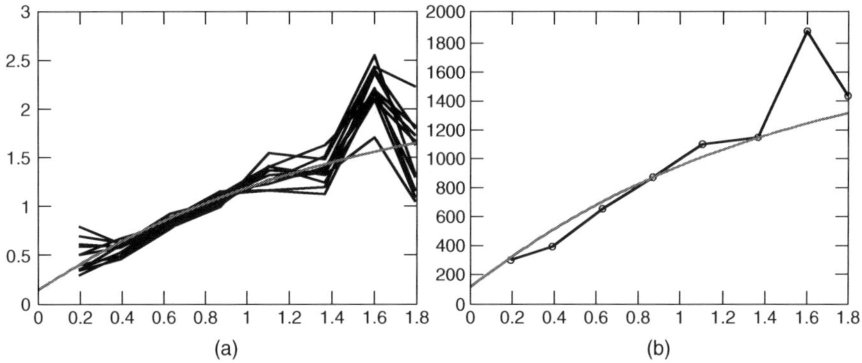

Figure 4.3 (a) Standardized empirical semivariograms for all of the 12 months and a fitted exponential semivariogram (grey). (b) Empirical semivariogram for the mean of rain residuals and a fitted exponential semivariogram (grey).

- for i=1:12
  plot(empvarioresrainmeanmonth{i}.lagh,empvarioresrainmeanmonth{i}.v
  /empvarioresrainmeanmonth{i}.var)
  hold on
  end
  hold on
  plot(0:0.05:1.8,2.3-(2.166666*exp(-3*(0:0.05:1.8)/4.5)))
  delta0=[0.1333,2.166666,4.5,0,0];

Next, let us calculate for each station the mean of rain residuals, the empirical semivariogram corresponding to these means and a fitted exponential semivariogram (Figure 4.3):

- meanresrainmeanmonth=mean(resrainmeanmonth')';
- empvario=empvariogram(xyelevation(:,1),xyelevation(:,2),
  meanresrainmeanmonth,0:0.25:4);
  figure
  plot(empvario.lagh,empvario.v,'o-')
  var0=var(meanresrainmeanmonth);
  hold on
  plot(0:0.05:1.8,var0*(2.3-(2.166666*exp(-3*(0:0.05:1.8)/4.5))))
- delta0rain(1:2)=var0*delta0(1:2); % parameters of the exponential
  delta0rain(3:5)=[4.5,0,0]; % semivariogram in Figure 4.3(b)

The fact that the standardized semivariograms are almost the same for all months means that the space-time random field is separable and that we can use one and the same semivariogram [the grey one in Figure 4.3(b)] for doing spatial sampling design for each month. In the next step we calculate the polar spectral distribution function corresponding to this semivariogram. Obviously, this spectral distribution function almost attains its maximum of 1735.2 at $w = 47$. We now select the frequencies $w_i, i = 1, 2, \ldots, 34$, calculate an approximation to the spectral distribution function via a step function (the steps are the a_i^2) and check whether this approximation to the spectral distribution function provides a good fit to the original covariance function (Figure 4.4):

- load wscaled
- plotspectraldist(0:0.5:47,delta0rain)
- w=wscaled*47;
- [wrain,deltarain]=expstep(w,delta0rain(3),delta0rain(2)); %the discrete spectrum
- hold on, approxspectraldist(wrain,deltarain)
- plotcovarianceapprox(wrain,45,deltarain,delta0rain,12.5:0.05:17,73,12.5, 12.5,15,70.8,73); %the worst approximating covariance function

A look at the approximating covariance function in Figure 4.4 shows that at the origin the difference between the true covariance function and the approximating covariance function is 20. This is small scale variation that the approximating covariance function does not take into account. Later in spatial sampling design we will add this value of 20 to the nugget effect 106.8 of the true covariance function. Thus, $20 + 106.8$ is the variance of the uncorrelated error process $\epsilon_0(x)$ in our BSLM (4.9).

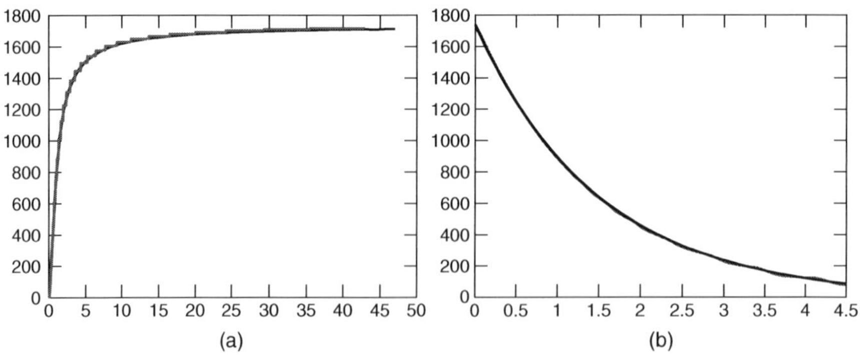

Figure 4.4 (a) Polar spectral distribution function and its approximation (grey). (b) True covariance function (black) and its worst approximation (grey). Note, the covariance function and its approximation are almost indistinguishable. The same is true also for the spectral distribution function and its approximation.

We now have all the quantities that we need in order to do spatial sampling design on the basis of our BSLM (4.9), corresponding to the assumption of Gaussianity of observations.

4.8.2 Optimal design for the BSLM

We consider adding 14 additional sampling locations from the complete design region **X** to the available grid of 36 sampling locations (Figure 4.5).

- [xadd,yadd,avglengthpredint]=optimally_add_n_locations_from_poolcomplete(
  {},..% no external drift
  [],..% no need to specify the matrix **U**
  xyelevation(:,1),xyelevation(:,2),ones(36,1),..% the available data locations
  20+delta0rain(1),.....................% the variance of the uncorrelated error
  ....% process $\epsilon_0(x)$
  [1000000, 0, 0; 0, 0.0000001, 0; 0, 0, 0.0000001],..% the a priori variance of
  ....% the constant trend must be given
  ....% almost infinite variance; no linear drift
  wrain,.......................................% the frequencies of the Bessel harmonics
  45,...% the largest angular frequency
  deltarain,...................................% the a_i^2
  delta0rain,.................................% the parameters of the
  ....% exponential covariance function
  1,...% the Box-Cox transformation parameter
  ....% (no transformation, Gaussian kriging)
  14,...% we want to add 14 samples
  12.5,15,70.8,73,.......................% the size of the design region
  10,...% maximally iterate 10 times in the exchange
  ....% step
  boundary,..................................% the polygonal design region
  17,17,.......................................% discretization in Easting and Northing
  ....% when considering new samples
  'z',...% we apply the Smith and Zhu (2004) design
  ....% criterion to ordinary kriging
  0..% no graphical output
  );

Figure 4.5 shows the optimal 8- and 14-point designs. Obviously, certain locations have been selected with multiplicities larger than 1. The reason is that the Smith and Zhu design criterion does not only take account of best prediction but also of covariance estimation; in order to get the nugget effect and the behaviour of the covariance function close to its origin well estimated locations are needed in the optimal design which are close to each other. Figure 4.6 plots the decrease of the average of the expected lenghts of the 95% predictive intervals when adding up to 14 design locations.

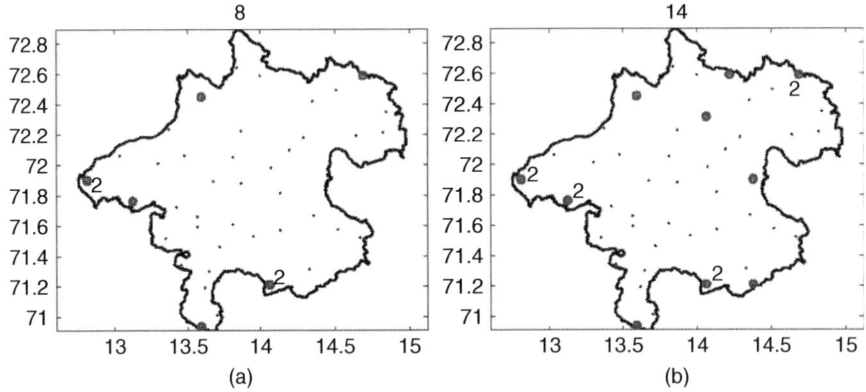

Figure 4.5 (a) Optimal 8-point design for Gaussian kriging. (b) Optimal 14-point design for Gaussian kriging. Certain locations have been selected with multiplicities larger than 1.

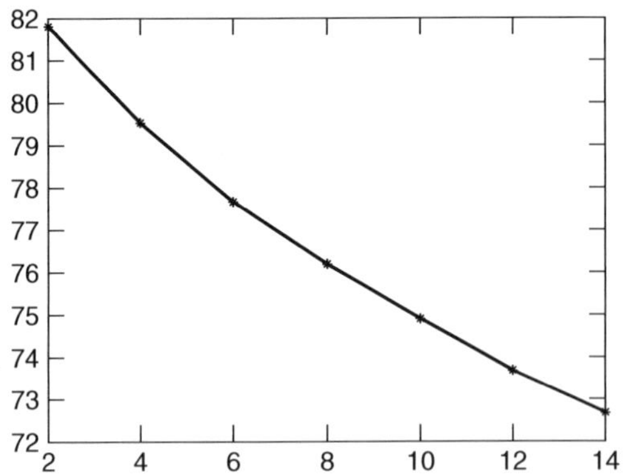

Figure 4.6 Average of the expected lengths of 95% predictive intervals, when adding up to 14 design locations.

The calculation of the optimal 14-point design takes about 1 week, on an Intel i7 8 Core CPU and a NVIDIA 580 GPU with 1.6 Gb RAM. Similar calculations for the simpler design functional (4.26) take, without NVIDIA Cuda support, 1 day.

4.8.3 Design for the trans-Gaussian kriging

In the above example session we have assumed the data to be Gaussian and have intended to use ordinary kriging for prediction, although, as is visible from Figure 4.7, the data are not Gaussian. Thus, we will consider now the assumption that the rainfall residuals can be transformed to Gaussianity by means of a Box–Cox transformation. Because the Box–Cox transformation works only for positive valued data we have to add a positive offset to the 12

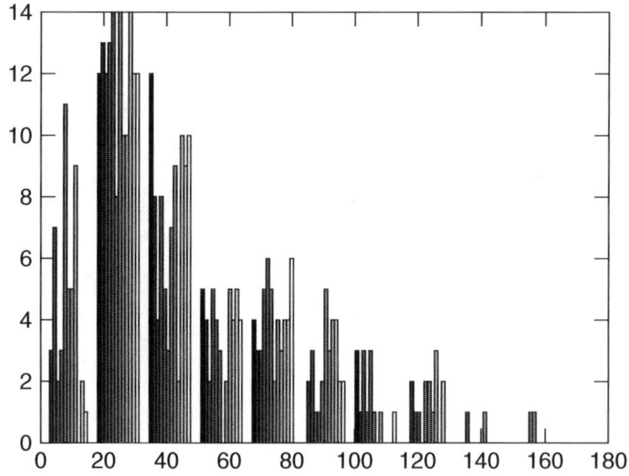

Figure 4.7 The 12 histograms of the rainfall residuals+53.

monthly sets of rainfall residuals. To identify the appropriate offset and optimal Box–Cox transformation parameter λ_0 we perform a sequence of Lilliefors tests for Gaussianity on the transformed rainfall residuals. We then retain this offset and the corresponding Box–Cox transformation parameter λ_0, where the sum of the 12 p-values from the Lilliefors tests attains its maximum. The following code demonstrates this idea:

- ```
  k=0;
  offset=46.19:100;
  nj=length(offset);
  lam=-1:0.05:1;
  ```
- ```
  for j=1:nj
  for l=1:41
  k=k+1;
  for i=1:12
  [ h(i),p(i) ] = lillietest(transform(offset(j)+resrainmeanmonth(:,i),
  lam(l)),0.1);
  end
  H(k)=sum(h);
  grid.x(k)=offset(j);
  grid.y(k)=lam(l);
  P(k)=sum(p);
  end
  end
  ```
- ```
  [xi,yi] = meshgrid(46.19:0.05:100,-1:0.05:1);
  ```
- ```
  figure
  PP = griddata(grid.x',grid.y',P',xi,yi,'cubic');
  ```

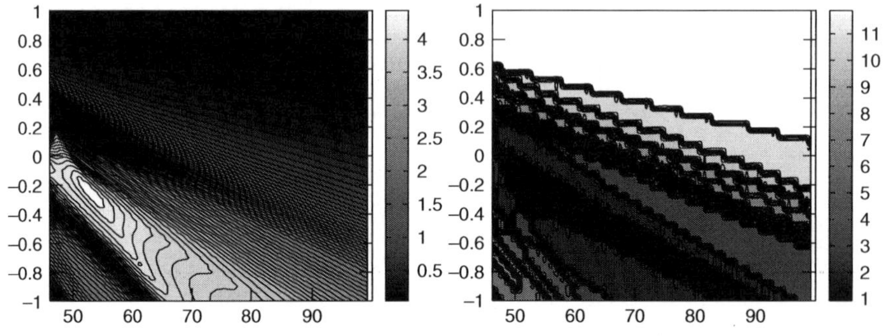

Figure 4.8 (a) Sum of the p*-values, depending on the transformation parameter λ and the offset. (b) Number of rejected hypotheses of Gaussianity at 10% significance level, depending on the transformation parameter λ and the offset.*

```
contourf(xi,yi,PP,60)
colorbar
```

- ```
  figure
  HH = griddata(grid.x',grid.y',H',xi,yi,'cubic');
  contourf(xi,yi,HH,60)
  colorbar
  ```

Figure 4.8 gives the corresponding surfaces of the sum of p-values and number of rejected hypotheses for Gaussianity at the 10% significance level. According to these figures the optimal parameters are chosen as:

- offset=53
- $\lambda_0 = -0.25$

Obviously, for these parameters only one hypothesis of Gaussianity is rejected at the 10% significance level.

We now proceed as in the Gaussian case. Figure 4.9 shows standardized empirical semivariograms of the Box–Cox transformed rain residuals and the semivariogram of the Box–Cox transformed means of rain residuals. Interestingly, the fitted semivariograms in this figure are just scaled versions of the fitted semivariograms from Figure 4.3. The parameters of the semivariogram in Figure 4.9(b) are delta0Box = [0.0047, 0.0768, 4.5000, 0, 0]. By means of

- ```
  load wscaled
  figure
  plotspectraldist(0:0.5:50,delta0Box)
  wBox=wscaled*50;
  [wBox,deltaBox]=expstep(wBox,delta0Box(3),delta0Box(2));
  hold on, approxspectraldist(wBox,deltaBox)
  ```

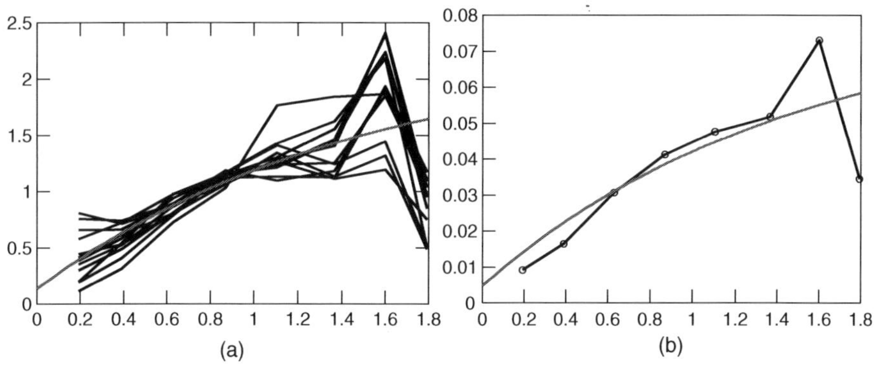

Figure 4.9 (a) Standardized empirical semivariograms of the transformed rain residuals for all of the 12 months and a fitted exponential semivariogram (grey). (b) Empirical semivariogram for the transformed mean of rain residuals and a fitted exponential semivariogram (grey).

```
figure
plotcovarianceapprox(wBox,45,deltaBox,delta0Box,12.5:0.05:17,73,12.5,
12.5,15,70.8,73)
```

we go on to calculate the spectral distribution function and its step-wise approximation and the worst approximating covariance function. We note that close to the origin $h = 0$ the difference between the true covariance function and its approximation is 0.001. Finally, the function call

```
• [xxresrainBox,yyresrainBox,avglengthpredintBox]=
  optimally_add_n_locations_from_poolcomplete({},[],xyelevation(:,1),
  xyelevation(:,2),meanresrainmeanmonth,delta0Box(1)+0.001,
  [10000000000000,0,0;0,0.00000000000001,0;0,0,0.00000000000001],
  wBox,45,deltaBox,delta0Box,lambda0,14,12.5,15,70.8,73,
  10,boundary,17,17,'z',0);
```

calculates the optimal 14-point design. Figure 4.10 visualizes optimal 8- and 14-point designs for trans-Gaussian kriging. Figure 4.11 gives the expected lengths of 95% predictive intervals. Obviously, the designs for trans-Gaussian kriging in Figure 4.10 are completely different from the designs for Gaussian kriging in Figure 4.5. Whereas the designs in Figure 4.5 are much more space-filling the design locations in Figure 4.10 have been selected in areas that have high average value of rainfall. This fact becomes more clear when we compare Figure 4.10 with Figure 4.12, where the median rainfall from the predictive distributions of trans-Gaussian kriging is visualized. Obviously, in areas with high average rainfall the average expected length of 95% predictive intervals can be most decreased. This fact results from a fundamental difference between designs for Gaussian random fields and designs for trans-Gaussian

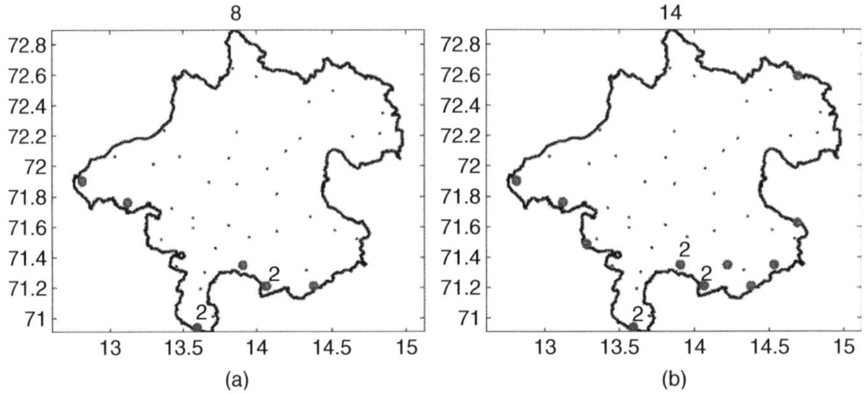

Figure 4.10 (a) Optimal 8-point design for trans-Gaussian kriging. (b) Optimal 14-point design for trans-Gaussian kriging. Certain locations have been selected with multiplicities larger than 1.

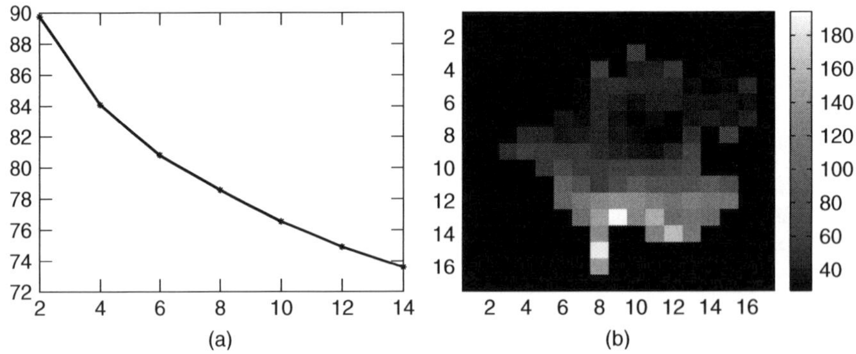

Figure 4.11 (a) Decrease of the average expected lengths of 95% predictive intervals, when adding up to 14 design locations. (b) Expected lengths of 95% predictive intervals corresponding to the optimal 14-point design.

kriging: designs for trans-Gaussian kriging are dependent also on the data, through the ordinary kriging predictor in formula (4.22).

4.9 Conclusions

Our approach to spatial sampling design is motivated by the fact that to date most sampling design implementations use stochastic search algorithms, e.g. simulated annealing. The approach to spatial sampling design proposed in this Chapter makes use of the mathematical structure of the investigated design functionals. Based on the polar spectral approximation of the isotropic random field we are

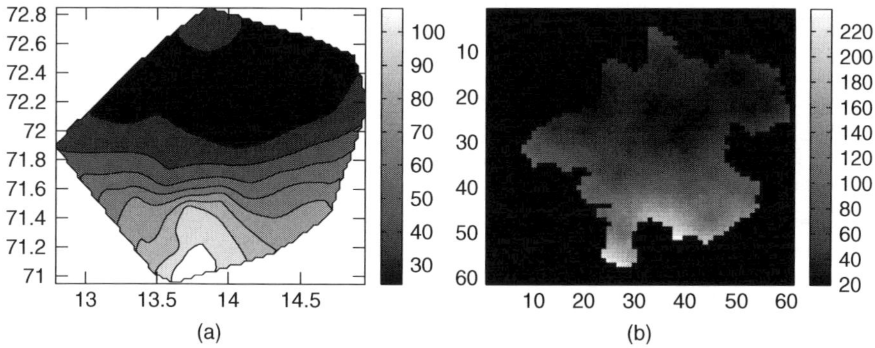

Figure 4.12 *(a) The median rainfall field+53 calculated from the predictive distributions of trans-Gaussian kriging applied to the 36 means of rainfall residuals. (b) Expected lengths of 95% predictive intervals corresponding to the trans-Gaussian kriging from (a). (Please see plate section for color version of the figure.)*

able to use classical experimental design theory for the calculation of spatial sampling designs. To date, a missing link for the Smith and Zhu design criterion and for the design criterion of trans-Gaussian kriging is the fact that we were not yet able to show convexity properties of the corresponding design functionals. This is surely a topic for future research.

We have demonstrated that designs for trans-Gaussian kriging are fundamentally different from designs for Gaussian kriging. Designs for trans-Gaussian kriging are dependent also on data values.

A further topic for future research is to extend the proposed approach also to nonstationary random fields. Some work towards this direction has already been done in Spöck and Pilz (2008). The approach discussed therein locally maintains isotropy but is globally nonstationary, and is based also on the approximation of a so-called locally isotropic random field by means of a cosine-sine-Bessel surface-harmonics regression model with random amplitudes.

References

Aarts, E. H. and Korst, J. (1989). *Simulated Annealing and Boltzman Machines*, John Wiley & Sons, Ltd, New York, 284 p.

Baume, O. P., Gebhardt, A., Gebhardt, C., Heuvelink, G. B. M. and Pilz, J. (2011). Network optimization algorithms and scenarios in the context of automatic mapping, *Computers & Geosciences*, 37, pp. 289–294.

Brown, P. J., Le, N. D. and Zidek, J. V. (1994). Multivariate spatial interpolation and exposure to air pollutants, *The Canadian Journal of Statistics*, 2, pp. 489–509.

Brus, D. J. and de Gruijter, J. J. (1997). Random sampling or geostatistical modeling? Chosing between design-based and model-based sampling strategies for soil (with discussion), *Geoderma*, 80, pp. 1–44.

Brus, D. J. and Heuvelink, G. B. M. (2007). Optimization of sample patterns for universal kriging of environmental variables, *Geoderma*, 138, pp. 86–95.

Caeiro, S., Painho, M., Goovaerts, P., Costa, H. and Sousa, S. (2003). Spatial sampling design for sediment quality assessment in estuaries, *Environmental Modelling and Software*, 18, pp. 853–859.

Chen, V. C. P., Tsui, K. L., Barton, R. R. and Meckesheimer, M. (2006). A review on design, modeling and applications of computer experiments, *IIE Transactions*, 38, pp. 273–291.

Cox Jr, L. A. (1999). Adaptive spatial sampling of contaminated soils, *Risk Analysis*, 19, pp. 1059–1069.

Delmelle E. M. and Goovaerts, P. (2009). Second-phase sampling designs for non-stationary spatial variables, *Geoderma*, doi:10.1016/j.geoderma.2009.08.007.

Diggle, P. and Lophaven, S. (2006). Bayesian Geostatistical Design, *Scandinavian Journal of Statistics*, 33, pp. 53–64.

Dobbie, M. J., Henderson, B. L. and Stevens Jr, D. L. (2008). Sparse sampling: spatial design for monitoring stream networks, *Statistics Surveys*, 2, pp. 113–153.

Fedorov, V. V. (1972). *Theory of Optimal Experiments*, transl. and ed. by W. J. Studden and E. M. Klimko, Academic Press, New York (Russian original: Nauka, Moscow, 1971), 292 p.

Fedorov, V. V. (1996). Design of spatial experiments: model fitting and prediction. In Gosh, S. and Rao, C. R., editors, *Handbook of Statistics*, 13. Elsevier, Amsterdam, pp. 515–553.

Fedorov, V. V. and Flanagan, D. (1997). Optimal monitoring network design based on Mercer's expansion of the covariance kernel, *Journal of Combinatorics, Information and System Sciences*, 23, pp. 237–250.

Fritz, J., Nowak, W. and Neuweiler, I. (2009). Application of FFT-based algorithms for large-scale universal kriging problems. *Mathematical Geosciences*, 41, pp. 509–533.

Fuentes, M., Chaudhuri, A. and Holland, D. M. (2007). Bayesian entropy for spatial sampling design of environmental data, *Journal of Environmental and Ecological Statistics*, 14, pp. 323–340.

Gebhardt, C. (2003). *Bayesian Methods for Geostatistical Design*, PhD Thesis, University of Klagenfurt, Austria, 139 p.

Gramacy, R. B. (2007). tgp: An R package for Bayesian nonstationary semiparametric nonlinear regression and design by treed Gaussian Process models, *Journal of Statistical Software*, 19, 47 p.

Gramacy, R. B. and Lee, H. K. H. (2010). Optimization under unknown constraints. In Bernardo, J. M. *et al.*, editors, *Bayesian Statistics*, 9. Oxford University Press, Oxford, 18 p.

Hengl, T., Rossiter, D. G. and Stein, A. (2003). Soil sampling strategies for spatial prediction by correlation with auxiliary maps, *Australian Journal of Soil Research*, 41, pp. 1403–1422.

Kleijnen, J. P. C. and van Beers, W. C. M. (2004). Application driven sequential design for simulation experiments: kriging metamodeling, *Journal of the Operational Research Society*, 55, pp. 876–893.

Lark, R. M. (2002). Optimized spatial sampling of soil for estimation of the variogram by maximum likelihood, *Geoderma*, 105, pp. 49–80.

Le, N. D. and Zidek, J. V. (2006). *Statistical Analysis of Environmental Space-Time Processes*, Springer, New York, 356 p.

Lim, Y. B., Sacks, J. and Studden, W. J. (2002). Design and analysis of computer experiments when the output is highly correlated over the input space, *Canadian Journal of Statistics*, 30, pp. 109–126.

Matheron, G. (1962). *Traite de geostatistique appliquee*, Tome I. Memoires du bureau de recherches geologiques et minieres principles, 14. Editions Technip, Paris, 333 p.

Mitchell, T. J. and Morris, M. D. (1992). Bayesian design and analysis of computer experiments: two examples, *Statistica Sinica*, 2, pp. 359–379.

Morris, M. D., Mitchell, T. J. and Ylvisaker, D. (1993). Bayesian design and analysis of computer experiments: use of derivatives in surface prediction, *Technometrics*, 35, pp. 243–255.

Müller, P., Sanso, B. and De Iorio, M. (2004). Optimal Bayesian design by inhomogeneous Markov chain simulation, *Journal of the American Statistical Association*, 99, pp. 788–798.

Müller, W. G. (2005). A comparison of spatial design methods for correlated observations, *Environmetrics*, 16, pp. 495–505.

Müller, W. G. and Pazman, A. (1998). Design measures and approximate information matrices for experiments without replications, *Journal of Statistical Planning and Inference*, 71, pp. 349–362.

Müller, W. G and Pazman, A. (1999). An algorithm for the computation of optimum designs under a given covariance structure, *Journal of Computational Statistics*, 14, pp. 197–211.

Müller, W. G. and Zimmerman, D. L. (1999). Optimal designs for variogram estimation. *Environmetrics*, 10, pp. 23–37.

Omre, H. (1987). Bayesian kriging – merging observations and qualified guess in kriging, *Mathematical Geology*, 19, pp. 25–39.

Omre, H. and Halvorsen, K. (1989). The Bayesian bridge between simple and universal kriging, *Mathematical Geology*, 21, pp. 767–786.

Pazman, A. and Müller, W. G. (2001). Optimal design of experiments subject to correlated errors, *Statistics and Probability Letters*, 52, pp. 29–34.

Pilz, J. (1991). *Bayesian Estimation and Experimental Design in Linear Regression Models*, John Wiley & Sons, Ltd, New York, 306 p.

Schonlau, M. (1997). *Computer Experiments and Global Optimization*, Dissertation, University of Waterloo, Canada, 130 p.

Smith, R. L. and Zhu, Z. (2004). Asymptotic theory for kriging with estimated parameters and its application to network design, www.stat.unc.edu/postscript/rs/supp5.pdf, 21 p.

Spöck, G. and Pilz, J. (2008). Non-stationary spatial modeling using harmonic analysis. In Ortiz, J. M. and Emery, X., editors, *Proceedings of the Eight International Geostatistics Congress*. Gecamin, Chile, pp. 389–398.

Spöck, G. and Pilz, J. (2010). Spatial sampling design and covariance-robust minimax prediction based on convex design ideas, *Stochastic Environmental Research and Risk Assessment*, 24, pp. 463–482.

Spöck, G., Zhu, Z. and Pilz, J. (2012). Simplifying objective functions and avoiding stochastic search algorithms in spatial sampling design, *Stochastic Environmental Research and Risk Assessment*, in press.

Trujillo-Ventura, A. and Ellis, J. H. (1991). Multiobjective air pollution monitoring network design, *Atmospheric Environment*, 25, pp. 469–479.

US-EPA (2002). *Guidance on choosing a sampling design for environmental data collection*, EPA QA/G-5S, United States Environmental Protection Agency, 70 p.

van Groenigen, J. W., Siderius, W. and Stein, A. (1999). Constrained optimisation of soil sampling for minimisation of the kriging variance, *Geoderma*, 87, pp. 239–259.

Yaglom, A. M. (1987). *Correlation Theory of Stationary and Related Random Functions I*, Springer-Verlag, New York, 526 p.

Zhu, Z. and Stein, M. L. (2006). Spatial sampling design for prediction with estimated parameters, *Journal of Agricultural, Biological and Environmental Statistics*, 11, pp. 24–49.

Zimmerman, D. L. and Homer, K. E. (1991). A network design criterion for estimating selected attributes of the semivariogram, *Environmetrics*, 2, pp. 425–441.

5

Entropy-based network design using hierarchical Bayesian kriging

Baisuo Jin[1], Yuehua Wu[2] and Baiqi Miao[1]

[1]*School of Management, University of Science and Technology of China, Hefei, People's Republic of China*

[2]*Department of Mathematics and Statistics, York University, Toronto, Canada*

5.1 Introduction

Data that vary spatially and temporally are spatio-temporal data. Typical spatio-temporal data include temperature, precipitation, atmospheric pressure, ozone concentration, personal income, infection prevalence, mosquito populations, among others. Assessing the changes both spatially and temporally of data is very important for monitoring these processes. Consider ozone concentrations for example. Assessing their changes both spatially and temporally can help in air quality change assessment, which has a great impact on the environment, society and the economy.

Some spatio-temporal data such as temperature, precipitation, atmospheric pressure and ozone concentration are collected from monitoring stations. However, the selection of these stations is often influenced by administrative, political,

Spatio-temporal design: Advances in efficient data acquisition, First Edition.
Edited by Jorge Mateu and Werner G. Müller.

and other pragmatic considerations (Zidek and Zimmerman 2009). Thus it is important to evaluate if an existing station is statistically unnecessary. It is also crucial to find a location such that a new monitoring station upon it can greatly improve the data modeling. These are considered to be environmental network design problems. From Müller (2005), Le and Zidek (2006) and Dobbie *et al.* (2008), most approaches to tackle this type of problem can be classified into the following three categories: (1) geometry-based; (2) probability-based; and (3) model-based. In this chapter, attention will be paid to the entropy-based design in Category 3. By defining an objective function using entropy, a solution for network design was proposed by Caselton and Husain (1980), Caselton and Zidek (1984), Shewry and Wynn (1987), and Sebastiani and Wynn (2000) among others. The entropy approach to design was implemented by Caselton *et al.* (1992) to obtain a method of ranking stations for possible elimination from an existing network; refinements were added by Wu and Zidek (1992). Later, Zidek *et al.* (2000) proposed a method for incorporating costs. More discussion can be found in Le and Zidek (2006) and Zidek and Zimmerman (2009) among others.

The main objective of this chapter is to show how to tackle the network design problems by the entropy approach. For demonstration, we consider the precipitation data from the Upper Austria region. The data vary over space and time with complicated spatial structure, temporal structure and spatio-temporal interactions. The traditional spatio-temporal approaches may not be suitable in consideration of the needs to specify these space, time and space-time interaction components of variation. Furthermore, missing data at some monitoring (gauge) stations also make the data modeling task more difficult. Thus, if a traditional spatio-temporal approach was used, it would be necessary to make potentially unrealistic simplifying assumptions in consideration of the needs to specify these space, time and space-time interaction components of variation and to deal with missing data, which may not be acceptable. As stated in Wilke *et al.* (1998), although we cannot escape the 'curse of dimensionality', we can take advantage of recent increases in computational speed and numerical advances (e.g., Markov chain Monte Carlo) that allow us to implement Bayesian space-time dynamical models in a hierarchical framework. Such specifications provide simple strategies for incorporating complicated space-time interactions at different stages of the model's hierarchy, and the models are feasible to implement in high dimensions. Two popular hierarchical Bayesian spatio-temporal modeling methods can be found in Wilke *et al.* (1998) and Le *et al.* (2001) among others. The latter will be used in this chapter.

To proceed, we first formulate the hierarchical spatio-temporal model to be employed based on observed precipitation data. We then fill in some missing observations such that the data have the staircase structure that is defined in the next section. Then, we estimate hyperparameters and obtain the spatial predictive distribution. Thus we are able to estimate the precipitation amount for those 445 areas located in the 18 districts of Upper Austria in which yearly data on greenland usage are available for 6 years. To decide if a new gauge station needs to be added at the center of these 445 areas or an existing station can be closed down, we solve this environmental network design problem using the principle of maximum entropy. In addition the relationship between the greenland usage, yearly precipitation and the geographic characteristics is modeled as a by-product.

Throughout this chapter, an indicator function for a set $\mathcal{A}$ is denoted by $I_{\mathcal{A}}(\cdot)$ or $I(\mathcal{A})$, a $p \times p$ identity matrix is written as I_p and the transpose of a matrix A is denoted by A^T. For a vector $\boldsymbol{c}$, $\boldsymbol{c}^T$ is its transpose, $\boldsymbol{c}(j)$ is its jth component, $|\boldsymbol{c}|$ and $\|\boldsymbol{c}\|$ are, respectively, its L_1-norm and L_2-norm (Euclidean norm). If $\mathcal{A}$ is a set, its complement and its cardinalities are denoted by $\mathcal{A}^c$ and $\sharp\mathcal{A}$, respectively. Let $\boldsymbol{a} = (a_1, \ldots, a_p)^T$ be a $p \times 1$ vector, $A = (a_{ij}) = (\boldsymbol{a}_1, \ldots, \boldsymbol{a}_p)$ be a $q \times p$ matrix, $\mathcal{B} = \{j_1, \ldots, j_k\}$ be an index set with $1 \leq j_1 < \cdots < j_k \leq p$, $\mathcal{C} = \{i_1, \ldots, i_\ell\}$ be an index set with $1 \leq i_1 < \cdots < i_\ell \leq q$, then denote $\boldsymbol{a}_{\mathcal{B}} = (a_{j_1}, \ldots, a_{j_k})^T$, $A_{\mathcal{B}} = (\boldsymbol{a}_{j_1}, \ldots, \boldsymbol{a}_{j_k})$, and denote $A_{\mathcal{C},\mathcal{B}} = (a_{ij}, i \in \mathcal{C}; j \in \mathcal{B})$. The relative complement of B in A is denoted by $A - B$ and the relative complement of B denoted by B^c. The length of a vector $\boldsymbol{x}$ is denoted by $\dim(\boldsymbol{x})$, $N(\cdot, \cdot)$ refers to a Gaussian distribution, $GIW(\cdot, \cdot)$ denotes the generalized inverted Wishart distribution, $IW(\cdot, \cdot)$ stands for the inverted Wishart distribution, and $\otimes$ represents the Kronecker product.

5.2 Entropy-based network design using hierarchical Bayesian kriging

Suppose that Y has a density function f. The total uncertainty about Y may be expressed by the entropy of its distribution; i.e., $\mathfrak{H}(Y) = E[\log f(Y)/h(Y)]$, where $h(\cdot)$ is a not necessarily integrable reference density (Jaynes 1963).

Let $ADD_m = \{add_m$, all sets of m locations$\}$. In light of Le and Zidek (2006), to add m gauge stations, the set of locations, $\widehat{add}_m$, is selected to maximize the corresponding entropy, i.e.,

$$\widehat{add}_m = \arg \max_{add_m \in ADD_m} \left(\mathfrak{H}(Y^{[u]} | Y^{[g]})\right) \tag{5.1}$$

where $Y^{[u]}$ denotes the unobserved responses at unobserved sites while $Y^{[g]}$ denotes the missing (unmeasured) and measured responses at gauge stations.

To compute the $\widehat{add}_m$, a predictive distribution of $Y^{[u]} | Y^{[g]}$ needs to be derived. Assume that there are missing observations at some gauge stations but the data can be arranged to have a staircase structure. Thus hierarchical Bayesian spatio-temporal modeling for the data with a staircase structure (Chapter 10 of Le and Zidek 2006) may be employed here, which is briefly introduced in the following.

Define

n = number of time points (e.g., number of months);
u = number of locations without gauge stations;
g = number of locations with gauge stations.

The gauge stations are organized into k blocks where the g_j $(j = 1, 2, \ldots, k)$ stations in the jth block have the same number of m_j timepoints at which no measurements are taken. These blocks are numbered so that the measurements correspond to a monotone data pattern or a staircase structure, i.e., ,

$$m_1 > m_2 > \cdots > m_k \geq 0.$$

The response variables can accordingly be organized as

$$\mathsf{Y} = [Y^{[u]}, Y^{[g]}].$$

Here $Y^{[u]}$: $n \times u$ denotes the unobserved responses at unobserved sites while $Y^{[g]}$: $n \times g$ is given by

$$Y^{[g]} = \left[Y^{[g_1]}, \ldots, Y^{[g_k]}\right] = \left[\begin{pmatrix} Y^{[g_1^m]} \\ Y^{[g_1^o]} \end{pmatrix}, \ldots, \begin{pmatrix} Y^{[g_k^m]} \\ Y^{[g_k^o]} \end{pmatrix}\right],$$

the missing (unmeasured) and measured responses at gauge stations, in which $Y^{[g_j^m]}$: $m_j \times g_j$ is the matrix of missing responses at the g_j gauge stations for the m_j timepoints; and $Y^{[g_k^o]}$: $(n - m_j) \times g_j$ is the matrix of measurements at the g_j gauge stations for the $(n - m_j)$ timepoints.

Denote $Y^{[g^o]} = \{Y^{[g_1^o]}, \ldots, Y^{[g_k^o]}\}$.

Assume the response matrix Y follows the hierarchical Bayesian model specified by

$$\begin{cases} \mathsf{Y}|\mathsf{B}, \boldsymbol{\Sigma} & \sim & N(X\mathsf{B}, I_n \otimes \boldsymbol{\Sigma}), \\ \mathsf{B}|\boldsymbol{\Sigma}, B_0 & \sim & N(B_0 W_a, V_B \otimes \boldsymbol{\Sigma}), \\ \boldsymbol{\Sigma} & \sim & GIW(\Theta, \boldsymbol{\delta}). \end{cases} \tag{5.2}$$

Here $GIW(\Theta, \boldsymbol{\delta})$ is defined in (c) of Appendix 5.1, X: $n \times l$ is the matrix of covariates, $\mathsf{B} = (B^{[u]}, B^{[g]})$ is the regression coefficient matrix with $B^{[u]}$: $l \times u$ and $B^{[g]}$: $l \times g$, $B_0 : l \times d$ is the hyperparameter matrix, $W_a = (w_{i,j})_{d\times(u+g)}$ is a constant matrix, $V_B > 0$: $l \times l$ is the variance components of B between its l rows, and

$$\boldsymbol{\Sigma} = \begin{pmatrix} \Sigma^{[u,u]} & \Sigma^{[u,g]} \\ \Sigma^{[g,u]} & \Sigma^{[g,g]} \end{pmatrix}$$

is the spatial covariance matrix, where $\Sigma^{[u,u]}$ is a $u \times u$ matrix for the unobserved sites and $\Sigma^{[u,u]}$ is a $g \times g$ matrix for the gauge stations. Hereafter the superscript $[g]$ in both $Y^{[g]}$ and $B^{[g]}$ and the superscript $[g, g]$ in $\Sigma^{[g,g]}$ will be suppressed if there is no confusion.

Let $\mathcal{H}$ be the set of hyperparameter values. Write $\mathcal{H} = [\mathcal{H}_u, \mathcal{H}_g]$. Here $\mathcal{H}_g$ represent the hyperparameters involved in the marginal distribution of $Y^{[g^o]}$ and $\mathcal{H}_u$ is the set of remaining parameters in $\mathcal{H}$. See (c) and (d) of Appendix 5.1 for details. By using hierarchical Bayesian kriging , the following is a general procedure for solving the network design problem for data having a staircase structure.

Step 1. Properly choose a covariate matrix X.

Step 2. Compute the hyperparameter values that maximize the marginal distribution $f(Y^{[g^o]}|\mathcal{H}_g)$ using an empirical Bayesian approach. Note that although $f(Y^{[g^o]}|\mathcal{H}_g)$ can be written as a matric-t distribution as in (5.19), a direct maximization of this marginal density presents a challenge. In this chapter, the expectation–maximization (EM) algorithm will be used to obtain $\hat{\mathcal{H}}_g$.

Step 3. Obtain the predictive distributions $f(Y^{[g_k^m]}|Y^{[g^o]}, \hat{\mathcal{H}}_g)$ of missing data as in (e) of Appendix 5.1. Fill in the missing observations by using the predictive distributions .

Step 4. Obtain the estimate $\hat{\Sigma}^{[g,g]}$ from the estimate of $\hat{\mathcal{H}}_g$. In terms of $\hat{\Sigma}^{[g,g]}$, select an appropriate spatial correlation function and obtain its estimate. Then use the thin-plate spline approach given in (f) of Appendix 5.1 to estimate the diagonal elements of $\Sigma^{[u,u]}$, which, jointly with the estimated spatial correlation function, yields $\hat{\Sigma}^{[u,g]}$ and $\hat{\Sigma}^{[u,u]}$.

Step 5. Estimate the hyperparameters $\hat{\mathcal{H}}_u$ and then obtain the predictive distribution $f(Y^{[u]}|Y^{[g]}, \hat{\mathcal{H}})$. See (g) of Appendix 5.1 for details.

Step 6. Carry out the network design by using (5.1).

Remark. *Note that if the data can not be arranged to have a staircase structure due to some missing observations, these observations need to be filled first so that the new data have the structure in order to use the procedure above, which will be demonstrated later via the example in this chapter.*

5.3 The data

There are 37 gauge stations located in the region of Upper Austria as shown in Figure 5.1, where the Gauss–Krüger coordinate system is used in this chapter. Each gauge station has 192 monthly precipitation data ranging from January 1994 to December 2009. All stations (with the exception of 10 stations) have missing data as shown in Figure 5.2. Moreover, four of them have over 100 missing data. In addition, yearly data on the greenland usage of 445 areas located in the 18 districts for the years 1995, 1999, 2003, 2005, 2007 and 2008 are available.

5.4 Spatio-temporal modeling

Denote the monthly precipitation data from 37 gauge stations over 192 months by a 192×37 data matrix Z. Note that the precipitation data are positive and highly skewed. A logarithmic transformation is applied to the data and the transformed data $Y = (\log(z_{ij})) = (\boldsymbol{y}_1, \ldots, \boldsymbol{y}_{37})$ are approximately normally distributed. In view of the fact that Z has a period of 12 consecutive months, the temporal

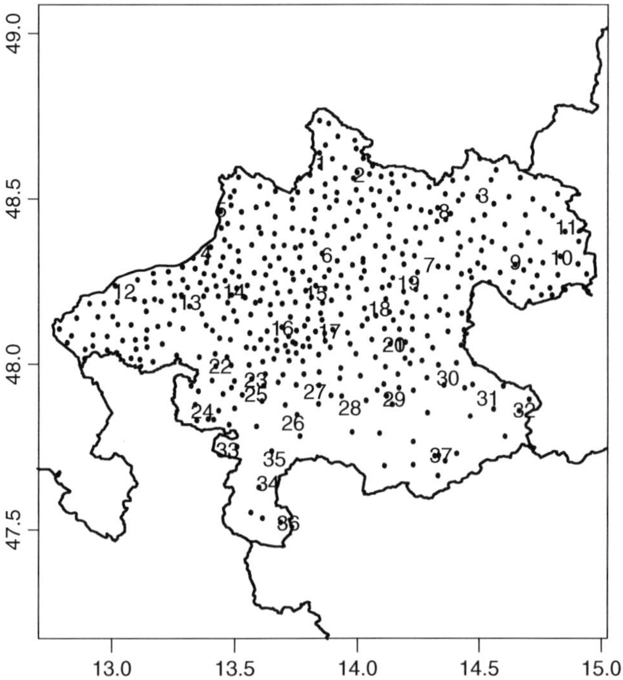

Figure 5.1 Gauge stations (indicated by numbers) and 445 areas (indicated by circles) in the 18 districts in Upper Austria.

linear model for each station is assumed to have the following form:

$$\boldsymbol{y}_j \sim N(X\boldsymbol{\beta}_j, \sigma_j^2 I_{192}), \quad j = 1, \ldots, 37, \tag{5.3}$$

where

$$X = \begin{pmatrix} 1 & \sin(1 \times 2\pi/12) & \cos(1 \times 2\pi/12) \\ 1 & \sin(2 \times 2\pi/12) & \cos(2 \times 2\pi/12) \\ \vdots & \vdots & \vdots \\ 1 & \sin(192 \times 2\pi/12) & \cos(192 \times 2\pi/12) \end{pmatrix} \triangleq \begin{pmatrix} \boldsymbol{x}_1^T \\ \boldsymbol{x}_2^T \\ \vdots \\ \boldsymbol{x}_{192}^T \end{pmatrix}. \tag{5.4}$$

For each station, we fit the model (5.3) to the observed precipitation data from that station. Table 5.1 lists some fitting results. From this table, it can be seen that all R^2s are larger than 0.97 and almost all the regression coefficients are significant. Thus the temporal model (5.3) is appropriate in terms of fitting precipitation observations for each station.

Even though the temporal linear model (5.3) is able to fit the data well for each individual station, by nature, the precipitation observations from the region of Upper Austria are supposed to interact with each other through a certain spatial mechanism. To pursue it, let us first transform the latitude–longitude

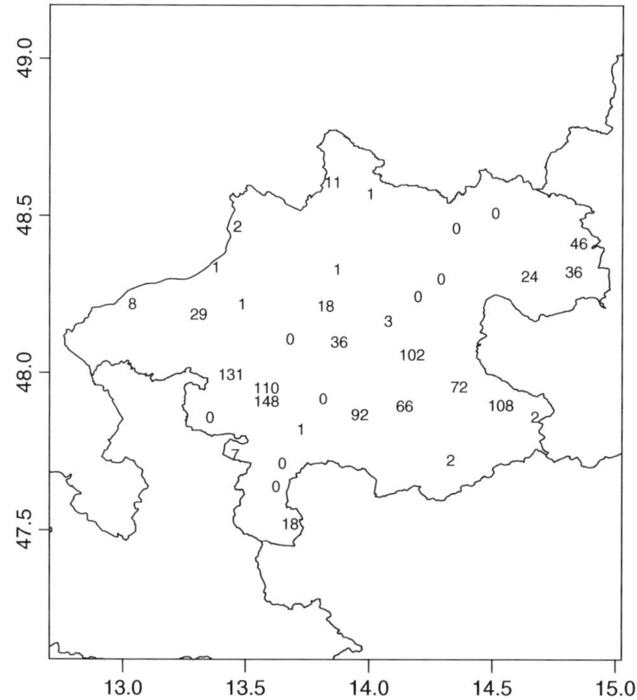

Figure 5.2 Number of missing observations at each gauge station.

Table 5.1 Regression coefficient estimates and R^2 for all the stations without any missing data.

ID	Station 3 Station 21	Station 7 Station 24	Station 8 Station 27	Station 16 Station 34	Station 19 Station 35
$\hat{\beta}_1$	3.97***	4.12***	4.24***	4.21***	4.05***
	4.30***	4.73***	4.50***	4.71***	4.83***
$\hat{\beta}_2$	−0.47***	−0.28***	−0.19**	−0.28***	−0.34***
	−0.30***	−0.26***	−0.36***	−0.29***	−0.28***
$\hat{\beta}_3$	−0.21**	−0.11†	−0.05	−0.19**	−0.13*
	−0.17**	−0.12†	−0.15*	−0.17**	−0.17**
R^2	0.977	0.977	0.976	0.976	0.981
	0.981	0.984	0.984	0.984	0.985

Significance: ***0.001; **0.01; *0.05; †0.1.

coordinates of a gauge station j to a rectangular coordinate system $(\mathsf{x}_j, \mathsf{y}_j)$. Denote the elevation of the station j by z_j. Further put $\tilde{\mathsf{x}}_j = \mathsf{x}_j/10^4, \tilde{\mathsf{y}}_j = \mathsf{y}_j/10^5$ and $\tilde{\mathsf{z}}_j = \mathsf{z}_j/10^2$ in order to have appropriate scales for these geographic data. Let $(\boldsymbol{b}_1^o, \boldsymbol{b}_2^o, \boldsymbol{b}_3^o)^T = (\hat{\boldsymbol{\beta}}_1^o, \ldots, \hat{\boldsymbol{\beta}}_{37}^o)$, where $\hat{\boldsymbol{\beta}}_j^o = (\hat{\beta}_{1,j}^o, \hat{\beta}_{2,j}^o, \hat{\beta}_{3,j}^o)^T$, $j = 1, \ldots, 37$, are the estimates of $\boldsymbol{\beta}_j$s obtained by fitting the observed precipitation data from the station j to the model (5.3). We seek for the spatial relationship among the elements of $\boldsymbol{b}_i^o$ for $i = 1, 2, 3$. Put

$$W_0 = \begin{pmatrix} 1 & \tilde{\mathsf{x}}_1 & \tilde{\mathsf{y}}_1 & \tilde{\mathsf{z}}_1 \\ 1 & \tilde{\mathsf{x}}_2 & \tilde{\mathsf{y}}_2 & \tilde{\mathsf{z}}_2 \\ \vdots & \vdots & \vdots & \vdots \\ 1 & \tilde{\mathsf{x}}_{37} & \tilde{\mathsf{y}}_{37} & \tilde{\mathsf{z}}_{37} \end{pmatrix}^T. \tag{5.5}$$

For $i = 1, 2, 3$, consider the linear model

$$\boldsymbol{b}_i = W_0^T \boldsymbol{\alpha}_i + \boldsymbol{\epsilon}_i, \tag{5.6}$$

where $\boldsymbol{\epsilon}_i \sim N(\boldsymbol{0}, \sigma_i^2 I_{37})$.

Fit $\boldsymbol{b}_i^o$ to the model (5.6) by the ordinary least squares for $i = 1, 2, 3$. The estimates of $\boldsymbol{\alpha}_i$ along with the R^2s are given in Table 5.2. Remove the terms with insignificant coefficients from the model (5.6) and re-estimate the regression coefficients. These estimates denoted by $\tilde{\boldsymbol{b}}_1^o$, $\tilde{\boldsymbol{b}}_2^o$, and $\tilde{\boldsymbol{b}}_3^o$ are reported in Table 5.2 along with the R^2s, which are larger than 0.9. Thus the relationships of $\hat{\beta}_{i,j}^o$ with $\tilde{\mathsf{x}}_j, \tilde{\mathsf{y}}_j, \tilde{\mathsf{z}}_j$ for $i = 1, 2, 3$ are well modeled by

$$\begin{cases} \hat{\beta}_{1,j}^o &= 6.349 - 0.662\tilde{\mathsf{y}}_j + 0.04\tilde{\mathsf{z}}_j + 0.12\epsilon_{1,j}, \\ \hat{\beta}_{2,j}^o &= -0.284 + 0.08\epsilon_{2,j}, \\ \hat{\beta}_{3,j}^o &= -0.15 + 0.05\epsilon_{3,j}, \end{cases} \tag{5.7}$$

where each of $\{\epsilon_{1,j}\}$, $\{\epsilon_{2,j}\}$, and $\{\epsilon_{2,j}\}$ is a sequence of independently and identically distributed (i.i.d.) normal random variables with 0 mean and variance 1.

Table 5.2 Estimates of regression coefficients and standard deviations, and R^2.

$\boldsymbol{\alpha}_i$	$\boldsymbol{b}_1^o$	$\boldsymbol{b}_2^o$	$\boldsymbol{b}_3^o$	$\tilde{\boldsymbol{b}}_1^o$	$\tilde{\boldsymbol{b}}_2^o$	$\tilde{\boldsymbol{b}}_3^o$
$\hat{\alpha}_{i,1}$	6.349***	−0.545**	−0.330**	6.349***	−0.284***	−0.150***
$\hat{\alpha}_{i,2}$	−0.006	2×10^{-6}	0.004	0	0	0
$\hat{\alpha}_{i,3}$	−0.656***	0.061	0.044	−0.662***	0	0
$\hat{\alpha}_{i,4}$	0.040***	0.012†	0.004	0.040***	0	0
R^2	0.999	0.944	0.922	0.999	0.930	0.905
$\hat{\sigma}_i$	0.122	0.077	0.049	0.120	0.080	0.050

Significance: ***0.001; **0.01; *0.05; †0.1.

Thus (5.7) can be utilized to characterize the spatial structures of the monthly precipitation data under consideration.

In view of Table 5.1 and Table 5.2, we will employ the following spatio-temporal model in a hierarchical framework:

$$\begin{cases} Y|B,\Sigma & \sim & N(XB, I_{192}\otimes\Sigma), \\ B|\Sigma, B_0 & \sim & N(B_0W, V_B\otimes\Sigma), \\ \Sigma & \sim & GIW(\Theta,\delta). \end{cases} \tag{5.8}$$

Here X is given in (5.4), B is the 3×37 regression coefficient matrix, Σ is the 37×37 covariance matrix, W consists of all but the second column of W_0,

$$B_0 = \begin{pmatrix} \alpha_0 & \eta_1 & \eta_2 \\ \zeta_1 & 0 & 0 \\ \zeta_2 & 0 & 0 \end{pmatrix},$$

$V_B > 0$: 3×3, the variance components of B between its rows, $\{\Theta,\delta\}$ is a set of model parameters [see (c) of Appendix 5.1 for details].

5.5 Obtaining a staircase data structure

By examining the data, it appears that if we can fill in some missing observations, the responses will have a staircase structure. We are then able to employ the hierarchical Baysian spatio-temporal modeling (or kriging) technique (see Chapter 10 of Le and Zidek 2006 or the review in Appendix 5.1) to model the data.

There are 13 gauge stations that have no more than 12 missing data . For such a station, say Station j, we replace a missing observation, say the kth element of $\boldsymbol{y}_j$, by its least squares prediction plus error based on the temporal linear model $\tilde{y}_{k,j} = \boldsymbol{x}_k^T\hat{\boldsymbol{\beta}}_j^o + \varepsilon_j$ with $\varepsilon_j \sim N(0,\hat{\sigma}_j^2)$, where $\hat{\boldsymbol{\beta}}_j^o$ is the estimate of $\boldsymbol{\beta}_j$ obtained by fitting the observed precipitation data from Station j to the temporal model (5.3) by least squares. In this way, we can fill in the missing data for all stations with no more than 12 missing observations. Denote the resulting dataset by $Y^{[1]} = (y_{i_1,i_2}^{[1]})$.

Let S_0 be the set of identification numbers (IDs) of the 23 stations that have no missing observations in the data matrix Y_1 and S_1 be the set of IDs of the remaining 14 stations. Table 5.3 lists the number of missing observations at the stations with IDs in S_1, where $\mathcal{M}_{t,j}$ denotes the set of time points in which there are no observations from the station with ID j (or simply Station j), $\mathcal{M}_{e,j}$ denotes the set of consecutive time points located at the end of the time period under consideration in which observations are not available from the same station j, and $\mathcal{M}_{r,j} = \mathcal{M}_{t,j} - \mathcal{M}_{e,j}$.

In order to have a staircase data structure, we fill in the missing observations for the stations with IDs $\{11, 13, 28, 22\}$. First, we replace the three missing observations for Station 11 with the time points in $\mathcal{M}_{r,11}$ by their least squares

Table 5.3 The number of missing observations at the stations with IDs in S_1.

ID	$\tilde{x}$	$\tilde{y}$	$\tilde{z}$	$\sharp\mathcal{M}_{t,ID}$	$\sharp\mathcal{M}_{e,ID}$	$\sharp\mathcal{M}_{r,ID}$
9	9.69	3.52	5.42	24	24	0
10	11.02	3.54	5.95	36	36	0
11	11.15	3.64	6.08	46	43	3
13	−0.25	3.38	4.40	29	0	29
15	3.56	3.41	3.68	18	18	0
17	3.98	3.28	3.60	36	36	0
22	0.68	3.17	5.71	131	2	129
23	1.74	3.12	5.01	110	110	0
25	1.75	3.07	4.69	148	148	0
28	4.62	3.03	5.40	92	0	92
29	5.98	3.06	4.52	66	66	0
30	7.62	3.13	3.50	72	72	0
31	8.86	3.06	3.79	108	108	0
36	2.76	2.64	20.50	18	18	0

predictions plus errors based on the temporal linear model (5.3) and obtain a new dataset $Y^{[2]} = (y^{[2]}_{i_1,i_2})_{192\times 37}$.

Secondly, we fill in all 29 missing data for Station 13 with the time points in $\mathcal{M}_{t,13}$ using all the data in the stations with IDs in S_0. Put $\tilde{Y} = Y^{[2]}_{\mathcal{M}_{t,13}\bigcup\mathcal{M}^c_{t,13},\ \{13\}\bigcup S_0}$, $\tilde{X}^T = (X^T)_{\mathcal{M}_{t,13}\bigcup\mathcal{M}^c_{t,13}}$ and $\tilde{W} = W_{\{13\}\bigcup S_0}$. To use the model (5.2), let $\tilde{Y}$, $\tilde{X}$ and $\tilde{W}$ substitute for Y, X and W_a, respectively, in (5.2). It can be seen that

$$d = 3,\ l = 3,\ n = 192,\ k = 2,\ m_1 = 29,\ m_2 = 0,\ g_1 = 1,\ g_2 = 23 \qquad (5.9)$$

from the structures and dimensions of $\tilde{Y}$ and $\tilde{X}$. We estimate hyperparameters of the distribution (5.19) by the EM algorithm, and then use the estimated predictive distribution (5.21) to fill in the missing data for Station 13 and obtain a new dataset $Y^{[3]} = (y^{[3]}_{i_1,i_2})_{192\times 37}$.

Thirdly, we fill in all 92 missing observations for Station 28 with the time points in $\mathcal{M}_{t,28}$ using the dataset $Y^{[3]}_{\{13\}\bigcup S_0}$. Put $\widetilde{Y} = Y^{[3]}_{\mathcal{M}_{t,28}\bigcup\mathcal{M}^c_{t,28},\ \{28,13\}\bigcup S_0}$, $\widetilde{X}^T = (X^T)_{\mathcal{M}_{t,28}\bigcup\mathcal{M}^c_{t,28}}$ and $\widetilde{W} = W_{\{28,13\}\bigcup S_0}$. To use the model (5.2), let $\widetilde{Y}$, $\widetilde{X}$ and $\widetilde{W}$ substitute for Y, X and W_a, respectively, in (5.2). At this time, d, l, n, k, m_2 and g_1 are the same as in (5.9) but $m_1 = 92$ and $g_2 = 24$ from the structures and dimensions of $\widetilde{Y}$ and $\widetilde{X}$. We estimate hyperparameters of the distribution (5.19) by the EM algorithm, and then use the estimated predictive distribution (5.21) to fill in the missing data for Station 28 and obtain a new dataset $Y^{[4]} = (y^{[4]}_{i_1,i_2})_{192\times 37}$.

Finally, we deal with the missing data for Station 22 with the time points in $\mathcal{M}_{t,22}$. Since there are only two observations missing at the end of the data sequence, we fill in the total 131 missing observations for Station 22 using the dataset $Y^{[4]}_{\{28,13\} \bigcup S_0}$ instead of filling the missing observations with the time points in $\mathcal{M}_{r,22}$. Let $\check{Y} = Y^{[4]}_{\mathcal{M}_{t,22} \bigcup \mathcal{M}^c_{t,22},\ \{22,28,13\} \bigcup S_0}$, $\check{X}^T = (X^T)_{\mathcal{M}_{t,22} \bigcup \mathcal{M}^c_{t,22}}$ and $\check{W} = W_{\{22,28,13\} \bigcup S_0}$. To use the model (5.2), let $\check{Y}$, $\check{X}$ and $\check{W}$ substitute for Y, X and W_a, respectively, in (5.2). Here, d, l, n, k, m_2 and g_1 are the same as in (5.9) but $m_1 = 92$ and $g_2 = 25$. We also estimate hyperparameters of the distribution (5.19) by the EM algorithm, and then use the estimated predictive distribution (5.21) to fill in the missing data for Station 22 and obtain a new dataset $Y^{[5]} = (y^{[5]}_{i_1,i_2})_{192\times 37}$, which has the staircase structure.

5.6 Estimating the hyperparameters $\mathcal{H}_g$ and the spatial correlations between gauge stations

Let $C_1 = \{192, 191, \ldots, 2, 1\}$ and $C_2 = S_2 \bigcup S_0$, where

$$S_2 = \{25, 23, 31, 30, 29, 22, 10, 17, 9, 15, 36, 22, 28, 13\}.$$

Table 5.4 lists the number of missing observations at the stations with IDs in S_2.

Table 5.4 The number of missing observations in $Y^{[5]}$ at the stations with IDs in S_2.

ID	$\tilde{x}$	$\tilde{y}$	$\tilde{z}$	$\sharp\mathcal{M}_{e,ID}$	ID	$\tilde{x}$	$\tilde{y}$	$\tilde{z}$	$\sharp\mathcal{M}_{e,ID}$
25	1.75	3.07	4.69	148	17	3.98	3.28	3.60	36
23	1.74	3.12	5.01	110	9	9.69	3.52	5.42	24
31	8.86	3.06	3.79	108	15	3.56	3.41	3.68	18
30	7.62	3.13	3.50	72	36	2.76	2.64	20.50	18
29	5.98	3.06	4.52	66	22	0.68	3.17	5.71	0
11	11.15	3.64	6.08	43	28	4.62	3.03	5.40	0
10	11.02	3.54	5.95	36	13	−0.25	3.38	4.40	0

Let $Y^* = Y^{[5]}_{C_1,\, C_2}$, $X^* = ((X^T)_{C_1})^T$ and $W^* = W_{C_2}$ and then Y^* has a staircase data structure. To use the model (5.2), let Y^*, X^* and W^* substitute for Y, X and W_a, respectively, in (5.2). Here, d, l and n are the same as in (5.9) but $k = 10$, $m_1 = 148$, $m_2 = 110$, $m_3 = 108$, $m_4 = 72$, $m_5 = 66$, $m_6 = 43$, $m_7 = 36$, $m_8 = 24$, $m_9 = 18$, $m_{10} = 0$, $g_1 = g_2 = \cdots = g_6 = 1$, $g_7 = 2$, $g_8 = 1$, $g_9 = 2$ and $g_{10} = 26$. We proceed to estimate the hyperparameters $\mathcal{H}_g = \{V_B, B_0, (\Upsilon_{0,1}, H_1, \Lambda_1, \delta_1), \ldots, (\Upsilon_{0,k-1}, H_{k-1}, \Lambda_{k-1}, \delta_{k-1}), (\Lambda_{10}, \delta_{10})\}$, defined in (d) of Appendix 5.1, by the EM algorithm . The estimates of B_0 are given in Table 5.5 and the estimate of V_B is given as follows:

$$\hat{V}_B = \begin{pmatrix} 0.1632 & 0.0479 & 0.0260 \\ 0.0479 & 0.0579 & 0.0296 \\ 0.0260 & 0.0296 & 0.0227 \end{pmatrix}.$$

- OPTIMALLY_IMPROVE_POOLDELETE_FROM_POOLCOMPLETE.m
- OPTIMALLY_IMPROVE_POOLDELETE_FROM_POOLADD.m

The names of these functions are self-explanatory: 'Pooldelete' is the discrete pool of locations that are allowed to be deleted from the design. 'Poolcomplete' is the complete compact area of points from **X** allowed to be added to the design. 'Pooladd' is the discrete pool of locations that are allowed to be added to the design. 'Improve' means the exchange algorithm, where locations from 'Pooldelete' may either be exchanged by locations from 'Poolcomplete' or from 'Pooladd' and the total number of sampling locations remains constant.

4.8 An example session

The purpose of this section is to demonstrate some capabilities of the spatDesign toolbox. The most important Matlab function calls related to sampling design with the Smith and Zhu design criterion are given. The dataset considered is the rainfall dataset from Upper Austria. The monitoring network comprises 36 locations. Average monthly rainfall has been measured at each location starting in January 1994 and ending in December 2009.

4.8.1 Preparatory calculations

The following Matlab code plots the 36 data locations and the 36 time series of rainfall at each station (Figure 4.1.):

- load designupperaustriaGPU
- plot(xyelevation(:,1),xyelevation(:,2),'o')
  hold on
  plot(boundary.x,boundary.y)
  axis equal
- raints=timeseries(rain');
  plot(raints)

In Figure 4.1 we see that there are obviously areas in the design region that look very empty, having no sampling locations. Next let us calculate from the above time series for each station the mean rainfall over the years, as well as the residual rainfall, for each of the 12 months (Figure 4.2):

- for i=1:12
  rainmeanmonth(:,i)=mean(raints(i:12:end),'MissingData','remove');
  end
- figure
  plot(rainmeanmonth')
  title('monthly mean of rain')

Table 5.5 The coefficient estimates for the spatio-temporal model (5.8).

$\hat{\alpha}_0$	$\hat{\eta}_1$	$\hat{\eta}_2$	$\hat{\zeta}_2$	$\hat{\zeta}_3$
5.911	−0.491	0.016	−0.554	−0.222

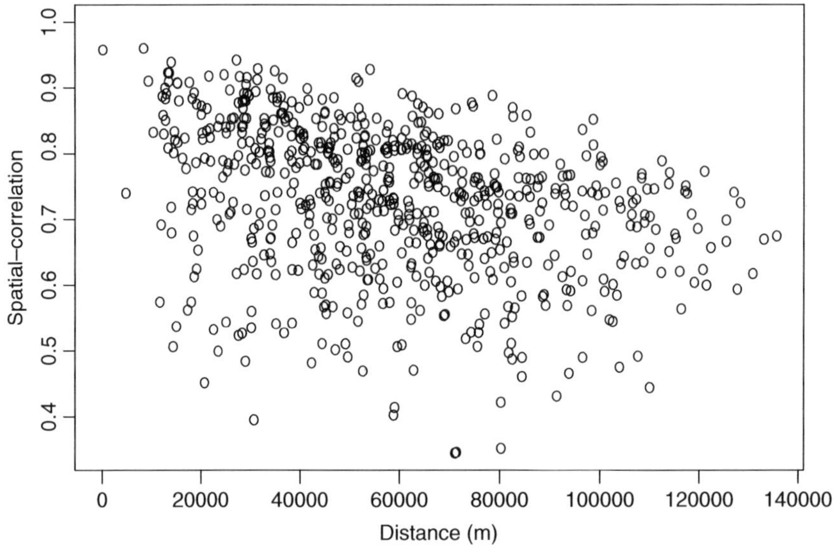

Figure 5.3 Estimated spatial correlations based on the hierarchical model (5.8).

The estimated spatial correlations between all gauge stations are displayed in Figure 5.3.

As shown in Figure 5.3, the estimated spatial correlation coefficients versus the distance between sites reveal the existence of heterogeneity in distribution. Thus, to model the relationship of the spatial correlation of any two locations, say $\boldsymbol{s}_i = (\mathsf{x}_i, \mathsf{y}_i)$ and $\boldsymbol{s}_j = (\mathsf{x}_j, \mathsf{y}_j)$, with the distance between these two locations, we use the following piecewise linear function:

$$\sum_{k=1}^{14} a_k I\{\|\boldsymbol{s}_i - \boldsymbol{s}_j\| \in ((k-1)10000, k10000]\}$$
$$+ a_{15}|\mathsf{x}_i - \mathsf{x}_j| + a_{16}|\mathsf{y}_i - \mathsf{y}_j| + a_{17}|\mathsf{z}_i - \mathsf{z}_j| \tag{5.10}$$

to approximate $Corr(Y(\boldsymbol{s}_i), Y(\boldsymbol{s}_j))$, where $\boldsymbol{s}_1 \neq \boldsymbol{s}_2$, $\|\boldsymbol{s}_1 - \boldsymbol{s}_2\| \leq 140000$, z_i and z_j are elevations, and $Y(\boldsymbol{s})$ denotes the $\log(Z(\boldsymbol{s}))$ with $Z(\boldsymbol{s})$ being the precipitation variable at the location $\boldsymbol{s}$. Note that (5.10) has 14 categorical coefficients $\{a_1, \ldots, a_{14}\}$. If $a_i = a_{i+1}$, for $1 \leq i \leq 13$, no more than 13 categorical coefficients are needed actually. For this reason, we re-express (5.10) via Helmert contrasts, which compare the first level of the factor with all later levels, the second level with all later levels, the third level with all later levels, and so on. The corresponding Helmert matrix, $\mathbf{H}_{14\times 14}$, is given by

$$\mathbf{H} = \begin{pmatrix} 1 & -1 & -1 & -1 & \cdots & -1 & -1 \\ 1 & 1 & -1 & -1 & \cdots & -1 & -1 \\ 1 & 0 & 2 & -1 & \cdots & -1 & -1 \\ \vdots & \vdots & \vdots & \vdots & \cdots & \vdots & \vdots \\ 1 & 0 & 0 & 0 & \cdots & 0 & 13 \end{pmatrix}.$$

Let the new parameters $\kappa_i, i = 1, 2, \ldots, 17$, satisfy $(a_1, \ldots, a_{14})^T = \mathbf{H}(\kappa_1, \ldots, \kappa_{14})^T$ and $\kappa_i = a_i$, $i = 15, 16, 17$. Write $\mathbf{H} = (\boldsymbol{h}_1, \ldots, \boldsymbol{h}_{14})^T$. Thus the piecewise linear function (5.10) will be re-expressed by the following

$$Corr(Y(\boldsymbol{s}_i), Y(\boldsymbol{s}_j)) = X_{\boldsymbol{s}_i,\boldsymbol{s}_j}\boldsymbol{\kappa} + \text{error}, \tag{5.11}$$

where $\boldsymbol{\kappa} = (\kappa_1, \ldots, \kappa_{17})^T$ and $X_{\boldsymbol{s}_i,\boldsymbol{s}_j} = (\boldsymbol{h}_k^T, |\mathsf{x}_i - \mathsf{x}_j|, |\mathsf{y}_i - \mathsf{y}_j|, |\mathsf{z}_i - \mathsf{z}_j|)$ if $\|\boldsymbol{s}_i - \boldsymbol{s}_j\| \in ((k-1)10000, k10000]$. There is a similarity between (5.11) and a statistical linear model. Actually, it is wise to treat it as a statistical linear model and hence extensive research tools for statistical linear models can be used for estimating the coefficients in (5.11). For this reason, we will treat an equation like (5.11) as a statistical linear model hereafter.

Note that if all $\kappa_i, i = 1, \ldots, 14$, are significant, then $a_i, i = 1, \ldots, 14$, are significantly different from each other. Further, if the R^2 of (5.11) is large, then it is appropriate to use the piecewise linear function (5.10). However there are 17 parameters in (5.11), and hence it is necessary to carry out variable selection for this model. To perform the variable selection, we can use the modern penalized model selection methods including the SCAD (Fan and Li 2001) and the MCP (Zhang 2010) that threshold small estimates to zeros. In light of Zhang (2010), such an estimate $\hat{\boldsymbol{\kappa}}$ of $\boldsymbol{\kappa}$ is given by

$$\hat{\boldsymbol{\kappa}} = \arg\min_{\boldsymbol{\kappa}} \left\{ \sum_{i=1}^{36} \sum_{j=i+1}^{37} (Corr(Y(\boldsymbol{s}_i), Y(\boldsymbol{s}_j)) - X_{\boldsymbol{s}_i,\boldsymbol{s}_j}\boldsymbol{\kappa})^2/666 \right.$$
$$\left. + \sum_{i=1}^{17} p_{\lambda,\gamma}(|\kappa_i|) \right\}, \tag{5.12}$$

where $666 = \sum_{i=1}^{36} \sum_{j=i+1}^{37} 1$, $\lambda > 0$, $\gamma > 0$, and $p_{\lambda,\gamma}(|\cdot|)$ is the penalty function. For demonstration, we consider the MCP in this chapter and $p_{\lambda,\gamma}(|u|)$ is given as follows for $u \in [0, \infty)$:

$$p_{\lambda,\gamma}(u) = \lambda u - \frac{u^2}{2\gamma} I_{[0,\,\gamma\lambda]}(u) + \frac{1}{2}\gamma\lambda^2 I_{(\gamma\lambda,\,\infty)}(u).$$

The estimate $\hat{\gamma}$ is set as 2.4 and $\hat{\lambda}$ is chosen by the Bayesian information criterion (BIC) as follows:

$$\hat{\lambda} = \arg\min_{\lambda}\{\log(666)DF_{\lambda} + 666\log(RSS_{\lambda}/666)\},$$

where DF_{λ} is the number of $\hat{\kappa}_i(\lambda) \neq 0$, $i = 1, \ldots, 17$, and

$$RSS_{\lambda} = \sum_{i=1}^{36}\sum_{j=i+1}^{37}(Corr(Y(s_i), Y(s_j)) - X_{s_i,s_j}\hat{\boldsymbol{\kappa}}(\lambda))^2$$

with $\hat{\boldsymbol{\kappa}}(\lambda)$ denoting $\hat{\boldsymbol{\kappa}}$ with λ used in the penalty function. Compute $\hat{\boldsymbol{\kappa}}$ by (5.12). Let $\hat{A} = \{i : \hat{\kappa}_i \neq 0, i = 1, \ldots, 17\}$. It is found that

$$\hat{A} = \{k_1 < k_2 \cdots < k_9\} = \{1, 5, 6, 8, 9, 12, 15, 16, 17\}.$$

Thus the model chosen by MCP is

$$\begin{aligned} Corr(Y(s_i), Y(s_j)) = &\sum_{t=1}^{5}\tilde{a}_t I(\|s_i - s_j\| \in ((k_t - 1)10000, (k_{t+1} - 1)10000]) \\ &+ \tilde{a}_6 I(\|s_i - s_j\| \in ((k_6 - 1)10000, 140000]) \\ &+ \tilde{a}_7|\mathsf{x}_i - \mathsf{x}_j| + \tilde{a}_8|\mathsf{y}_i - \mathsf{y}_j| + \tilde{a}_9|\mathsf{z}_i - \mathsf{z}_j| + \text{error}. \end{aligned} \quad (5.13)$$

We again use the Helmert contrasts and re-express (5.13) similar to what is done above. We apply the first step of backward elimination model selection to (5.13) to see if a coefficient is not significant. If all coefficients are significant, the model (5.13) is kept. If one of $\tilde{a}_7$, $\tilde{a}_8$ and $\tilde{a}_9$ is selected to be removed, we perform the next step of the backward elimination model selection. If a categorical coefficient is selected to be removed, we once more use the Helmert contrasts and re-express the model as done previously for the model (5.13). We again apply the first step of the backward elimination model selection to the new model to see if a term needs to be removed from the model. We repeat doing it till all coefficients are significant. In this way, we reach the following piecewise linear function

$$\begin{aligned} &a_1^* I(\|s_i - s_j\| \leq 40000) + a_2^* I(\|s_i - s_j\| \in (40000, 80000]) \\ &+ a_3^* I(\|s_i - s_j\| > 80000)) + a_4^*|\mathsf{y}_1 - \mathsf{y}_2| \end{aligned} \quad (5.14)$$

which is an optimal approximation to $Corr(Y(s_i), Y(s_j))$. Let

$$\mathbf{H}_{3\times3} = \begin{pmatrix} 1 & -1 & -1 \\ 1 & 1 & -1 \\ 1 & 0 & 2 \end{pmatrix},$$

Table 5.6 Spatial correlation function estimation.

Parameter	Estimate	Standard error	t-value	$Pr(>\lvert t\rvert)$
κ_1^*	0.760	0.009	89.302	<2e-16
κ_2^*	−0.019	0.005	−3.730	2e-4
κ_3^*	−0.015	0.004	−3.908	1e-4
κ_4^*	−9e-07	2e-07	−4.696	3e-6
$R^2 = 0.980$		$\hat{\sigma} = 0.105$		

κ_i^*, $i = 1, 2, 3$, satisfy $(a_1^*, a_2^*, a_3^*)^T = \mathbf{H}_{3\times 3}(\kappa_1^*, \kappa_2^*, \kappa_3^*)^T$ and $\kappa_4^* = a_4^*$. The estimates of $\kappa_i^*, i = 1, 2, 3, 4$, are displayed in Table 5.6. By transforming $\hat{\boldsymbol{\kappa}}^*$ back to $\hat{\boldsymbol{a}}^*$, we obtain the following approximation of $Corr(Y(\boldsymbol{s}_i), Y(\boldsymbol{s}_j))$:

$$\begin{aligned}\widehat{Corr}(Y(\boldsymbol{s}_i), Y(\boldsymbol{s}_j)) &= 0.794I(\|\boldsymbol{s}_i - \boldsymbol{s}_j\| \le 40000)\\ &\quad + 0.756I(\|\boldsymbol{s}_i - \boldsymbol{s}_j\| \in (40000, 80000])\\ &\quad + 0.73I(\|\boldsymbol{s}_i - \boldsymbol{s}_j\| > 80000)) - 9|\mathsf{y}_i - \mathsf{y}_j|/10^7. \end{aligned} \tag{5.15}$$

Note that a correlation matrix is positive definite and its elements have values between 0 and 1. If the above method fails to produce spatial correlation function estimation, one may try some semi-variogram models such as exponential models and Gaussian models (Journel and Huijbregts 1978; Wachernagel 2003; Le and Zidek 2006) or use other methods including the one given in Barry and Ver Hoef (1996). Since all negative eigenvalues of the correlation matrix estimate have small absolute values in this data example, we have thus replaced them by 0.001 and the resulting matrix becomes positive definite.

5.7 Spatial predictive distribution over the 445 areas located in the 18 districts of Upper Austria

By utilizing what we obtained above, we will derive spatial predictive distribution over the 445 areas located in the 18 districts of Upper Austria in this section. Note that latitude–longitude coordinates of center points $\{\boldsymbol{u}_i, i = 1, \ldots, 445\}$ of these areas are given in the data. For convenience, the ith area with center point $\boldsymbol{u}_i$ is denoted by u_i, which may also stand for the ith center point for convenience. Combining the latitude–longitude coordinates of center points $\{\boldsymbol{u}_i, i = 1, \ldots, 445\}$, spatial correlation function (5.15), estimate of $\mathcal{H}_g$, and (f) of Appendix 5.1, we can estimate the hyperparameters Λ_0, Υ_{00}, H_0 and δ_0 by (g) of Appendix 5.1. Their estimates are given in Appendix 5.2.

By now all the hyperparameters have been estimated, and we can thus predict the monthly precipitation at all the points $\{\boldsymbol{u}_i, i = 1, \ldots, 445\}$ by generating samples from the predictive distribution for the time period under consideration.

For examining the performance of the monthly precipitation predictions over these center points, we generate 200 samples from the predictive distribution for each month in the years 1995, 1999, 2003, 2005, 2007 and 2008, in which yearly data on the greenland usage of 445 areas located in the 18 districts is available. Figures 5.4 and 5.5 display the contours of the sample means and standard deviations for January 1995 for demonstration purposes.

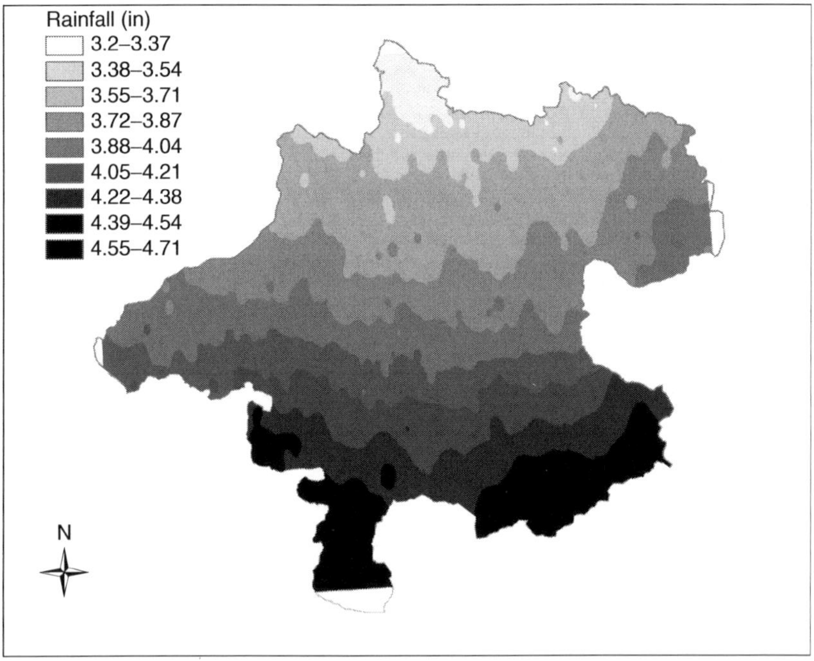

Figure 5.4 Contour of sample means for January 1995.

We now consider using the yearly predictive precipitation and the elevation as primary cofactors and build a model for the greenland usage. For the tth year in $\{1995, 1999, 2003, 2005, 2007, 2008\}$ and the area ℓ, let $G_{t,\ell}$ denote the logarithm of the ratio of the two areas, i.e., area of arable land and area of greenland, and $\tilde{Z}_{t,\ell}$ denote the average of the 12 monthly sample means from the 200 precipitation predictions generated above. For the same reason given in Section 15.3, we put $\tilde{\mathsf{x}}_\ell = \mathsf{x}_\ell/10^4$, $\tilde{\mathsf{y}}_\ell = \mathsf{y}_\ell/10^5$ and $\tilde{\mathsf{z}}_\ell = \mathsf{z}_\ell/10^2$, where $(\mathsf{x}_\ell, \mathsf{y}_\ell)$ is the rectangular coordinates and z_ℓ is the elevation of the area ℓ. In order to model the relationship between the greenland usage, yearly precipitation and the geographic characteristics well, we fit $\{G_{t,\ell}\}$ to $\{\tilde{Z}_{t,\ell}, \tilde{\mathsf{x}}_\ell, \tilde{\mathsf{y}}_\ell, \tilde{\mathsf{z}}_\ell\}$ by a polynomial regression of a higher order. Here the order is set to be 6 and the polynomial regression is given by

$$E[\exp(G_{t,\ell})] = \sum_{i=1}^{6} \left[\vartheta_i \tilde{Z}_{t,\ell}^i + \omega_{1,i}\tilde{\mathsf{x}}_\ell^i + \omega_{2,i}\tilde{\mathsf{y}}_\ell^i + \omega_{3,i}\tilde{\mathsf{z}}_\ell^i\right].$$

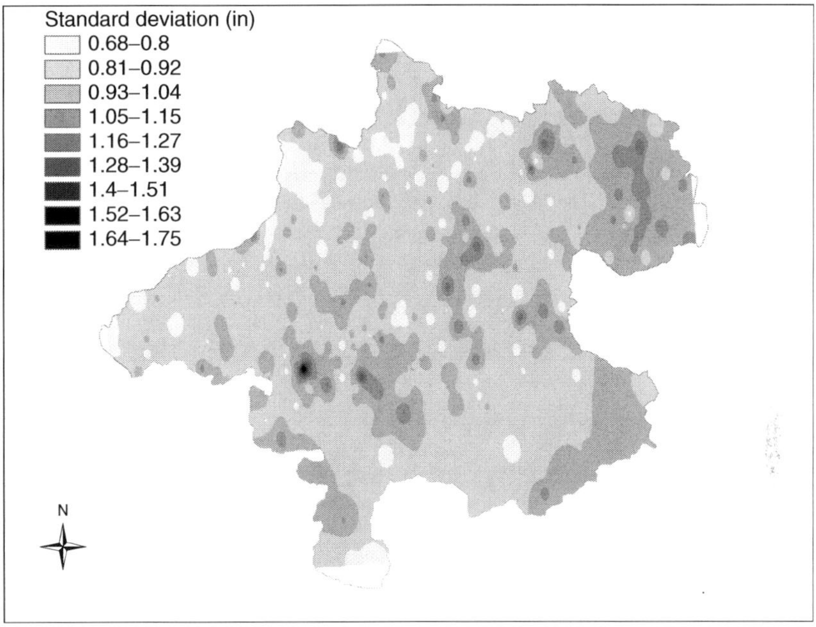

Figure 5.5 Contour of sample standard deviations for January 1995.

Applying the backward elimination model selection, the following model with R^2 larger than 0.83 is selected

$$E[\exp(G_{t,\ell})] = \sum_{i=1}^{5} \vartheta_i \tilde{Z}^i_{t,\ell} + \sum_{i\neq 2,4}^{6} \omega_{1,i}\tilde{\mathsf{x}}^i_\ell + \sum_{i=1}^{6} \left[\omega_{2,i}\tilde{\mathsf{y}}^i_\ell + \omega_{3,i}\tilde{\mathsf{z}}^i_\ell\right],$$

and the estimates of regression coefficients are displayed in Table 5.7. It can be seen that all these regression coefficients are significant. It may also be observed that the regression coefficient estimates of $\omega_{2,i}$ and $\omega_{3,i}$ have the same signs for $i = 1, \ldots, 6$, and these signs are opposite to the signs of the regression coefficient estimates of ϑ_i for $i = 1, \ldots, 5$. In view that $(10^4\tilde{\mathsf{x}}_j, 10^5\tilde{\mathsf{y}}_j)$ and $10^2\tilde{\mathsf{z}}_j$ are, respectively, the rectangular coordinates and elevation of a gauge station j, the above phenomena are attributed by the significantly negative correlation between the precipitation amount and the latitude (Figure 5.4), and the higher elevation in the northern region of Upper Austria. In addition, the positivity of the estimates of $\omega_{\ell,1}$, $\ell = 1, 2, 3$, the respective coefficients of $\tilde{\mathsf{x}}_j$, $\tilde{\mathsf{y}}_j$ and $\tilde{\mathsf{z}}_j$, indicates the positive effects of the longitude, latitude and elevation on $\exp(G_{t,\ell})$, while the negativity of the estimate of ϑ_1, the coefficient of $\tilde{Z}_{t,\ell}$, shows the negative influence of the precipitation on $\exp(G_{t,\ell})$.

Table 5.7 Regression coefficient estimates.

Coefficients	Estimate	Standard error	t-value	$Pr(>\|t\|)$
ϑ_1	−1.159e+05	1.539e+04	−7.531	6.88e-14
ϑ_2	5.314e+04	7.076e+03	7.509	8.08e-14
ϑ_3	−1.217e+04	1.628e+03	−7.471	1.07e-13
ϑ_4	1.391e+03	1.875e+02	7.417	1.60e-13
ϑ_5	−6.352e+01	8.644e+00	−7.348	2.66e-13
$\omega_{1,1}$	7.133e-02	1.449e-02	4.923	9.03e-07
$\omega_{2,1}$	1 1.882e+05	2.453e+04	7.672	2.36e-14
$\omega_{3,1}$	1 8.139e+01	1.233e+01	6.599	4.97e-11
$\omega_{2,2}$	−1.456e+05	1.866e+04	−7.801	8.73e-15
$\omega_{3,2}$	−3.655e+01	6.052e+00	−6.039	1.77e-09
$\omega_{1,3}$	3.258e-03	8.661e-04	3.762	0.000172
$\omega_{2,3}$	5.972e+04	7.544e+03	7.917	3.55e-15
$\omega_{3,3}$	8.281e+00	1.526e+00	5.425	6.31e-08
$\omega_{2,4}$	−1.371e+04	1.710e+03	−8.017	1.60e-15
$\omega_{3,4}$	−1.009e+00	2.091e-01	−4.822	1.50e-06
$\omega_{1,5}$	−9.739e-05	2.100e-05	−4.636	3.72e-06
$\omega_{2,5}$	1.669e+03	2.060e+02	8.103	8.09e-16
$\omega_{3,5}$	6.298e-02	1.479e-02	4.257	2.14e-05
$\omega_{1,6}$	5.968e-06	1.343e-06	4.445	9.14e-06
$\omega_{2,6}$	−8.426e+01	1.031e+01	−8.174	4.57e-16
$\omega_{3,6}$	−1.581e-03	4.227e-04	−3.740	0.000188
	$R^2 = 0.8312$	$\hat{\sigma} = 0.6983$		

5.8 Adding gauge stations over the 445 areas located in the 18 districts of Upper Austria

The total entropy $\mathfrak{H}(Y^{[u]}|Y^{[g]})$ can be expressed, using the predictive distribution (5.22), as

$$\mathfrak{H}(Y^{[u]}|Y^{[g]}) = \frac{1}{2}\log|\Psi^{[u|g]}| + c, \tag{5.16}$$

where c is a constant. From (5.1), to add m gauge stations, $\widehat{add}_m$ is selected to maximize the corresponding entropy, i.e.,

$$\widehat{add}_m = \arg \max_{add_m \in ADD_m} \left(\frac{1}{2}\log\left|\Psi^{[u_{add_m}|g]}\right|\right), \tag{5.17}$$

where $\sharp add_m = m$ and $u_{aad_m} = \{u_i, i \in add_m\}$.

Recall the 445 center points $\{\boldsymbol{u}_i, 1 \leq i \leq 445\}$ defined previously. We consider adding some gauge stations over these center points. Let $m = 1$ and we obtain $\log\left|\Psi^{[u_{i_1}|g]}\right| \geq \log\left|\Psi^{[u_{i_2}|g]}\right| \geq \cdots \geq \log\left|\Psi^{[u_{i_{445}}|g]}\right|$. The first 20 areas are ranked as follows:

$$v_1 = u_{197}, v_2 = u_{149}, v_3 = u_{203}, v_4 = u_{98}, v_5 = u_{139}, v_6 = u_{139}, v_7 = u_{224},$$
$$v_8 = u_{15}, v_9 = u_{214}, v_{10} = u_{81}, v_{11} = u_{60}, v_{12} = u_{64}, v_{13} = u_{122}, v_{14} = u_{164},$$
$$v_{15} = u_{11}, v_{16} = u_{40}, v_{17} = u_1, v_{18} = u_8, v_{19} = u_{172}, v_{20} = u_2.$$

Thus if we only add one gauge station, it will be in the area u_{197}. Now consider adding m gauge stations in these 20 locations. Let $C_m = \log|\Psi^{[v_{\widetilde{add}_m}|g]}|$, $m = 1, \ldots, 19$, where the subscript m indicates that m gauge stations from $\{v_1, \ldots, v_{20}\}$ are planned to be added and

$$\widetilde{add}_m = \arg \max_{add_m \in \{1,\ldots,20\}} \left(\frac{1}{2}\log\left|\Psi^{[v_{add_m}|g]}\right|\right).$$

Set $D(m) = C_m - C_{m-1} - (C_{m-1} - C_{m-2})$, $m = 3, \ldots, 19$, which is plotted in Figure 5.6. Hence we choose $m \leq 5$, i.e., the number of added gauge stations is limited to no more than 5. Table 5.8 presents the results of adding m gauge stations with $m \leq 5$. For $m = 5$, the chosen areas are displayed in Figure 5.7.

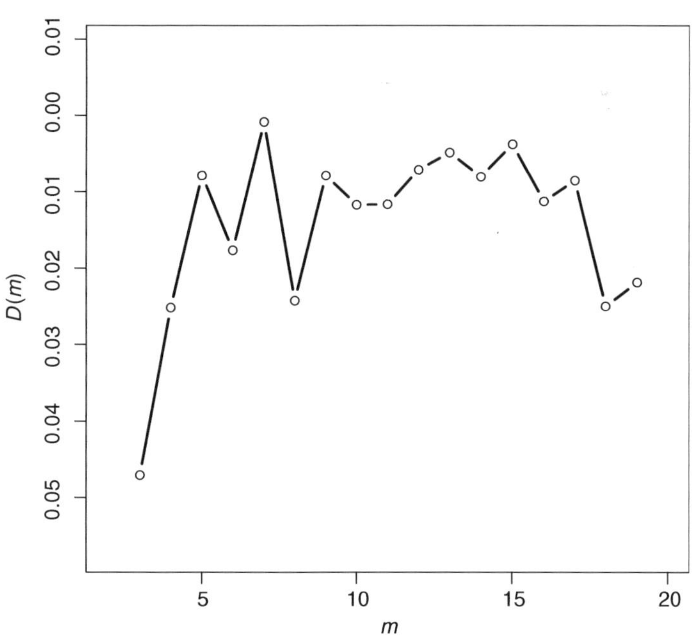

Figure 5.6 Plot of D(m) *versus m.*

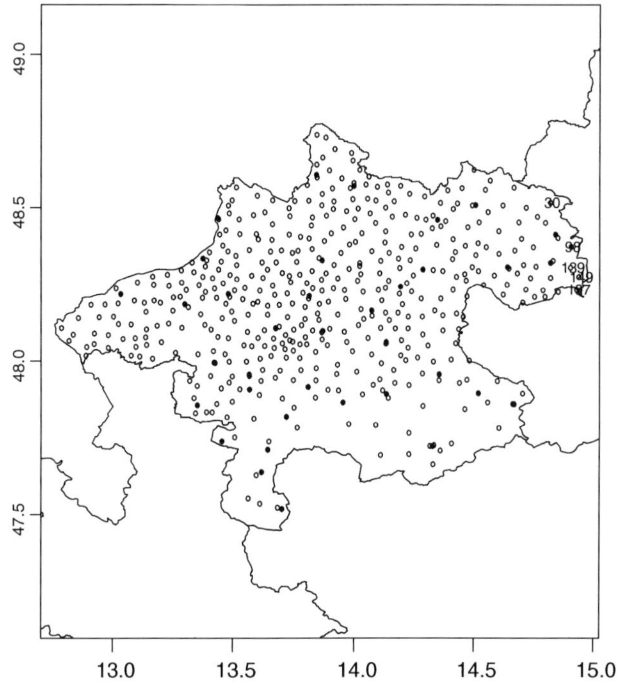

Figure 5.7 Additon of five monitoring stations. The 445 areas are indicated by small circles, the gauge stations are indicated by larger black circles and the gauge stations to add are indicated by numbers.

Table 5.8 Adding m gauge stations.

m	$u^*_{i_1}$	$u^*_{i_2}$	$u^*_{i_3}$	$u^*_{i_4}$	$u^*_{i_5}$	C_m
1	u_{197}					−4.204
2	u_{197}	u_{149}				−8.462
3	u_{197}	u_{149}	u_{98}			−12.768
4	u_{197}	u_{149}	u_{98}	u_{30}		−17.098
5	u_{197}	u_{149}	u_{203}	u_{98}	u_{30}	−21.436

5.9 Closing down an existing gauge station

For each gauge station, we first find the predictive distribution as done above with all available data excluding the observations from it, and then compute the entropy for this station using (5.16). The entropies for all stations are listed in the last column of Table 5.9. It can be seen that Station 15 has the minimum entropy. Thus if a station is to be closed down, it seems an optimal choice to close down Station 15. As a matter of fact, Station 15 is located in the center of the Upper Austria region, so closing it down seems to have little impact.

Table 5.9 Relative differences, standard deviations and entropies.

ID	Longitude	Latitude	Elevation	Relative difference	Standard deviation	Entropy
25	17563.774	307566.6	469	−0.012653208	0.1280227	−1.696983
23	17486.991	312415.5	501	0.079931694	0.1245034	−1.708437
31	88628.749	306765.1	379	−0.045404916	0.1262025	−1.708766
30	76283.855	313385.4	350	−0.040100606	0.1288432	−1.712841
29	59841.583	306157.0	452	−0.058423180	0.1326789	−1.713300
11	111547.573	364692.9	608	0.037744325	0.1298020	−1.714720
10	110274.548	354471.5	595	0.047187196	0.1076588	−2.282720
17	39825.614	328803.4	360	−0.008343848	0.1074783	−2.282720
9	96972.725	352654.1	542	0.090594293	0.1145361	−1.715623
15	35609.379	341409.2	368	−0.018733851	0.1132170	−2.283435
36	27673.096	264273.5	2050	−0.047956094	0.0901443	−2.283433
20	59611.685	324402.4	383	0.020656584	0.1095231	−2.246988
22	6816.502	317057.4	571	−0.050626721	0.1062921	−2.246985
1	37405.985	385314.9	725	−0.021530171	0.1149397	−2.246984
2	49046.355	381341.5	602	−0.103781051	0.1053635	−2.246989
3	86718.983	374888.0	549	0.152313152	0.1065431	−2.246988
4	3004.210	355045.6	350	0.057342220	0.1037514	−2.246984
5	7537.145	369567.2	307	0.007920166	0.1104956	−2.246985
6	39453.064	354392.3	400	0.038152576	0.1069542	−2.246987
7	70710.843	351341.7	262	0.063620965	0.1152221	−2.246984
8	74997.527	369391.6	685	−0.039791648	0.1096464	−2.246999
12	−22235.211	342085.5	382	0.058101602	0.1023377	−2.246986
13	−2533.843	338350.8	440	−0.002877260	0.1023881	−2.246986
14	10895.801	342081.9	435	−0.007224642	0.1093014	−2.246993
16	25430.782	329741.5	634	−0.018140433	0.1125951	−2.246984
18	54962.654	336310.5	309	0.045263177	0.1066169	−2.246986
19	63771.874	345112.7	298	0.060725647	0.1090552	−2.246986
21	59529.176	324370.7	382	0.013316904	0.1149015	−2.246984
24	1290.660	301837.3	491	−0.097327386	0.1151995	−2.246984
26	28888.879	297758.9	1618	−0.122258022	0.1207841	−2.246990
27	35500.634	308574.9	424	−0.019561930	0.1136047	−2.261379
28	46280.080	303029.2	540	−0.124750363	0.1157572	−2.246987
32	99893.304	302963.2	428	−0.043489927	0.1103940	−2.246987
33	9015.492	288707.2	539	−0.011602014	0.1156496	−2.246988
34	21383.122	277558.6	510	0.066050797	0.1018765	−2.246985
35	23292.197	285729.8	469	−0.092800166	0.1080578	−2.246984
37	74542.610	287322.3	600	−0.049034678	0.1160656	−2.246986

5.10 Model evaluation

Define the relative difference between the real observation r and its prediction $\hat{r}$ by $|\hat{r} - r|/r$. For each gauge station, we generate 1000 precipitation predictions from the predictive distribution obtained in the previous section for a month. We then compute the sample means. We compare the real observations with these sample means to evaluate the validity of the proposed method by the relative differences between them. As an example, the relative differences and sample standard deviations for the month of January 1995 and for all stations are given in Table 5.9. From this table, it can be seen that almost all absolute values of relative differences are smaller than 0.1 and hence the proposed method works well in this example.

Appendix 5.1: Hierarchical Bayesian spatio-temporal modeling (or kriging)

Consider the model (5.2). The following is mainly based on Chapter 10 of Le and Zidek (2006).

(a) Parameter partitioning

Partition the matrix W_a: $d \times (u+g)$ corresponding to the l covariates in conformance with the block structure as

$$W_a = (W_a^{[u]}, W_a^{[g_1]}, \cdots, W_a^{[g_k]}).$$

Write $W_a^{[g_{j_1},\ldots,g_{j_r}]} = (W_a^{[g_{j_1}]}, \cdots, W_a^{[g_{j_r}]})$ and $W_a^{[g]} = (W_a^{[g_1]}, \cdots, W_a^{[g_k]})$.

Likewise, partition the covariance matrix $\boldsymbol{\Sigma} : (u+g) \times (u+g)$ over monitored and unobserved sites conformably as

$$\boldsymbol{\Sigma} = \begin{pmatrix} \Sigma^{[u,u]} & \Sigma^{[u,g]} \\ \Sigma^{[g,u]} & \Sigma^{[g,g]} \end{pmatrix},$$

where $\Sigma^{[u,u]}$ is a $u \times u$ matrix for the unobserved sites. Further partition the covariance matrix $\Sigma^{[g,g]} : g \times g$ for the gauge station blocks as follows:

$$\Sigma^{[g,g]} = \begin{pmatrix} \Sigma^{[g_1,g_1]} & \ldots & \Sigma^{[g_1,g_k]} \\ \vdots & \vdots & \vdots \\ \Sigma^{[g_k,g_1]} & \ldots & \Sigma^{[g_k,g_k]} \end{pmatrix}.$$

Similarly, for $j = 1, \ldots, k$, let

$$\Sigma^{[g_j,\cdots,g_k]} = \begin{pmatrix} \Sigma^{[g_j,g_j]} & \ldots & \Sigma^{[g_j,g_k]} \\ \vdots & \vdots & \vdots \\ \Sigma^{[g_k,g_j]} & \ldots & \Sigma^{[g_k,g_k]} \end{pmatrix}.$$

(b) The Bartlett transformation (Bartlett 1933)

Deriving the predictive distribution is facilitated by reparameterizing the matrix $\boldsymbol{\Sigma}$ through the recursive one-to-one Bartlett transformation for the two blocks:

$$\boldsymbol{\Sigma} = \begin{pmatrix} \Gamma^{[u]} + \left(\Upsilon^{[u]}\right)^T \Sigma^{[g,g]}\Upsilon^{[u]} & (\Upsilon^{[u]})^T \Sigma^{[g,g]} \\ \Sigma^{[g,g]}\Upsilon^{[u]} & \Sigma^{[g,g]} \end{pmatrix}$$

where $\Gamma^{[u]} = \Sigma^{[u,u]} - \Sigma^{[u,g]}(\Sigma^{[g,g]})^{-1}\Sigma^{[g,u]}$ and $\Upsilon^{[u]} = (\Sigma^{[g,g]})^{-1}\Sigma^{[u,g]}$. Similarly, applying the Bartlett decomposition, one can represent the (sub-)matrix $\Sigma^{[g_j,\cdots,g_k]}$, for $j = 1, \cdots, k-1$, as

$$\Sigma^{[g_j,\cdots,g_k]} = \begin{pmatrix} \Gamma_j + \Upsilon_j^T \Sigma^{[g_{j+1},\cdots,g_k]}\Upsilon_j & \Upsilon_j^T \Sigma^{[g_{j+1},\cdots,g_k]} \\ \Sigma^{[g_{j+1},\cdots,g_k]}\Upsilon_j & \Sigma^{[g_{j+1},\cdots,g_k]} \end{pmatrix}.$$

Here $\Gamma_k = \Sigma^{[g_k,g_k]}$, and for $j = 1, \ldots, k-1$,

$$\Gamma_j : g_j \times g_j = \Sigma^{[g_j,g_j]} - \Sigma^{[g_j,(g_{j+1},\ldots,g_k)]}\left(\Sigma^{[g_{j+1},\ldots,g_k]}\right)^{-1}\Sigma^{[(g_{j+1},\ldots,g_k),g_j]},$$

$$\Upsilon_j : (g_{j+1} + \ldots + g_k) \times g_j = \left(\Sigma^{[g_{j+1},\ldots,g_k]}\right)^{-1}\Sigma^{[(g_{j+1},\ldots,g_k),g_j]},$$

with

$$\Sigma^{[(g_{j+1},\ldots,g_k),g_j]} = \begin{pmatrix} \Sigma^{[g_{j+1},g_j]} \\ \vdots \\ \Sigma^{[g_k,g_j]} \end{pmatrix},$$

and $\Sigma^{[g_j,(g_{j+1},\ldots,g_k)]} = \left(\Sigma^{[(g_{j+1},\ldots,g_k),g_j]}\right)^T$ for $j = 1, \ldots, k-1$.

(c) The GIW prior $GIW(\Theta, \delta)$

The GIW prior distribution for $\boldsymbol{\Sigma}$ in (5.2) is equivalently defined in terms of $(\Gamma^{[u]}, \Upsilon^{[u]})$ and $\{(\Gamma_1, \Upsilon_1), \ldots, (\Gamma_{k-1}, \Upsilon_{k-1}), \Gamma_k\}$ as follows.

$$\begin{cases} \Upsilon^{[u]}|\Gamma^{[u]} & \sim & N(\Upsilon_{00}, H_0 \otimes \Gamma^{[u]}), \\ \Gamma^{[u]} & \sim & IW(\Lambda_0, \delta_0), \\ \Upsilon_j|\Gamma_j & \sim & N(\Upsilon_{0j}, H_j \otimes \Gamma_j), \quad j = \ldots, k-1, \\ \Gamma_j & \sim & IW(\Lambda_j, \delta_j), \quad j = 1, \ldots, k. \end{cases} \tag{5.18}$$

From (5.18), $\Upsilon^{[u]}$ is the slope of the optimal linear predictor of $Y^{[u]}$ based on $Y^{[g]}$ and $\Gamma^{[u]}$, the residual covariance of the optimal linear predictor. Similar interpretations apply to Υ_j and Γ_j, for $j = 1, \ldots, k-1$.

Let $\mathcal{H}$ be the set of the hyperparameters in (5.2)–(5.18), i.e., $\mathcal{H} = \{\Theta, \boldsymbol{\delta}, V_B, B_0\}$, where Θ labels the set of hyperparameters: $\Theta = \{(\Upsilon_{00}, H_0, \Lambda_0), (\Upsilon_{01}, H_1, \Lambda_1), \ldots, (\Upsilon_{0k-1}, H_{k-1}, \Lambda_{k-1}), \Lambda_k\}$ with degrees of freedom parameters $\boldsymbol{\delta} = (\delta_0, \delta_1, \ldots, \delta_k)^T$. The dimensions of H_j and Λ_j are $(g_{j+1} + \cdots + g_k) \times (g_{j+1} + \cdots + g_k)$ and $g_j \times g_j$, respectively.

(d) Estimation of $\mathcal{H}_g = \{V_B, B_0, (\Upsilon_{01}, H_1, \Lambda_1, \delta_1), \ldots, (\Upsilon_{0,k-1}, H_{k-1}, \Lambda_{k-1}, \delta_{k-1}), (\Lambda_k, \delta_k)\}$

As indicated in Le and Zidek (2006), the predictive distributions derived through the integrated framework developed above are completely characterized by their hyperparameters, which may be estimated by using an empirical Bayes approach. This means estimating them by maximizing the marginal likelihood of all the measured responses (conditional on those hyperparameters) evaluated at their observed values. This procedure is referred to as type-II maximum likelihood estimation (type-II MLE). To use it to estimate $\mathcal{H}_g$, the following procedure can be employed:

Compute the hyperparameter values that maximize the marginal distribution $f(Y^{[g^o]}|\mathcal{H}_g)$, where $\mathcal{H}_g = \{V_B, B_0, (\Upsilon_{01}, H_1, \Lambda_1, \delta_1), \ldots, (\Upsilon_{0,k-1}, H_{k-1}, \Lambda_{k-1}, \delta_{k-1}), (\Lambda_k, \delta_k)\}$ are the hyperparameters of interest, and $Y^{[g^o]} = \{Y^{[g_1^o]}, \ldots, ^{[g_k^o]}\}$. The subscript g indicates that not all the hyperparameters are involved in this marginal distribution. Assume the response matrix Y has the Gaussian-generalized inverted Wishart model specified by (5.2) and (5.18). Then, the marginal distribution $f(Y^{[g^o]}|\mathcal{H}_g)$ can be written as

$$Y^{[g^o]}|\mathcal{H}_g \sim \prod_{j=1}^{k} t_{(n-m_j)\times g_j}(\Xi_o^{[j]}, \Phi_o^{[j]} \otimes \Psi_o^{[j]}, \delta_o^{[j]}), \tag{5.19}$$

where $t_{a\times b}$ denotes a matric-t distribution and $\Xi_o^{[j]} = \Xi_{(2)}^{[j]}$, $\Phi_o^{[j]} = A_{22}^{[j]}$, $\Psi_o^{[j]} = \frac{\Lambda_j}{\delta_j - g_j + 1}$, $\delta_o^{[j]} = \delta_j - g_j + 1$ with $\tilde{\mathcal{E}}^{[g_{j+1},\ldots,g_k]} = Y^{[g_{j+1},\ldots,g_k]} - X B_0 W_a^{[g_{j+1},\ldots,g_k]}$, for $j = 1, \ldots, k-1$,

$$\begin{pmatrix} \Xi_{(1)}^{[j]} \\ \Xi_{(2)}^{[j]} \end{pmatrix} : \begin{pmatrix} m_j \times g_j \\ (n-m_j) \times g_j \end{pmatrix} = X B_0 W_a^{[g_j]} + \tilde{\mathcal{E}}^{[g_{j+1},\ldots,g_k]} \Upsilon_{0j},$$

$$\begin{pmatrix} A_{11}^{[j]} & A_{12}^{[j]} \\ A_{21}^{[j]} & A_{22}^{[j]} \end{pmatrix} : \begin{pmatrix} m_j \times m_j & m_j \times (n-m_j) \\ (n-m_j) \times m_j & (n-m_j) \times (n-m_j) \end{pmatrix}$$

$$= I_n + X V_B X^T + \tilde{\mathcal{E}}^{[g_{j+1},\ldots,g_k]} H_j \left(\tilde{\mathcal{E}}^{[g_{j+1},\ldots,g_k]}\right)^T.$$

Although $f(Y^{[g^o]}|\mathcal{H}_g)$ can be written as a product of matric-t distributions as in (5.19), direct maximization of this marginal density presents a challenge. In this chapter, the EM algorithm is used instead.

(e) Predictive distributions of missing data

From Theorem 10.1 of Le and Zidek (2006), it follows that

$$Y^{[g_k^m]}|Y^{[g^o]}, \mathcal{H}_g \sim t_{m_k \times g_k}\left(\Xi_{(m|g)}^{[k]}, \Phi_{(m|g)}^{[k]} \otimes \Psi_{(m|g)}^{[k]}, \delta_{(m|g)}^{[k]}\right) \tag{5.20}$$

$$Y^{[g_j^m]}|Y^{[g_{j+1}^m,\ldots,g_k^m]}, Y^{[g^o]}, \mathcal{H}_g \sim t_{m_k \times g_k}\left(\Xi_{(m|g)}^{[j]}, \Phi_{(m|g)}^{[j]} \otimes \Psi_{(m|g)}^{[j]}, \delta_{(m|g)}^{[j]}\right) \quad (5.21)$$

where

$$\Xi_{(m|g)}^{[j]} = \Xi_{(1)}^{[j]} + A_{12}^{[j]}(A_{22}^{[j]})^{-1}\left(Y^{[g_j^o]} - \Xi_{(2)}^{[j]}\right),$$

$$\Phi_{(m|g)}^{[j]} = \frac{\delta_j - g_j + 1}{\delta_j - g_j + n - m_j + 1}[A_{11}^{[j]} - A_{12}^{[j]}(A_{22}^{[j]})^{-1}A_{21}^{[j]}],$$

$$\Psi_{(m|g)}^{[j]} = \frac{1}{\delta_j - g_j + 1}\left[\Lambda_j + \left(Y^{[g_j^o]} - \Xi_{(2)}^{[j]}\right)^T (A_{22}^{[j]})^{-1}\left(Y^{[g_j^o]} - \Xi_{(2)}^{[j]}\right)\right],$$

and $\delta_{(m|g)}^{[j]} = \delta_j - g_j + n - m_j + 1$.

(f) Thin-plate spline approach

Let f be the smooth mapping from the original location $\boldsymbol{s}_i$ to z_i, $i = 1, \ldots, N$. Then

$$f(\boldsymbol{s}) = \alpha_0 + \alpha_1 s^{(1)} + \alpha_2 s^{(2)} + \sum_{i=1}^{N} \beta_i u_i(\boldsymbol{s})$$

with the convention that $u_i(\boldsymbol{s}_i) = 0$ for $1 \le i \le N$, where $u_i(\boldsymbol{s}) = \|\boldsymbol{s} - \boldsymbol{s}_i\|^2 \log \|\boldsymbol{s} - \boldsymbol{s}_i\|$ and $s^{(j)}$ indicates the jth coordinate of the location $\boldsymbol{s}$ for $j = 1, 2$. The parameters to be fitted are αs and βs. Compute the function f as a thin-plate spline (Wabba and Wendelberger 1980) for z_i. A smoothing parameter is incorporated into this second step, allowing users to choose a desirable level of smoothness.

For any specified value smoothing parameter λ, αs and βs are chosen to minimize

$$\sum_{i=1}^{N}(z_i - f(\boldsymbol{s}_i))^2 + \lambda[J_2(f_1)],$$

while J_2 measures the smoothness of the functions defined as

$$J_2(f) = \int_R \left[\left(\frac{\partial^2 f}{\partial x_1^2}\right)^2 + 2\left(\frac{\partial^2 f}{\partial x_1 \partial x_2}\right)^2 + \left(\frac{\partial^2 f}{\partial x_2^2}\right)^2\right] dx_1 dx_2.$$

This construction ensures that $\beta_i \to 0$, $1 \le i \le N$, when $\lambda \to \infty$; that is, in this case, it is a simple linear mapping.

(g) Spatial predictive distribution of unobserved sites

The spatial predictive distribution over these unobserved sites is given by

$$(Y^{[u]}|Y^{[g]}, \mathcal{H}) \sim t_{n \times up}\left(\Xi^{u|g}, \frac{\Phi^{[u|g]} \otimes \Psi^{[u|g]}}{\delta_0^*}, \delta_0^*\right), \quad (5.22)$$

where $\delta_0^* = \delta_0 - u + 1$, $\Psi^{[u|g]} = \Lambda_0$, $\Xi^{[u|g]} = XB_0W_a^{[u]} + (Y^{[g]} - XB_0W_a^{[g]})\Upsilon_{00}$ and

$$\Phi^{[u|g]} = I_n + XV_BX^T + (Y^{[g]} - XB_0W_a^{[g]})H_0(Y^{[g]} - XB_0W_a^{[g]})^T.$$

The estimates of B_0 and V_B have been obtained in (d).

Define $\mathcal{S} = \{\boldsymbol{s}_1, \ldots, \boldsymbol{s}_g\} \bigcup \{\boldsymbol{u}_1, \ldots, \boldsymbol{u}_u\}$. First estimate the spatial covariance between response variables at two different locations $\boldsymbol{l}_i \in \mathcal{S}$ and $\boldsymbol{l}_j \in \mathcal{S}$ such that at least one of them is not a location with a gauge station. Note that $\Sigma_{i,j} = \text{Cov}(Y(\boldsymbol{l}_i), Y(\boldsymbol{l}_j)) = \sigma_{Y(\boldsymbol{l}_i)}\sigma_{Y(\boldsymbol{l}_j)}Corr(Y(\boldsymbol{l}_i), Y(\boldsymbol{l}_j))$. Assume that $\widehat{Corr}(Y(\boldsymbol{l}_i), Y(\boldsymbol{l}_j))$, an optimal approximation of the spatial correlation between $Y(\boldsymbol{l}_i)$ and $Y(\boldsymbol{l}_j)$, have been obtained [e.g. (5.15)]. $\hat{\sigma}_{Y(\boldsymbol{l}_i)}$ and $\hat{\sigma}_{Y(\boldsymbol{l}_j)}$, respective estimates of $\sigma_{Y(\boldsymbol{l}_i)}$ and $\sigma_{Y(\boldsymbol{l}_i)}$, can then be obtained by the thin-plate spline approach [see (f) for details]. Thus $\Sigma_{i,j}$ can be estimated by

$$\hat{\Sigma}_{i,j} = \widehat{\text{Cov}}(Y(\boldsymbol{l}_i), Y(\boldsymbol{l}_j)) = \hat{\sigma}_{Y(\boldsymbol{l}_i)}\hat{\sigma}_{Y(\boldsymbol{l}_j)}\widehat{Corr}(Y(\boldsymbol{l}_i), Y(\boldsymbol{l}_j)).$$

Then in terms of $\hat{\mathcal{H}}_g$ obtained in (d), estimate the hyperparameters Λ_0, Υ_{00}, H_0 and δ_0 via

$$\hat{\delta}_0 = \frac{\hat{\delta}_1 + \cdots + \hat{\delta}_k}{k}, \quad \hat{H}_0 = \hat{\Lambda}^{[1,\ldots,k]}, \quad \hat{\Upsilon}_{00} = \left(\hat{\Sigma}^{[g,g]}\right)^{-1}\hat{\Sigma}^{[g,u]}, \tag{5.23}$$

$$\hat{\Lambda}_0 = \frac{\hat{\delta}_0 - u - 1}{1 + tr\left(\hat{\Sigma}^{[g,g]}\hat{H}_0\right)}\left(\hat{\Sigma}^{[u,u]} - \hat{\Upsilon}_{00}^T\hat{\Sigma}^{[g,g]}\hat{\Upsilon}_{00}\right), \tag{5.24}$$

where

$$\hat{\Lambda}^{[j,\ldots,k]} = \begin{pmatrix} \hat{\Lambda}_j + \hat{\Upsilon}_{0j}^T\hat{\Lambda}^{[j+1,\ldots,k]}\hat{\Upsilon}_{0j} & \hat{\Upsilon}_{0j}^T\hat{\Lambda}^{[j+1,\ldots,k]} \\ \hat{\Lambda}^{[j+1,\ldots,k]}\hat{\Upsilon}_{0j} & \hat{\Lambda}^{[j+1,\ldots,k]} \end{pmatrix}$$

for $j = 1, \ldots, k-1$, and $\hat{\Lambda}^{[k]} = \hat{\Lambda}_k$.

Appendix 5.2: Some estimated parameters

$\hat{\delta}_0 = 448.9954$,

$$\hat{H}_0 = \begin{pmatrix} 37.25 & -30.32 & -2.56 & \cdots & 3.09 & 0.93 & 5.44 \\ -30.32 & 48.06 & 7.09 & \cdots & -2.01 & -6.06 & -0.79 \\ \vdots & \vdots & \vdots & \vdots & \vdots & \vdots & \vdots \\ 0.93 & -6.06 & -1.09 & \cdots & -0.94 & 2.55 & 0.27 \\ 5.44 & -0.79 & -5.72 & \cdots & -0.51 & 0.27 & 5.52 \end{pmatrix}_{37\times37},$$

$$\hat{\Upsilon}_{00} = \begin{pmatrix} 0.005 & 0.003 & 0.001 & \cdots & 0.046 & 0.038 & 0.040 \\ 0.007 & 0.004 & 0.002 & \cdots & 0.065 & 0.062 & -0.063 \\ \vdots & \vdots & \vdots & \vdots & \vdots & \vdots & \vdots \\ 0.010 & 0.007 & 0.004 & \cdots & 0.037 & 0.013 & 0.015 \\ -0.027 & -0.022 & -0.018 & \cdots & -0.029 & -0.018 & -0.021 \end{pmatrix}_{37\times 445},$$

$$\hat{\Lambda}_0 =$$

$$\begin{pmatrix} 0.013 & 0.004 & 0.003 & \cdots & 1e-04 & -4e-05 & -2e-05 \\ 0.004 & 0.013 & 0.003 & \cdots & 6e-05 & -6e-05 & -5e-05 \\ \vdots & \vdots & \vdots & \vdots & \vdots & \vdots & \vdots \\ -4e-05 & -6e-05 & -7e-05 & \cdots & 0.001 & 0.011 & 0.001 \\ -2e-05 & -5e-05 & -6e-05 & \cdots & 0.001 & 0.001 & 0.011 \end{pmatrix}_{445\times 445}.$$

Acknowledgments

The work was partially supported by the Natural Sciences and Engineering Research Council of Canada. We thank Prof. Dr Werner G. Müller and an anonymous reviewer for helpful comments and suggestions.

References

Bartlett MS (1933). On the theory of statistical regression. *Proc. R. Soc. Edinburgh* **53**, 260–283.

Barry R, Ver Hoef J (1996). Blackbox kriging: spatial prediction without specifying variogram models. *J. Agric. Biol. Environ. Statist.* **1**, 297–322.

Caselton WF, Husain T (1980). Hydrologic networks: information transmission. *J. Water Res. Plann. Manage. Div. ASCE* **106**, 503–520.

Caselton WF, Zidek JV (1984). Optimal monitoring network designs. *Statist. Probab. Lett.* **2**, 223–227.

Caselton WF, Kan L, Zidek JV (1992). Quality data networks that minimize entropy. In: Walden A, Guttorp P (eds), *Statistics in Environmentali and Earth Sciences*. Griffin, London, 10–38.

Dobbie MJ, Herderson BL, Stevens Jr DL (2008). Sparse sampling: spatial design for monitoring stream networks. *Statist. Surv.* **2**, 113–153.

Fan J, Li R (2001). Variable selection via nonconcave penalized likelihood and its oracle properties. *J. Am Statist. Assoc.* **96**, 1348–1360.

Jaynes ET (1963). *Information Theory and Statistical Mechanics*. In: Ford K (ed.), *Statistical Physics*. Benjamin, New York, 181–218.

Journel AG, Huijbregts CJ (1978). *Mining Geostatistics.* Academic, London.

Le ND, Zidek JV (2006). *Statistical Analysis of Environmental Space-Time Processed*. Springer, New York.

Le ND, Sun W, Zidek JV (2001). Spatial prediction and temporal backcasting for environmental fields having monotone data patterns. *Can. J. Statist.* **29**, 529–554.

Müller WG (2005). A comparison of spatial design methods for correlated observations. *Environmetrics* **16**, 495–506.

Shewry M, Wynn H (1987). Maximum entropy sampling. *Appl. Statist.* **14**, 165–207.

Sebastiani P, Wynn HP (2000). Maximum entropy sampling and optimal Bayesian experimental design. *J. R. Statist. Soc. Ser. B* **62**, 145–157.

Wabba G, Wendelberger J (1980). Some new mathematical methods for variational objective analysis using splines and cross-validation. *Mon. Weather Rev.* **108**, 1122–1143.

Wachernagel H (2003). *Multivariate Geostatistics*. Springer, Berlin.

Wikle CK, Berliner LM, Cressie N (1998). Hierarchical Bayesian space-time models. *Environ. Ecol. Statist*. **5**, 117–154.

Wu S, Zidek JV (1992). An entropy-based analysis of data from selected NADP/NTN network sites for 1983-1986. *Atmos. Environ*. **26A**, 2089–2103.

Zhang C (2010). Nearly unbiased variable selection under minimax concave penalty. *Ann. Statist.* **38**, 894–942.

Zidek JV, Zimmerman D (2009). Monitoring network designs. In: Gelfand A, Diggle P, Fuentes M and Guttorp P (eds), *Handbook of Spatial Statistics*. Chapman & Hall/CRC Press, Boca Raton, FL, 131–148.

Zidek JV, Sun W, Le ND (2000). Designing and integrating composite networks for monitoring multivariate Gaussian pollution fields. *Appl. Statist.* **49**, 63–79.

6

Accounting for design in the analysis of spatial data

Brian J. Reich and Montserrat Fuentes
Department of Statistics, North Carolina State University, USA

6.1 Introduction

The fundamental objective of geostatistical data analysis is to estimate the continuous spatial process $\mu(\mathbf{s})$ over a domain of interest $\mathcal{D}$. Estimation is based on measurements $y_1, \ldots, y_n$ taken at locations $\mathbf{s}_1, \ldots, \mathbf{s}_n$. It is typically assumed that the measurement locations are fixed when conducting the analysis. However, in some situations this assumption is untenable. For example, the United States Environmental Protection Agency (US EPA) maintains a nationwide network of measurement stations to monitor ambient air quality. The locations of these stations were not selected at random. They were selected to best achieve the EPA's scientific objectives, including to intensely monitor regions that may be out of compliance with EPA air quality standards and to monitor air quality in areas with high population density. In both cases, monitors are placed in locations where the process of interest is likely to be high, and areas with good air quality are under-represented in this sample. This may result in a positive bias when interpolating from the measurement locations to locations without monitors.

While there is a vast literature on methods for designing networks to best meet the scientific objectives of a spatial analysis (e.g., Müller 2007), there has until recently been little work on accounting for these design factors in

Spatio-temporal design: Advances in efficient data acquisition, First Edition.
Edited by Jorge Mateu and Werner G. Müller.

the subsequent spatial analysis. We identify two settings in which the analysis should incorporate knowledge of the design process: informative missingness and informative sampling. Informative missingness occurs if the original design was fixed without regard to the spatial process of interest, but some observations are not recorded, and this missing data process depends on the spatial process of interest. Examples of informative missingness include:

1. Rain gauges that are more often missing on days with heavy precipitation.
2. Clouds obscuring gridded satellite measurements of environmental processes.

On the other hand, informative sampling refers to situations where there is no fixed set of locations with measurements thinned following a missing data process. Rather, the measurement locations are selected from the entire domain $\mathcal{D}$ in a way that depends on the spatial process to be measured. Examples of informative sampling include:

1. Sampling locations for ambient air quality monitoring devices.
2. Catch size of fishermen.
3. Residential locations of asthma patients.

In all three cases, the locations are selected by individuals with prior knowledge of the process of interest (air quality, fish density, and health risk for asthma patients, respectively), and it is natural to question whether these locations can be treated as a random sample or fixed in the subsequent analysis.

Figure 6.1 shows a one-dimensional simulated example to illustrate the pitfalls of ignoring informative sampling. In this example, the data are generated so that the density of sampling locations and the associated responses are both high around locations $s = 0.1$ and $s = 0.6$. The true spatial process is low at $s = 0.4$, but since there are no samples in this area the kriging estimate is pulled towards to the overall mean. Therefore, ignoring the relationship between the underlying spatial process and the sampling density results in poor spatial interpolation in this unsampled location.

One approach to accommodating both informative missingness and informative sampling is to jointly model the observed responses and the locations. In this shared variable approach, the underlying process of interest μ enters the model for both $\mathbf{s}_i$ and y_i, and thus both sources of data inform about the spatial process. The shared-variable approach attenuates prediction bias and provides valid inference for model parameters.

The remainder of the chapter proceeds as follows. Section 6.2 provides a review of spatial methods that account for the sampling design. Informative missingness is discussed in Section 6.2.1, while informative sampling is discussed in Section 6.2.2. The informative missingness model is applied to analyze yearly total precipitation in Upper Austria in Section 6.3 and Section 6.4 concludes.

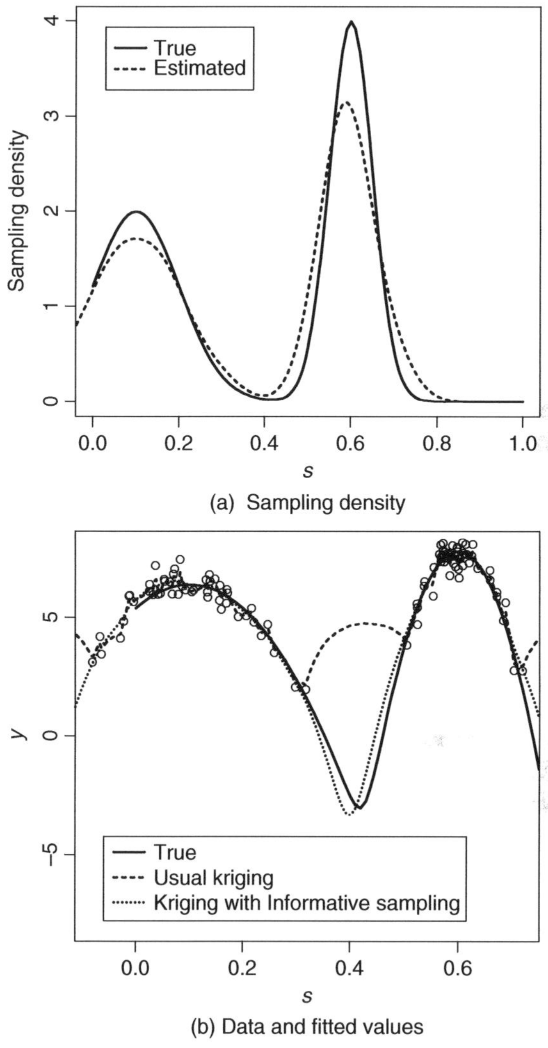

Figure 6.1 Simulated example with informative sampling. The true sampling density $\mathrm{p}(s)$ *in (a) is an equal mixture of* $N(0.1, 0.1^2)$ *and* $N(0.6, 0.05^2)$ *densities, and the estimated density is a kernel density estimate. The* $\mathrm{n} = 100$ *sampling locations in (b) are drawn from* $\mathrm{p}(\mathbf{s})$*, and the responses are generated independently as* $\mathrm{y_i} = N(5 + 2\log[\mathrm{p}(\mathbf{s}_\mathrm{i})], 0.5^2)$*. The two fits are the kriging estimates with and without the estimated log sampling intensity as a linear predictor.*

6.2 Modeling approaches

Let $\mu(\mathbf{s})$ be a spatial process of interest on the domain $\mathcal{D} \subset \mathcal{R}^2$. The data consist of n spatial locations $\mathcal{S} = \{\mathbf{s}_1, \ldots, \mathbf{s}_n\} \subset \mathcal{D}$ and observations $\boldsymbol{y} = (y_1, \ldots, y_n)^T$, with y_i corresponding to location $\mathbf{s}_i$, $i = 1, \ldots, n$. In the shared variable approach, we model $\boldsymbol{y}$ and $\mathcal{S}$ simultaneously conditioned on the entire spatial process μ. Given the sequential nature of the sampling, it is convenient to express this joint distribution as $p(\boldsymbol{y}, \mathcal{S}|\mu) = p(\boldsymbol{y}|\mathcal{S}, \mu)p(\mathcal{S}|\mu)$. We will assume throughout that $\boldsymbol{y}|\mathcal{S}, \mu$ is multivariate normal with mean $\mathrm{E}(y_i) = \mu(\mathbf{s}_i)$ and $\mathrm{Cov}(\boldsymbol{y}) = \sigma^2 I_n$.

For the prior, we assume that μ is a Gaussian process with mean $\mathrm{E}[\mu(\mathbf{s})] = \mathbf{x}(\mathbf{s})^T\boldsymbol{\beta}_1$ and covariance $\mathrm{Cov}[\mu(\mathbf{s}), \mu(\mathbf{s}')] = \tau_1^2\rho_1(||\mathbf{s} - \mathbf{s}'||)$, where $\mathbf{x}(\mathbf{s})$ is a p-vector of spatial covariates and $\boldsymbol{\beta}_1$ is the corresponding coefficient vector. For simplicity, we assume stationary covariances and Gaussian responses throughout, although more complex models may be embedded in this Bayesian hierarchical framework. Denote $p(\mu)$ as the process prior for μ. Then its posterior is

$$p(\mu|\mathcal{S}, \boldsymbol{y}) \propto p(\boldsymbol{y}|\mathcal{S}, \mu)p(\mathcal{S}|\mu)p(\mu). \tag{6.1}$$

Clearly, if $\mathcal{S}$ is independent of μ, then inference on μ can proceed as if the sampling locations are fixed and thus $p(\mu|\mathcal{S}, \boldsymbol{y}) \propto p(\boldsymbol{y}|\mathcal{S}, \mu)p(\mu)$, as usual. If $\mathcal{S}$ depends on μ, then $p(\mathcal{S}|\mu)$ remains in the posterior, i.e., the sampling locations are informative about μ and should be accounted for in the analysis.

The remainder of this section considers different models for $\mathcal{S}|\mu$ that are appropriate for different designs. We begin with the case of informative missingness in Section 6.2.1, since this is conceptually and computationally simpler than the informative sampling model described in Section 6.2.2. We also describe a two-stage approach in Section 6.2.3, which attempts to account for informative sampling and missingness by including the proper covariate in the mean of μ.

6.2.1 Informative missingness

Assume a set of N potential measurement locations $\mathcal{T} = \{\mathbf{t}_1, \ldots, \mathbf{t}_N\}$ is selected without regard to the latent process μ. For each potential location, denote

$$\delta_j = \begin{cases} 1 & \text{the observation at } \mathbf{t}_j \text{ is missing} \\ 0 & \text{otherwise} \end{cases} \tag{6.2}$$

so that $\mathcal{S} = \{\mathbf{t}_j | \delta_j = 0\}$. Following Reich and Bandyopadhyay (2010), we model the missing data process as

$$g[P(\delta_j = 1)] = \theta(\mathbf{t}_j) + a\mu(\mathbf{t}_j), \tag{6.3}$$

where g is a monotonically increasing link function and θ is a Gaussian process with mean $\mathrm{E}[\theta(\mathbf{t})] = \mathbf{x}(\mathbf{t})^T\boldsymbol{\beta}_2$ and covariance $\mathrm{Cov}[\theta(\mathbf{t}), \theta(\mathbf{t}')] = \tau_2^2\phi_2(||\mathbf{t} - \mathbf{t}'||)$

that controls spatial patterns in the missing data process that are not captured by μ.

The parameter a controls the degree of informative missingness. If $a = 0$, then the sampling locations $\mathcal{S}$ are independent of μ, and the missing data process can be ignored when estimating μ. In contrast, if $a > 0$ ($a < 0$), then locations with large (small) values of the spatial process are more likely to be missing. Therefore, the posterior of a can be used to test for the presence of informative missingness.

Reich and Bandyopadhyay (2010) use a probit link for g, although other links such as the logistic link are possible. Under the probit link, $P(\delta_j = 1) = \Phi[\theta(\mathbf{t}_j) + a\mu(\mathbf{t}_j)]$, where Φ is the standard normal cumulative distribution function. The probit model is equivalent to the auxiliary variable model $\delta_j = \mathbf{I}(z_j > 0)$, where $z_j \sim \mathrm{N}[\theta(\mathbf{t}_j) + a\mu(\mathbf{t}_j), 1]$, independent over j (Albert and Chib 1993). This gives the model at site $\mathbf{t}_j$

$$y_j \sim \mathrm{N}\left[\mu(\mathbf{t}_j), \sigma^2\right] \quad \text{and} \quad z_j \sim \mathrm{TN}_{A_j}\left[\theta(\mathbf{t}_j) + a\mu(\mathbf{t}_j), 1\right], \tag{6.4}$$

where y_j and z_j are conditionally (on μ) independent, and $\mathrm{TN}_A(m, s^2)$ denotes the normal density with mean m and standard deviation s, truncated to the region A. The truncation region is $A_j = (-\infty, 0)$ if $\delta_j = 0$ and $A_j = (0, \infty)$ if $\delta_j = 1$. The latent variables z_j are treated as unknown parameters in the Bayesian model. Therefore, Markov chain Monte Carlo (MCMC) sampling naturally marginalizes over the latent z_j, giving $P(\delta_j = 1) = P(z_j > 0) = \Phi[\theta(\mathbf{t}_j) + a\mu(\mathbf{t}_j)]$.

This auxiliary model is convenient for MCMC sampling, as z_j, $\boldsymbol{\mu} = [\mu(\mathbf{t}_1), \ldots, \mu(\mathbf{t}_N)]$, $\boldsymbol{\theta} = [\theta(\mathbf{t}_1), \ldots, \theta(\mathbf{t}_N)]$, and a have conjugate full conditional distributions (truncated normal for the z_j and Gaussian for $\boldsymbol{\mu}$, $\boldsymbol{\theta}$, and a), which permits Gibbs sampling and leads to straight forward coding and tuning. Also, the missing y_j at sites with $\delta_j = 1$ are simply updated from their likelihood, $y_j \sim \mathrm{N}\left[\mu(\mathbf{t}_j), \sigma^2\right]$.

6.2.2 Informative sampling

Informative sampling considers a more intricate design. Instead of a fixed and finite set, the measurement locations are drawn from the measurable set $\mathcal{D}$. In this case, it is impossible to model the presence or absence of measurements at each individual location. In the seminal paper by Diggle *et al.* (2010), the authors propose modeling the sampling locations as a realization of an inhomogeneous Poisson process with intensity proportional to $\exp[\theta(\mathbf{t}) + a\mu(\mathbf{t})]$. Conditioned on the number of samples, n, the sampling locations $s_1, \ldots, s_n$ are independent with logit-Gaussian density

$$p(\mathbf{s}) = \frac{\exp[\theta(\mathbf{s}) + a\mu(\mathbf{s})]}{\int_{\mathcal{D}} \exp[\theta(\mathbf{t}) + a\mu(\mathbf{t})]\, dt}. \tag{6.5}$$

The interpretation of θ and a are similar to the interpretations given in Section 6.2.1; θ controls spatial patterns in the sampling intensity not explained by μ, and a determines the strength of informative sampling.

Pati *et al.* (2011) study the theoretical properties of a Bayesian implementation of this model. Assuming a powered-exponential spatial correlation function and proper, in some cases bounded, priors for $\boldsymbol{\beta}_j$, τ_j^2, σ^2, and spatial correlation parameters, they show that the posterior of a is proper even if its prior is the flat improper prior. That is, the data are informative about this important parameter. Also, assuming a Matérn correlation function and infill asymptotics (Stein 1999), they establish weak posterior consistency for θ, μ, and a.

Computation for this model is not straight forward because of the integral in the denominator of (6.5). Evaluating this integral requires the Gaussian processes μ and θ at every location in $\mathcal{D}$, which is not feasible. Therefore, an approximation to the likelihood of $\mathcal{S}$ is required. Let the points $\boldsymbol{u}_1, \ldots, \boldsymbol{u}_M$ be a dense grid of points covering $\mathcal{D}$, and define the approximate likelihood

$$\tilde{p}(\mathbf{s}) = \frac{\exp\left[\theta(\mathbf{s}) + a\mu(\mathbf{s})\right]}{\sum_j \exp\left[\theta(\boldsymbol{u}_j) + a\mu(\boldsymbol{u}_j)\right]}. \tag{6.6}$$

For large M, $\tilde{p}(\mathbf{s}) \approx p(\mathbf{s})$. Since $\tilde{p}(\mathbf{s})$ only requires the Gaussian processes at a finite collection of locations, it is possible to evaluate it in closed form. Diggle *et al.* (2010) use this approximate likelihood in a frequentist analysis using a Monte Carlo likelihood method. The Bayesian analysis in Pati *et al.* (2011) uses MCMC. Unlike the probit model in Section 6.2.1, the full conditionals are not conjugate and Metropolis sampling is required. Pati *et al.* (2011) further reduce the dimension of the model using an approximation (Higdon *et al.* 1998) for the Gaussian processes μ and θ.

6.2.3 A two-stage approach for informative sampling

To avoid the difficult computation that results from (6.5), Pati *et al.* (2011) consider a two-stage approach. This approach first makes a reparameterization from

$$y_i = \mu(\mathbf{s}_i) + \varepsilon_i \quad \text{and} \quad p(\mathbf{s}) = \frac{\exp\left[\theta(\mathbf{s}) + a\mu(\mathbf{s})\right]}{\int_{\mathcal{D}} \exp\left[\theta(\mathbf{t}) + a\mu(\mathbf{t})\right] dt} \tag{6.7}$$

for $\varepsilon_i \overset{iid}{\sim} \mathrm{N}(0, \sigma^2)$ to

$$y_i = f(\mathbf{s}_i) + bg(\mathbf{s}_i) + \varepsilon_i \quad \text{and} \quad p(\mathbf{s}) = \frac{\exp\left[g(\mathbf{s})\right]}{\int_{\mathcal{D}} \exp\left[g(\mathbf{t})\right] dt}. \tag{6.8}$$

As with μ and θ, the spatial processes f and g have Gaussian process priors. We note that although the span of the model under these two parameterizations is the same and there is a one-to-one correspondence between them, by specifying priors for f and g we are in fact considering a different Bayesian model than that presented in Section 6.2.2. However, this parameterization has useful properties, as described below.

In this parameterization, the mean of y_i is $\mu(\mathbf{s}_i) = f(\mathbf{s}_i) + bc + b\log[p(\mathbf{s}_i)]$, where $c = \log\{\int_{\mathcal{D}} \exp\left[g(\mathbf{t})\right] dt\}$. Therefore, the log sampling intensity enters the mean response as a linear predictor. This suggests an alternative to the shared variable approach. Let $\hat{\lambda}(\mathbf{s})$ be an estimate of $\log[p(\mathbf{s}_i)]$, perhaps a kernel density estimate based on $\mathcal{S}$. Then, absorbing bc into f's prior mean, we consider the model

$$y_i = f(\mathbf{s}_i) + b\hat{\lambda}(\mathbf{s}_i) + \varepsilon_i. \tag{6.9}$$

Since $\hat{\lambda}(\mathbf{s})$ is considered fixed, this approximate model is simply the usual geostatistical model without informative sampling, but with an additional linear predictor in the mean. Therefore, this model can be fitted by standard statistical software packages such as R and SAS. Also, more complex sampling mechanisms can be accounted for by allowing the mean of y_i to be a nonlinear function of $\hat{\lambda}(\mathbf{s}_i)$.

Figure 6.1 illustrates this two-stage approach on a one-dimensional simulated example. Figure 6.1(a) shows the kernel density estimate of the sampling density. Including the log of the estimated intensity as a linear predictor in Figure 6.1(b) clearly improves estimation of the underlying process.

Although computationally convenient, one would expect this two-stage approximation to underestimate posterior variability for μ and its hyperparameters since uncertainty regarding the log sampling density is not propagated through to μ's posterior. An extension would be to extract both an estimate and corresponding measure of uncertainty for the log sampling density from the first stage density estimate, and carry both of these quantities to the second stage. That is, let $\boldsymbol{\lambda} = [\lambda(\mathbf{s}_1), \ldots, \lambda(\mathbf{s}_n)]^T$ and $\hat{\boldsymbol{\lambda}} = [\hat{\lambda}(\mathbf{s}_1), \ldots, \hat{\lambda}(\mathbf{s}_n)]^T$ be, respectively, the true and estimated log sampling density at the sample locations. Then in the first stage we obtain the estimate $\hat{\boldsymbol{\lambda}} \sim \mathrm{N}(\boldsymbol{\lambda}, \Sigma_\lambda)$. Uncertainty in the estimated log sampling density could then be accounted for in the model

$$\mathbf{y} \sim \mathrm{N}\left[\mathbf{f} + b\boldsymbol{\lambda}, \sigma^2 I_n\right] \quad \text{and} \quad \boldsymbol{\lambda} \sim N(\hat{\boldsymbol{\lambda}}, \Sigma_\lambda), \tag{6.10}$$

where $\mathbf{f} = [f(\mathbf{s}_1), \ldots, f(\mathbf{s}_n)]^T$. Marginally over $\boldsymbol{\lambda}$,

$$\mathbf{y} \sim \mathrm{N}\left[\mathbf{f} + b\hat{\boldsymbol{\lambda}}, \sigma^2 I_n + b^2\Sigma_\lambda\right]. \tag{6.11}$$

Compared with the plug-in model (6.9), uncertainty in the sampling intensity propagates through to the response by adding $b^2\Sigma_\lambda$ to the covariance. Therefore, this hybrid approach accounts for uncertainty in the log sampling density, while maintaining a computationally tractable model with conjugate conditionals for $\mathbf{f}$ and $\boldsymbol{\lambda}$.

6.3 Analysis of the Austrian precipitation data

In this section, we analyze the precipitation data from Upper Austria using the informative missingness model described in Section 6.2.1. Data are available at $n = 37$ spatial locations for the years 1994–2009. For these data, 18% of the observations are missing, usually following a clearly nonrandom pattern, e.g., observations at a station may be missing for the first 5 years and observed for the remaining years.

Let $y_i(\mathbf{t})$ be the log of the total annual rainfall (mm) at location $\mathbf{t}$ for year $i = 1994, \ldots, 2009$. We model these data using an extension of the model in Section 6.2.1 to accommodate replication across years. Let

$$y_i(\mathbf{t}) \sim \mathrm{N}\left[\alpha(\mathbf{t}) + \mu_i(\mathbf{t}), \sigma^2\right] \tag{6.12}$$
$$z_i(\mathbf{t}) \sim \mathrm{N}\left[\beta(\mathbf{t}) + \theta_i(\mathbf{t}) + a\{\alpha(\mathbf{t}) + \mu_i(\mathbf{t})\}, 1\right],$$

where $y_i(\mathbf{t})$ is missing if $z_i(\mathbf{t}) > 0$. In this model, $\alpha(\mathbf{t})$ and $\beta(\mathbf{t})$ control the overall spatial trends in precipitation and missing data probabilities, respectively, and the Gaussian processes α, β, μ_i, and θ_i are all independent with exponential correlation functions and means $\mathrm{E}[\alpha(\mathbf{t})] = c$, $\mathrm{E}[\beta(\mathbf{t})] = d$, and $\mathrm{E}[\mu_i(\mathbf{t})] = \mathrm{E}[\theta_i(\mathbf{t})] = 0$. For priors, we assume $a, c, d \sim \mathrm{N}(0, 100^2)$ and $\sigma^{-2} \sim$ Gamma(0.1,0.1). For all Gaussian processes with exponential covariance of the form $\mathrm{Cov}[\alpha(\mathbf{t}), \alpha(\mathbf{t}')] = \tau^2 \exp(-||\mathbf{t} - \mathbf{t}'||/\rho)$, where $\rho > 0$ is the spatial range parameter which determines the rate of decay of the spatial correlation. We use InvGamma(0.1, 0.1) priors for the variances τ^2 and LogNormal$(0, 10^2)$ priors for the spatial ranges (where locations are given in degrees latitude/longitude). Implementing this spatio-temporal model is very similar to the spatial-only model described above. Given the spatial processes α and β, the model is independent across years following the spatial-only model. Updating α and β proceeds using the usual normal/normal conjugate updates.

There is no reason to believe that there is informative missingness in these data. In fact, an analysis using the full data show that the posterior 95% interval for a covers zero, suggesting the missing data mechanism is not associated with precipitation level. Therefore, to illustrate the modeling techniques discussed above, we create synthetic data with informative missingness. We first fit the model without informative missingness (i.e., with $a = 0$) to the full dataset to estimate $m_i(\mathbf{t}) = \alpha(\mathbf{t}) + \mu_i(\mathbf{t})$. After standardizing these values to have mean zero and variance one, we withhold observation $y_i(\mathbf{t})$ with probability $\Phi[m_i(\mathbf{t}) - 1]$. We then fit model (6.12) with informative missingness (IM) and with noninformative missingness (NIM) by setting $a = 0$.

Figures 6.2 and 6.3 plot the data and the results for 1994 and 2009, respectively. In both years we see a strong north/south trend in the fitted values due to the elevation. For the IM model, the posterior 95% interval for a is (2.54, 3.85). Therefore, we detect the presence of informative sampling using these data. Figures 6.2(c) and 6.3(c) show that the posterior mean of the missing data probability, i.e., $\Phi\{\beta(\mathbf{t}) + \theta_i(\mathbf{t}) + a[\alpha(\mathbf{t}) + \mu_i(\mathbf{t})]\}$, varies considerably across space, from 0.1 in the north to 0.8 in the south. As a result of accounting for IM, the posterior mean of $\alpha(\mathbf{t}) + \mu_i(\mathbf{t})$ increases by as much as 0.12 log mm in the south [Figures 6.2(d) and 6.3(d)].

Finally, we repeated this experiment 100 times, each time drawing a new set of missing values with probabilities $\Phi[m_i(\mathbf{t}) - 1]$. Comparing the performance of these two models averaged over the 100 datasets, we find that the root mean square prediction error for the withheld $y_i(\mathbf{t})$ was 0.086 (standard error 0.001) for the IM model compared with 0.100 (standard error 0.002) for the NIM model.

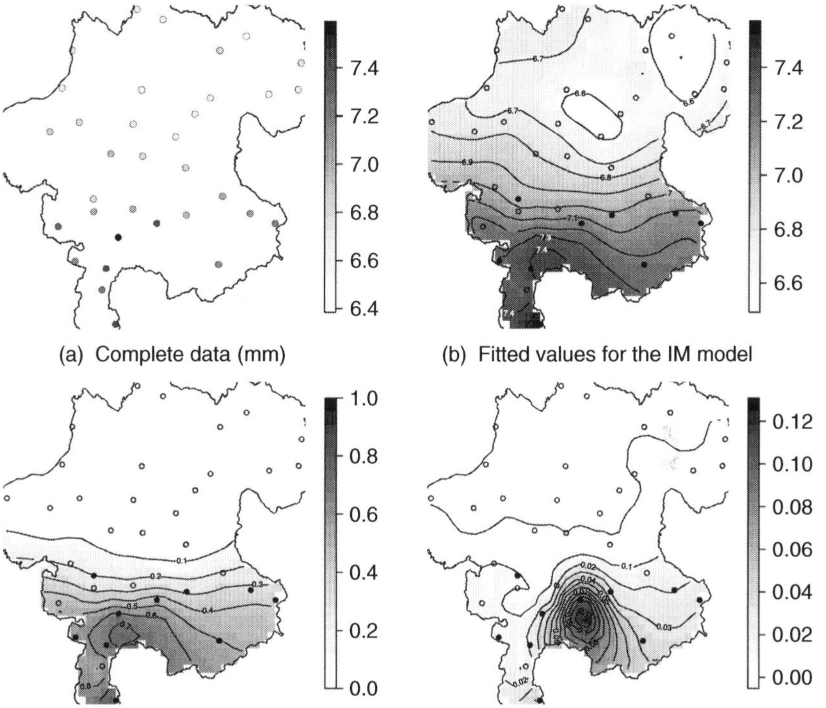

(a) Complete data (mm)

(b) Fitted values for the IM model

(c) Missingness probability for the IM model

(d) IM-NIM models, fitted values

Figure 6.2 Data and results for 1994. (a) shows the complete data (i.e., before simulating missing observations). (b) and (c) plot the posterior mean of $\alpha + \mu_{1994}$*, where* μ_{1994} *is the spatial process for 1994, and the missingness probability, respectively, for the informative missingness (IM) model. (d) plots the difference between the posterior mean of* $\alpha + \mu_{1994}$ *for the IM and noninformative missingness (NIM) models. In (b)–(d), open circles represent data locations and solid circles represent missing data locations.*

While both models have coverage near the nominal 95% level, the bias is more pronounced for the NIM model (-0.027, standard error 0.002) than the IM model (-0.002, standard error 0.001).

6.4 Discussion

In this chapter, we have reviewed several current approaches in the literature that account for design in a spatial analysis. We discuss separately informative missingness and informative sampling. Both problems are handled with a shared variable approach and MCMC computing. We also present a two-stage method for the computationally intensive shared variable approach to informative sampling.

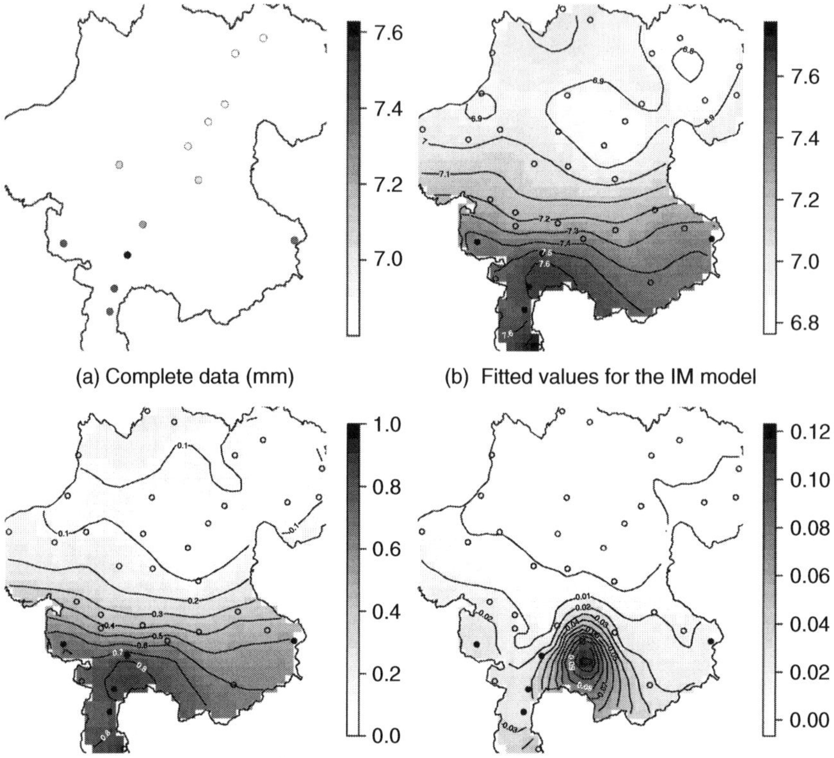

(a) Complete data (mm)

(b) Fitted values for the IM model

(c) Missingness probability for the IM model

(d) IM-NIM models, fitted values

Figure 6.3 Data and results for 2009. (a) shows the complete data (i.e., before simulating missing observations). (b) and (c) plot the posterior mean of $\alpha + \mu_{2009}$*, where* μ_{2009} *is the spatial process for 2009, and the missingness probability, respectively, for the informative missingness (IM) model. (d) plots the difference between the posterior mean of* $\alpha + \mu_{2009}$ *for the IM and noninformative missingness (NIM) models. In (b)–(d), open circles represent data locations and solid circles represent missing data locations.*

The shared variable approach requires several parametric assumptions that are difficult to verify in practice. For example, we assume normality, stationarity, and a log-linear relationship between the spatial process and the sampling intensity. It may be possible to expand the parametric model to account for each of these complex features (e.g., Gelfand *et al.* 2010). However, more flexible and robust models would certainly be desirable.

We have discussed only point-referenced spatial data. Accounting for design is also possible for areal and point pattern spatial data. Reich and Bandyopadhyay (2010) analyze informative missingness for areal spatial data using a Gaussian Markov random field (e.g., Gelfand *et al.* 2010) for the latent spatial process. Also, Chakraborty *et al.* (2010) discuss preferential sampling for point process data in the context of a study to estimate animal density where areas thought to have high density are monitored more intensely than other regions.

References

Albert JH and Chib S 1993. Bayesian analysis of binary and polychotomous response data. *Journal of the American Statistical Association*, **88**, 669–679.

Chakraborty A, Gelfand AE, Wilson AM, Latimer AM, and Silander JA 2010. Modeling large scale species abundance with latent spatial processes. *Annals of Applied Statistics*, **4**, 1403–1429.

Diggle P, Meneze R, and Su T 2010. Geostatistical inference under preferential sampling (with discussion). *Journal of the Royal Statistical Society: Series C (Applied Statistics)*, **59**, 191–232.

Gelfand AE, Diggle PJ, Fuentes M, and Guttorp P 2010. *Handbook of Spatial Statistics*. Chapman & Hall/CRC, Boca Raton, FL.

Higdon D, Swall J, and Kern J 1998. *Non-Stationary Spatial Modeling*. Bayesian Statistics 6–Proceedings of the Sixth Valencia Meeting. Bernardo JM, Berger JO, Dawid AP, and Smith AFM (editors). Clarendon Press, Oxford, 761–768.

Müller WG 2007. *Collecting spatial data: optimum design of experiment for random fields*. Springer, Heidelberg.

Pati D, Reich BJ, and Dunson DB 2011. Bayesian geostatistical modeling with informative sampling locations. *Biometrika*, **98**, 35–48.

Reich BJ and Bandyopadhyay D 2010. A latent factor model for spatial data with informative missingness. *Annals of Applied Statistics*, **4**, 439–459.

Stein ML 1999. *Interpolation of Spatial Data: Some Theory for Kriging*. Springer, New York.

7

Spatial design for knot selection in knot-based dimension reduction models

Alan E. Gelfand[1], **Sudipto Banerjee**[2] **and Andrew O. Finley**[3]

[1]*Department of Statistical Science, Duke University, Durham USA*
[2]*Division of Biostatistics, School of Public Health, University of Minnesota, Minneapolis, USA*
[3]*Department of Geography and Department of Forestry, Michigan State University, East Lansing, USA*

7.1 Introduction

This chapter has a different flavor from others in this volume. Here, the problem is not one of spatial design for data collection but, rather, one of spatial design to facilitate fitting of spatial models. That is, we have a post-data collection, pre-data analysis design problem. More precisely, we address the setting where there is need to specify knots in order to fit desired spatial models. That is, we seek to fit hierarchical models in order to enable full inference and to adequately capture uncertainty. However, we are unable to directly fit the model we would

Spatio-temporal design: Advances in efficient data acquisition, First Edition.
Edited by Jorge Mateu and Werner G. Müller.

like and so we adopt an *approximate* model to make computation feasible. Such a situation arises when we work with very large datasets in space (and in time).

We note that, with the advent of inexpensive high speed computing, statisticians today routinely encounter geographically referenced datasets containing multiple variables observed across thousands of locations, perhaps across thousands of time points. There is, by now, a rich literature in probabilistic/statistical modeling for location-referenced spatial data; see, for example, the books by Cressie (1993), Banerjee *et al.* (2004), and Schabenberger and Gotway (2004) as well as the very recent volumes from Gelfand *et al.* (2010) and Cressie and Wikle (2011) for broad discussion and applications. The need to model spatially referenced outcomes, perhaps vector-valued, across large spatial (and possibly temporal) domains is emerging in many fields including the geological and environmental sciences, ecological systems, digital terrain cartography, computer experiments, public health, and social and economic systems.

The general challenge of the proposed approximation is to express a surface over a spatial domain through the use of *knots*, often called low rank or reduced rank representations (Wahba 1990; Rasmussen and Williams 2006). Here, the problem takes a critical subdivision. Is the surface viewed as 'random', i.e., a realization of a stochastic process? Are there an uncountable number of variables over the spatial domain, say D? Or, is the surface viewed as 'fixed', i.e., a parametric function?[1]

In the former case, we write the surface as $\{w(\mathbf{s}); \mathbf{s} \in D\}$ where $w(\mathbf{s}) = \sum_{j=1}^{m} l(\mathbf{s}, \mathbf{s}_j^*)Z(\mathbf{s}_j^*)$. Thus, the surface/process realization is completely determined by the function $l(\cdot, \cdot)$ and the set of variables, $\{Z(\mathbf{s}_j^*), j = 1, 2, \ldots, m\}$. The collection of $\mathbf{s}_j^*$'s are the knots and their specification, both number and location, is the design issue of this chapter. Evidently, knot selection affects both $l(\cdot, \cdot)$ and the distribution of the Z's.

In the latter case, we think in terms of basis representations for the function, in particular, using splines. We might write the surface as $f(\mathbf{s}) = \sum_{j=1}^{m} b_j g_j(\mathbf{s})$ where the b_j are a set of coefficients and $\{g_j(\mathbf{s}), j = 1, 2, \ldots, m\}$ are a spline basis set (typically, but not necessarily orthonormal). The spline functions are defined with regard to a set of knots which, again, we can label as $\{\mathbf{s}_j^*, j = 1, 2, \ldots, m \in D\}$. The literature here is enormous and, in fact, is a subset of functional data analysis which is often applied over domains of dimension higher than 2 and is called multivariate function approximation in the applied math community, nonparametric regression in the statistics realm, and neural nets in the machine learning/computer science world.

In this literature, with interest in the function explicitly, typically, a knot is taken at the 'location' (covariate level) of each data point (Ramsay and Silverman 2005). In this case, there is no dimension reduction. Alternatively, knot selection is subsumed within the fitting problem. The fitting is algorithmic, e.g., an algorithm for fitting the curve to the observed data. It is usually done through

[1] Here, we mean fixed in the classical sense, i.e., a parametric form. Of course, in the Bayesian view, if the parametric function is unknown, it is taken to be random.

recursive partitioning as in CART and MARS (Breiman *et al.* 1984; Friedman 1991; Berk 2008; Hastie *et al.* 2009) or as L^p (usually $p = 1, 2$) optimization, introducing regularization through a roughness penalty (Wahba 1990; Gu 2002). Knot selection can also be achieved using Generalized Cross-Validation (GCV), which is a form of regularization that trades off goodness-of-fit against model complexity using a formula to approximate the error that would be determined by leave-one-out validation (Wahba 1990; Lin *et al.* 2000; Ruppert *et al.*, 2003). Bayesian analogs of the above include variants of Bayesian adaptive regression models as discussed, for example, in Lang and Brezger (2004), in the book by Denison *et al.* (2004), and in Chipman *et al.* (1998, 2010).

Returning to the stochastic process context, we find the situation to be very different if we are interested in model-based inference. With regard to a stochastic process, we think in terms of a covariance function, say $C(\boldsymbol{s}, \boldsymbol{s}')$ to measure the covariance between $w(\boldsymbol{s})$ and $w(\boldsymbol{s}')$, to capture the pairwise dependence between the variables in D. If we are going to build an approximate model for the process, we need to incorporate this dependence structure into the knot design.

First, let us clarify why we need to approximate the process. It arises primarily due to our desire to fit Bayesian hierarchical models (Banerjee *et al.* 2004; Gelman *et al.* 2004) which are recognized as versatile inferential tools for capturing the rich dependence structures underlying spatial data and offering full inference regarding the spatial processes. Hierarchical models implemented through Markov chain Monte Carlo (MCMC) methods have become especially popular for spatial model fitting, due to their ability to handle models that would be infeasible otherwise.

Under such modeling, covariance matrices arise; determinants and inverses for these matrices are required in order to work with the likelihoods. When the dimension, say n, of these matrices is large, fitting hierarchical spatial models involves expensive matrix computation of the order n^3. In a fully Bayesian paradigm, where one seeks formal inference on the 'variance' parameters, these matrix computations occur in each iteration of the MCMC. As modern data technologies acquire and exploit massive amounts of data, statisticians analyzing spatially referenced datasets confront settings where the number of geo-referenced locations is so large as to make hierarchical modeling fitting infeasible. The situation is further exacerbated in multivariate settings with several spatially dependent response variables, where the matrix dimensions increase by a factor of the number of spatially dependent variables being modeled (Gelfand and Banerjee 2010). It is also aggravated when data are collected at frequent time points and spatio-temporal process models are used, again increasing dimension by a factor of the number of time points.

So, we see the value of representations of the spatial process in lower-dimensional subspaces, representations that easily generalize to multivariate and/or spatio-temporal processes. As noted above, these are often referred to as low-rank or reduced-rank spatial models and have been explored in different contexts (Wikle and Cressie 1999; Lin *et al.* 2000; Higdon 2002; Kamman and Wand 2003; Ver Hoef *et al.* 2004; Xia and Gelfand 2006; Stein 2007, 2008;

Banerjee *et al.* 2008; Crainiceaniu *et al.* 2008; Cressie and Johannesson 2008). Many of these methods are variants of the so-called 'subset of regressors' methods used in Gaussian process regressions for large datasets in machine learning (Rasmussen and Williams 2006). The idea here is to assume, quite reasonably, that the spatial information available from the entire set of observed locations can be summarized in terms of a smaller, but representative, sets of locations, or 'knots'. That is, we still use all of the data but we represent the spatial structure through a dimension reduction. So, again, in implementing the reduction, we need to *design* the knots. Consideration of this issue forms the balance of the chapter.

The format of the chapter is as follows. Section 7.2 provides a brief review of different approaches to modeling large point-referenced datasets. Section 7.3 turns the focus to knot-based dimension reduction using low rank models/approximations and the predictive process in particular. Section 7.4 discusses some basic ideas for designing knots and connects this objective with existing knowledge on spatial designs. A specific strategy for knot-design for predictive process models is outlined in Section 7.4.2. Section 7.5 presents some numerical examples using simulated data as well as a forestry dataset and the Austria monitoring network dataset. We conclude the chapter with some final thoughts and an eye toward future work.

7.2 Handling large spatial datasets

Modeling for large point-referenced spatial datasets has been receiving increased attention. Here, we provide a brief review of this work. Vecchia (1988) proposed approximating the likelihood with a product of local conditional distributions (in the spirit of Markov random fields, see, e.g., Banerjee *et al.* 2004; Rue and Held 2005) to obtain maximum-likelihood estimates. Stein *et al.* (2004) adapt this idea effectively for restricted maximum likelihood estimation, pointing out the need to include some nonlocal locations in the approximation in order to better learn about the overall behavior of the parametric covariance function that has been assumed. Another possibility is to approximate the likelihood using spectral representations of the spatial process (Fuentes 2007). These likelihood approximations yield a joint distribution (but not a process) that facilitates spatial interpolation. However, approximation in the spectral domain, e.g., a periodogram estimate of the spectral density can be subtle. Expertise in tailoring and tuning is required and does not easily adapt to multivariate processes. As a result, concern arises regarding the adequacy of the resultant likelihood approximation. Also, the spectral density approaches seem best suited to stationary covariance functions. Yet another approach considers compactly supported correlation functions, captured through covariance tapering, which yields sparse correlation structures (Furrer *et al.*, 2006; Du *et al.* 2009; Kaufman *et al.*, 2009). More efficient computation using sparse solvers can then be employed for kriging and variance estimation, but the tapered structures may limit modeling flexibility. Also, full likelihood-based inference requires determinant computations that still may be problematic. Another approach either replaces the process (random field) model by a *Markov* random field (Cressie 1993) or else approximates the random field model by a Markov random field (Rue and Tjelmeland, 2002; Rue and Held, 2005). This approach is best suited for points on a regular grid. With irregular locations, realignment to a grid or a torus is required, done by an algorithm, possibly introducing unquantifiable errors in precision.

In recent work Rue *et al.* (2009) propose a promising `INLA` (Integrated Nested Laplace Approximation) algorithm as an alternative to MCMC that utilizes the sparser matrix structures to deliver fast and accurate posterior approximations. However, the method involves a mode-finding exercise for hyper-parameters that may be problematic when the number of hyperparameters is more than 10. Its effectiveness is unclear for multivariate models with different structured random effects and unknown covariance matrices as hyperparameters. Also, its inference is limited; we do not see the entire posterior for a parameter, we do not see the entire predictive distribution when we krige to a new location.

Adapting the above approaches to more complex hierarchical spatial models involving multivariate processes (Gelfand *et al.* 2004; Wackernagel 2006), spatiotemporal processes and spatially varying regressions (Gelfand *et al.* 2003) and nonstationary covariance structures (Paciorek and Schervish 2006) is potentially problematic. When retaining the richness and flexibility of hierarchical models is of primary interest, knot-based low rank models seem to be a preferred option.

7.3 Dimension reduction approaches

In Section 7.3.1 we review some of the properties of a dimension reduction representation. Then, in Section 7.3.2 we turn to the predictive process, a particular reduced rank approach with which we have the most experience in terms of knot design.

7.3.1 Basic properties of low rank models

Following Section 7.1, we write the reduced rank specification for $\{s \in D\}$ as

$$\tilde{w}(s) = \sum_{j=1}^{m} l(s, s_j^*)Z(s_j^*). \tag{7.1}$$

Again, the surface/process realization is completely determined by the function $l(\cdot,\cdot)$ and the set of variables, $\{Z(s_j^*), j = 1, 2, \ldots, m\}$. The collection of s_j^*'s are the knots. For a collection of locations, with associated vector denoted by $\tilde{\mathbf{w}} = (\tilde{w}(s_1), \tilde{w}(s_2), \ldots, \tilde{w}(s_n))^T$, we write

$$\tilde{\mathbf{w}} = \boldsymbol{L}z^*, \tag{7.2}$$

where $\boldsymbol{L}$ is the $n \times m$ matrix with (i, j)th element $l(\boldsymbol{s}_i, \boldsymbol{s}_j^*)$ and z^* is the $m \times 1$ vector with entries $Z(\boldsymbol{s}_j^*)$ with $m < n$.

From (7.2) we see that, despite there being n $\tilde{w}$'s, we will only have to work with m Z's. Since we anticipate $m \ll n$, the dimension reduction is evident and, since we will write the model in terms of the Z's [with the $\tilde{w}$'s being deterministic from the Z's, given $l(\cdot, \cdot)$], the associated matrices we work with will be $m \times m$. Evidently, the $\tilde{w}(\boldsymbol{s})$ process spans only an m-dimensional space; we create an uncountable number of variables through a finite number of variables. When $n > m$, the joint distribution of $\tilde{\mathbf{w}}$ is singular. However, we do create a valid stochastic process. In particular, the valid covariance function is

$$\text{cov}(\tilde{w}(\boldsymbol{s}), \tilde{w}(\boldsymbol{s}')) = \mathbf{l}(\boldsymbol{s})^T \Sigma_{\mathbf{Z}^*} \mathbf{l}(\boldsymbol{s}') \tag{7.3}$$

where $\mathbf{l}(\boldsymbol{s})$ is the $m \times 1$ vector with entries $l(\boldsymbol{s}, \boldsymbol{s}_j^*)$. From (7.3), we see that, even if $l(\cdot, \cdot)$ is stationary, i.e., of the form $l(\cdot - \cdot)$, the induced covariance function is not. Also, if the Z's are Gaussian, then $\tilde{w}(\boldsymbol{s})$ is a Gaussian process.

How do we view the Z's? Are they a collection of w's from a process of interest, whence the $\tilde{w}$'s provide an approximation to enable computational tractability? Or are they merely a specification to provide a spatial process model? Also, are the $\boldsymbol{s}_j^*$ a subset of the observed $\boldsymbol{s}$'s or chosen otherwise? We conclude here with a few more words about the first three questions. The last question takes us back to the design problem.

The most prevalent specification for the Z's is i.i.d. normal with mean 0 and variance σ^2, i.e., $Z(\boldsymbol{s})$ is a white noise process, whence (7.3) simplifies to $\sigma^2 \mathbf{l}(\boldsymbol{s})^T \mathbf{l}(\boldsymbol{s}')$. This form appears in Barry and Ver Hoef (1996) and in a series of papers by Higdon and collaborators (Higdon *et al.*, 1998; Higdon, 2002), the former calling it a 'moving average' model, the latter, 'kernel convolution'. In particular, a natural choice for l is a kernel function, say $K(\boldsymbol{s} - \boldsymbol{s}')$ which puts more weight on $\boldsymbol{s}'$ near $\boldsymbol{s}$. The kernel would have parameters (which induces a parametric covariance function) and might be spatially varying (Higdon 2002; Paciorek and Schervish 2006). The reduced rank form can be viewed as a discretization of a process specification of the form $\tilde{w}(\boldsymbol{s}) = \int_{R^2} K(\boldsymbol{s} - \boldsymbol{s}') Z(\boldsymbol{s}') d\boldsymbol{s}'$ (see Xia and Gelfand 2006 for discussion regarding this discrete approximation). Gaussian kernels are frequently used though they lead to Gaussian covariance functions which, typically, yield process realizations too smooth to be satisfactory in practice (Stein 1999; Paciorek and Schervish 2006). Moreover, the scope of processes that can be obtained through kernel convolution is limited; for instance the widely used exponential covariance function does not arise from kernel convolution.

A different approach to specification for the Z's is to endow them with a stochastic process model having a selected covariance function. Again, from (7.3), this will impart a covariance function to the $\tilde{w}$'s. Reversing the perspective, if we have a particular covariance function that we wish for the $\tilde{w}(\boldsymbol{s})$, what covariance function shall we choose for the Z's? We argue in the next subsection that, in some sense, the predictive process provides an *optimal* choice.

7.3.2 Predictive process models: A brief review

Detailed descriptions of hierarchical Gaussian predictive process models are given in Banerjee *et al.* (2008), Finley *et al.* (2009), and Eidsvik *et al.* (2011). Here, we offer a brief review.

For an $n \times 1$ vector of observed outcomes, $\boldsymbol{y} = (y(\boldsymbol{s}_1), y(\boldsymbol{s}_2), \ldots, y(\boldsymbol{s}_n))^T$ with a first stage conditionally independent Gaussian specification and associated priors, we consider the Bayesian hierarchical model

$$p(\boldsymbol{\theta}, \boldsymbol{\beta}, \boldsymbol{w} \mid \boldsymbol{y}) \propto p(\boldsymbol{\theta}) \times N(\boldsymbol{\beta} \mid \boldsymbol{\mu}_\beta, \Sigma_\beta) \\ \times N(\boldsymbol{w} \mid \boldsymbol{0}, \boldsymbol{C}(\boldsymbol{\theta}_1)) \times \prod_{i=1}^{n} N(y(\boldsymbol{s}_i) \mid \boldsymbol{x}(\boldsymbol{s}_i)^T \boldsymbol{\beta} + w(\boldsymbol{s}_i), \tau^2)\,, \tag{7.4}$$

where $\boldsymbol{x}(\boldsymbol{s}_i)^T$ is a $1 \times p$ vector of explanatory variables and $\boldsymbol{\theta} = \{\boldsymbol{\theta}_1, \tau^2\}$. The parameter τ^2 is called the *nugget* and captures unstructured noise that may arise in the form of measurement error or microscale variability. Customarily, either a flat or a multivariate Gaussian prior is assigned to $\boldsymbol{\beta}$. Zhang (2004) demonstrated, rather remarkably, that the process parameters $\boldsymbol{\theta}_1$ were not consistently estimable (in the classical sense) for a rather general class of covariance functions. This has consequence in the Bayesian framework since it implies that the impact of the prior on inference is not overwhelmed with increasing sample size. Hence, informative priors will be needed for some parameters in order to identify all of the process unknowns.

Fitting (7.4) customarily proceeds using MCMC methods (Robert and Casella, 2004). With Gaussian likelihoods, we can integrate out the spatial effects $\boldsymbol{w}$. This marginalization results in $N\left(\boldsymbol{y} \mid \boldsymbol{X}\boldsymbol{\beta}, \boldsymbol{C}(\boldsymbol{\theta}_1) + \tau^2 \boldsymbol{I}_n\right)$, where $\boldsymbol{X}$ is $n \times p$ with $\boldsymbol{x}(\boldsymbol{s}_i)^T$ as the ith row. In any case, as noted above, MCMC fitting involves matrix decompositions of cubic order in the number of locations, which becomes infeasible for large n. Evidently, multivariate and spatial-temporal settings aggravate the situation.

Reduced rank spatial process models use a fixed set of 'knots' $\mathcal{S}^* = (\boldsymbol{s}_1^*, \ldots, \boldsymbol{s}_m^*)$ with $m \ll n$, which, as suggested above, need not be a subset of $\mathcal{S}$. An optimal projection of the process $w(\boldsymbol{s})$ at location $\boldsymbol{s}$, based upon its realization over $\mathcal{S}^*$, is given by the 'kriging equation', $\tilde{w}(\boldsymbol{s}) = \mathrm{E}\{w(\boldsymbol{s}) \mid \boldsymbol{w}^*\}$, where $\boldsymbol{w}^* = (w(\boldsymbol{s}_1^*), w(\boldsymbol{s}_2^*), \ldots, w(\boldsymbol{s}_m^*))^T$. We refer to $\tilde{w}(\boldsymbol{s})$ as the *predictive process* derived from the *parent process* $w(\boldsymbol{s})$.

For a zero-centered Gaussian process with covariance function $C(\boldsymbol{s}_1, \boldsymbol{s}_2; \boldsymbol{\theta}_1)$ as the parent, the predictive process is $\tilde{w}(\boldsymbol{s}) = \mathrm{E}\{w(\boldsymbol{s}) \mid \boldsymbol{w}^*\} = \boldsymbol{c}(\boldsymbol{s}, \boldsymbol{\theta}_1)^T \boldsymbol{C}^*(\boldsymbol{\theta}_1)^{-1} \boldsymbol{w}^*$, where $\boldsymbol{c}(\boldsymbol{s}; \boldsymbol{\theta}_1)^T$ is the $1 \times m$ vector whose jth element is $C(\boldsymbol{s}, \boldsymbol{s}_j^*; \boldsymbol{\theta}_1)$ and $\boldsymbol{C}^*(\boldsymbol{\theta}_1)$ is the $m \times m$ covariance matrix with elements $C(\boldsymbol{s}_i^*, \boldsymbol{s}_j^*; \boldsymbol{\theta}_1)$. The product $\boldsymbol{c}(\boldsymbol{s}, \boldsymbol{\theta}_1)^T \boldsymbol{C}^*(\boldsymbol{\theta}_1)^{-1}$ provides the $l(\boldsymbol{s}, \boldsymbol{s}_j^*)$ entries that appear in the general expression (7.1) for the reduced rank form in Section 7.3.1. Since $\boldsymbol{w}^*$ is multivariate normal with zero mean and $m \times m$

dispersion matrix $\boldsymbol{C}^*(\boldsymbol{\theta}_1)$, $\tilde{w}(\boldsymbol{s})$ is itself a nonstationary Gaussian process arising from a spatially adaptive linear transformation of the parent process over the set of knots. Immediately, $\text{cov}(\tilde{w}(\boldsymbol{s}), \tilde{w}(\boldsymbol{s}')) = \boldsymbol{c}(\boldsymbol{s}, \boldsymbol{\theta}_1)^T \boldsymbol{C}^{*-1}(\boldsymbol{\theta}_1)\boldsymbol{c}(\boldsymbol{s}', \boldsymbol{\theta}_1)$ and, thus, $\text{var}(\tilde{w}(\boldsymbol{s})) = \boldsymbol{c}(\boldsymbol{s}, \boldsymbol{\theta}_1)^T \boldsymbol{C}^{*-1}(\boldsymbol{\theta}_1)\boldsymbol{c}(\boldsymbol{s}, \boldsymbol{\theta}_1)$. Because of the dimension reduction, $\tilde{w}(\boldsymbol{s})$ *lives* in a lower dimensional space than $w(\boldsymbol{s})$. Thus, $\text{var}(\tilde{w}(\boldsymbol{s})) < \text{var}(w(\boldsymbol{s}))$, as seen from $\text{var}(\tilde{w}(\boldsymbol{s})|\boldsymbol{w}^*; \boldsymbol{\theta}_1) = \boldsymbol{C}(\boldsymbol{s}, \boldsymbol{s}; \boldsymbol{\theta}_1) - \boldsymbol{c}(\boldsymbol{s}, \boldsymbol{\theta}_1)^T \boldsymbol{C}^{*-1}(\boldsymbol{\theta}_1)\boldsymbol{c}(\boldsymbol{s}, \boldsymbol{\theta}_1) > 0$.

In the spirit of the previous section, replacing $w(\boldsymbol{s})$ with $\tilde{w}(\boldsymbol{s})$ in (7.4), yields its predictive process counterpart,

$$p(\boldsymbol{\theta}, \boldsymbol{\beta}, \boldsymbol{w}^* \mid \boldsymbol{y}) \propto p(\boldsymbol{\theta}) \times N(\boldsymbol{\beta} \mid \boldsymbol{\mu}_\beta, \Sigma_\beta) \times N(\boldsymbol{w}^* \mid \boldsymbol{0}, \boldsymbol{C}^*(\boldsymbol{\theta}_1)) \times \prod_{i=1}^{n} N(y(\boldsymbol{s}_i) \mid \boldsymbol{x}(\boldsymbol{s}_i)^T \boldsymbol{\beta} + \tilde{w}(\boldsymbol{s}_i), \tau^2). \tag{7.5}$$

Computational gains are achieved since matrix computations now involve the $m \times m$ matrix $\boldsymbol{C}^*(\boldsymbol{\theta}_1)$, with m much smaller than n. Unlike some other knot-based approaches, the predictive process does not introduce additional parameters nor does it involve projecting data onto a grid. Thus, it avoids identifiability issues or spurious decrease in uncertainty (Banerjee *et al.* 2008). Indeed, the attraction of predictive process models is that they are directly induced by the parent process (adopting its associated covariance function) without requiring choices of basis functions or kernels or alignment algorithms for the locations.

7.4 Some basic knot design ideas

In Section 7.4.1 we offer a short review of potential knot design approaches that we could use, settling upon an approach of Diggle and Lophaven (2006) for illustration. We elaborate this approach and the associated optimality criterion in Section 7.4.2.

7.4.1 A brief review of spatial design

Selection of knots is a challenging problem, evidently more difficult in two dimensions than in one. Suppose, for the moment, that m is given. As noted in Section 7.1, in the spline smoothing literature (and in most of the literature on functional data/regression modeling using basis representations), it is customary to place knots at every data point (Ramsay and Silverman 2005). This is not an option for us but raises the question of whether to use a subset of the observed spatial locations or a disjoint set of locations (perhaps, even a combination of both). If we use a subset of the sampled locations, should we draw this set at random? If we do not use a subset then we are dealing with a design problem, with the difference being that we already have samples at n locations.

The question to ask is whether and how the fact that we have already collected observations at $(\boldsymbol{s}_1, \boldsymbol{s}_2, \ldots, \boldsymbol{s}_n)$ would affect the choice of knots for the MCMC model fitting. Hence, we depart from the well-studied pre-posterior analysis (Bayesian design) literature. *Cost* also takes on a different role. Customarily, it is the cost associated with data collection, e.g., erecting a monitoring site, maintaining a monitoring site, collecting the data from that site, etc. For us, it is computational cost in terms of run times for MCMC model fitting. Also, from a Bayesian (and a likelihood perspective, as well), we can not revise our model to include a new prior which is the updated posterior resulting from the data collection and then embark upon knot selection. We need to select the knots in order to *fit* the model.

There is a rich literature in spatial and spatio-temporal design, i.e., design for data with structured dependence. Indeed, this volume provides the current state of the art, as well as full referencing. In our context, we only mention a very small portion of this work. One approach would be a so-called space-filling knot selection following the design ideas of Nychka and Saltzman (1998). Such designs are based upon geometric criteria, measures of how well a given set of points covers the study region, independent of the assumed covariance function. A variation on this would be a probability based approach which includes widely used simple random sampling without replacement. In our case, this would amount to placing a uniform distribution over the region D and choosing m points at random from this distribution. Either of these strategies will tend to be robust in that they make no population assumptions regarding, for example, mean structure or dependence structure.

Opportunities for more interesting design arise when the problem is cast as one of optimal spatial sampling design, assuming a particular spatial model. Model-based design has followed either a regression model path or a random process model path. Under regression modeling with independent data, optimality is defined with regard to efficiency of the estimates of the regression coefficients. An optimization criterion that is a function of the design matrix is specified and then the 'best' design optimizes this criterion over all design matrices. See Pukelsheim (1993) or Müller (2001) for details. This theory is not directly extensible to spatial design but approximately optimal solutions based upon information-theoretic measures have emerged, most notably employing the recursion in Brimkulov *et al.* (1986) (also see Fedorov 1996). Xia *et al.* (2006) consider algorithms such as sequential selection, block selection and stochastic search. Recent work by Zhu and Stein (2005) focuses on designs based upon optimization using the likelihood. They suggest working with the Fisher information as a measure in the form of a ratio of determinants and implement the optimization using a simulated annealing algorithm.

Model-based design, motivated by a random process specification, has been promoted in Le and Zidek (1992) and Zidek *et al.* (2000) as well as references therein. The proposal is an entropy-based design where the selection of the next site to be added will be the one with the largest entropy where entropy can be viewed as uncertainty. Under a Gaussian process assumption, the criterion that emerges is the conditional variance of an observation at a new location given the locations already selected. This conditional variance depends only upon the previously selected locations but not on the data already collected at those locations. The site with the most uncertainty is the one with the largest conditional variance given the selected sites. Extension to multivariate data at a location converts the criterion to a conditional covariance matrix. This approach has no interest in mean structure. In fact, quoting Zidek *et al.* (2000, p. 66), '[I]t avoids the need to specify objectives like parameter estimation.'

In the multiparameter case (almost certainly the case of interest in applications), both the information and entropy criteria emerge as matrices. So, to achieve a single number summary for a design, one has to summarize the resulting matrix either through a determinant or a (possibly weighted) trace.

Recently, Diggle and Lophaven (2006) discuss spatial designs through modification of designs on regular grids. They explore a Bayesian design criterion which minimizes the spatially averaged prediction variance. Their Bayesian design approach naturally combines the goal of efficient spatial prediction while allowing for uncertainty in the values of model parameters. These designs augment the lattice with close pairs or infill. We examine such designs for knot selection in our simulation example.

7.4.2 A strategy for selecting knots

For a given set of observations, Finley *et al.* (2009) proposed a knot selection strategy designed to improve the induced predictive process as an approximation to the parent process. For a selected set of knots, $\tilde{w}(\boldsymbol{s}) = \mathrm{E}[w(\boldsymbol{s}) \mid \boldsymbol{w}^*]$ is considered as an approximation to the parent process. Given $\boldsymbol{\theta}_1$, the associated predictive variance of $w(\boldsymbol{s})$ conditional on the predictive process $\boldsymbol{w}^*$ on $\mathcal{S}^*$ is

$$\begin{aligned} V_{\boldsymbol{\theta}_1}(\boldsymbol{s}, \mathcal{S}^*) &= \mathrm{var}[w(\boldsymbol{s}) \mid \boldsymbol{w}(\cdot), \mathcal{S}^*, \boldsymbol{\theta}_1] \\ &= \boldsymbol{C}(\boldsymbol{s}, \boldsymbol{s}; \boldsymbol{\theta}_1) - \boldsymbol{c}(\boldsymbol{s}, \boldsymbol{\theta}_1)^T \boldsymbol{C}^{*-1}(\boldsymbol{\theta}_1)\boldsymbol{c}(\boldsymbol{s}, \boldsymbol{\theta}_1) \end{aligned} \tag{7.6}$$

which measures how well we approximate $w(\boldsymbol{s})$ by the predictive process $\tilde{w}(\boldsymbol{s})$. This measure is in the spirit of the foregoing work of Zidek and colleagues (Le and Zidek 1992; Zidek *et al.* 2000), i.e., measuring knot *value* in terms of conditional variance. There, the best knot selection maximizes conditional variance given the previously selected knots. Here, we measure the effectiveness of a given collection of selected knots through small conditional variance.

In particular, the knot selection criterion is then defined as a function of $V_{\boldsymbol{\theta}_1}(\boldsymbol{s}, \mathcal{S}^*)$. One commonly used criterion is:

$$V_{\boldsymbol{\theta}_1}(\mathcal{S}^*) = \int_D V_{\boldsymbol{\theta}_1}(\boldsymbol{s}, \mathcal{S}^*) g(\boldsymbol{s}) d\boldsymbol{s} = \int_D \mathrm{var}[w(\boldsymbol{s}) \mid \boldsymbol{w}^*, \boldsymbol{\theta}_1] g(\boldsymbol{s}) d\boldsymbol{s} \tag{7.7}$$

where $g(\boldsymbol{s})$, integrable over D, is the weight assigned to location $\boldsymbol{s}$ (Zidek *et al.* 2000; Diggle and Lophaven 2006). Here, we only consider the simple case for which $g(\boldsymbol{s}) \equiv 1$. $V_{\boldsymbol{\theta}_1}(\mathcal{S}^*)$ can be regarded as a spatially averaged predictive variance. The integral in (7.7) is analytically intractable and discrete approximations such as numerical quadrature or Monte Carlo integration will be required. We use the discrete approximation which computes the spatially averaged prediction variance over all the observed locations,

$$V_{\boldsymbol{\theta}_1}(\mathcal{S}^*) \approx \frac{\sum_{i=1}^n \mathrm{var}[w(\boldsymbol{s}_i) \mid \boldsymbol{w}^*, \boldsymbol{\theta}_1]}{n} \tag{7.8}$$

We ultimately reduce the problem of knot performance to the minimization of a design criterion, which is the function $V_{\boldsymbol{\theta}_1}(\mathcal{S}^*)$.

The following facts are easily verified for $V_{\boldsymbol{\theta}_1}(\mathcal{S}^*)$ defined in (7.7):

- $V_{\boldsymbol{\theta}_1}(\{\mathcal{S}^*, \boldsymbol{s}_0\}) - V_{\boldsymbol{\theta}_1}(\mathcal{S}^*) < 0$ for a new site $\boldsymbol{s}_0$;
- $V_{\boldsymbol{\theta}_1}(\{\mathcal{S}^*, \boldsymbol{s}_0\}) - V_{\boldsymbol{\theta}_1}(\mathcal{S}^*) \longrightarrow 0$ when $\|\boldsymbol{s}_0 - \boldsymbol{s}_i^*\| \longrightarrow 0$, where $\boldsymbol{s}_i^*$ is any member of $\mathcal{S}^*$;
- $V_{\boldsymbol{\theta}_1}(\mathcal{S}) = 0$, where $\mathcal{S} = \{\boldsymbol{s}_1, \ldots, \boldsymbol{s}_n\}$ are the original observed locations.

The variance–covariance matrix under the parent process model is $\Sigma_{\boldsymbol{Y}} = \boldsymbol{C}(\boldsymbol{\theta}_1) + \tau^2 \boldsymbol{I}$, while that from the corresponding predictive process is given by $\tilde{\Sigma}_{\boldsymbol{Y}} = \mathcal{C}(\boldsymbol{\theta}_1)^T \boldsymbol{C}^{*-1}(\boldsymbol{\theta}_1)\mathcal{C}(\boldsymbol{\theta}_1) + \tau^2 \boldsymbol{I}$. The Frobenius norm between $\Sigma_{\boldsymbol{Y}}$ and $\tilde{\Sigma}_{\boldsymbol{Y}}$ is $\|\Sigma_{\boldsymbol{Y}} - \tilde{\Sigma}_{\boldsymbol{Y}}\|_F \equiv \mathrm{tr}\left([\boldsymbol{C}(\boldsymbol{\theta}_1) - \mathcal{C}(\boldsymbol{\theta}_1)^T \boldsymbol{C}^{*-1}(\boldsymbol{\theta}_1)\mathcal{C}(\boldsymbol{\theta}_1)]^2\right)$. Since $\boldsymbol{C}(\boldsymbol{\theta}_1) - \mathcal{C}(\boldsymbol{\theta}_1)^T \boldsymbol{C}^{*-1}(\boldsymbol{\theta}_1)\mathcal{C}(\boldsymbol{\theta}_1)$ is positive definite, the Frobenius norm $\|\Sigma_{\boldsymbol{Y}} - \tilde{\Sigma}_{\boldsymbol{Y}}\|_F \equiv \sum \lambda_i^2$, where λ_i is the ith eigenvalue of $\Sigma_{\boldsymbol{Y}} - \tilde{\Sigma}_{\boldsymbol{Y}}$. Also, the averaged predictive variance is given by $\bar{V} = \frac{1}{n}\mathrm{tr}(\Sigma_{\boldsymbol{Y}} - \tilde{\Sigma}_{\boldsymbol{Y}}) = \frac{1}{n}\sum \lambda_i$.

Note that, even after discretization, we can not evaluate $V_{\boldsymbol{\theta}_1}(\mathcal{S}^*)$ since it depends upon the unknown $\boldsymbol{\theta}_1$. Available options to accommodate this include obtaining parameter estimates by using a subset of the original data or more fully Bayesian strategies which place a prior on $\boldsymbol{\theta}_1$ and then minimizes $E_{\boldsymbol{\theta}_1}(V_{\boldsymbol{\theta}_1}(\mathcal{S}^*))$ (Diggle and Lophaven 2006). In fact, we might naturally use the same prior as we would use to fit the model.

Regardless of which of these strategies we adopt, how shall we proceed to find a good $\mathcal{S}^*$? Suppose the values of the parameters and the knot size m are given. The following sequential search algorithm finds an approximately optimal design:

- Initialization. As in all cases where the domain is continuous, for implementation of an optimal design, we need to reduce the possible sampling locations to a finite set. Natural choices include a fine grid set, the observed set of locations or the union of these two sets.
- Specify an initial set of locations of size $m_0 \ll m$ as starting points for knot selection; possible choices include a coarse grid, or a subset of the observed locations, chosen randomly or deterministically.
- At step $t + 1$,
  - For each sample point $\boldsymbol{s}_i$ in the allowable sample set, evaluate $V(\{\mathcal{S}^{*(t)}, \boldsymbol{s}_i\})$.

- Remove the sample point with maximum decrease in $\bar{V}$ from the allowable sample set and add it to the knot set.

- Repeat the above procedure until we obtain m knots.

The sequential evaluation of $\bar{V}$ is achieved using a very efficient routine incorporating block matrix computation. Utilizing this, we have successfully implemented the sequential algorithm in a simulation study shown in Section 7.5.2. We remark that the sequential algorithm does not necessarily achieve the global optimization solution. Alternative computational approaches are available to us in finding approximately optimal designs such as stochastic search and block selection (Xia *et al.* 2006).

As to the choice of m, the obvious answer is 'as large as possible'. Evidently, this is governed by computational cost and sensitivity to choice. So, for the former, we will have to implement the analysis over different choices of m to consider run time. For the latter, we look for stability of predictive inference as m increases. We measure this by the value of minimized $\bar{V}$ under different choices of m. Unlike more formal sampling design contexts, our goal here is to achieve 'good' knot selection to enable model fitting. We find coarse progression in m to be adequate. Finally, we can implement the analysis in two steps by combining this knot selection procedure with the modified predictive process in the obvious way: (1) choose a set of knots to minimize the averaged predictive variances; (2) then use the modified process in the model fitting.

7.5 Illustrations

7.5.1 A simulation example

A basic illustrative model employed in Banerjee *et al.* (2008) simulated the response $y(\boldsymbol{s})$ using

$$y(\boldsymbol{s}) = \beta_0 + w(\boldsymbol{s}) + \epsilon(\boldsymbol{s}) \ , \tag{7.9}$$

where β_0 is an intercept, while the spatial process $w(\boldsymbol{s})$ was generated using a stationary *anisotropic* Matérn covariance function given by $C(\boldsymbol{s}_1, \boldsymbol{s}_2; \boldsymbol{\theta}) = \sigma^2/\left(\Gamma(\nu)2^{\nu-1}\right)\left(2\sqrt{\nu d(\boldsymbol{s}_1, \boldsymbol{s}_2)}\right)^{\nu} \kappa_\nu\left(2\sqrt{\nu d(\boldsymbol{s}_1, \boldsymbol{s}_2)}\right)$, where we use the Mahalanobis distance $d(\boldsymbol{s}_1, \boldsymbol{s}_2) = \left(\boldsymbol{s}_1 - \boldsymbol{s}_2\right)^T \Sigma^{-1}\left(\boldsymbol{s}_1 - \boldsymbol{s}_2\right)$. We further parameterize $\Sigma = G(\psi)\Lambda^2 G(\psi)^T$ where $G(\psi)$ is a rotation matrix with angle ψ and Λ is a diagonal matrix with positive diagonal elements, say λ's. The vector $\boldsymbol{\theta} = (\nu, \psi, \Lambda)$ denotes the spatial parameters; ν controls the smoothness, while the rate of spatial decay is controlled by the λ's.

We generated $y(\boldsymbol{s}_i)$'s using 3000 irregularly scattered locations over a 1000×1000 domain. In this case, though very slow, we can fit (7.9) without resorting to the predictive process; comparison with various choices of m along with knot design can be made. Parameter values generating the simulated process are given in the second column in Table 7.1. Figure 7.1(a) clearly shows the dominant

Table 7.1 Parameter credible intervals, 50% (2.5% 97.5%) and predictive validation for the predictive process models using a regular grid of knots. Bold entries indicate where the true value is missed. The last column shows results for the parent model.

Parameter	True	144	256	529	3000
β	1.0	0.94 (0.56, 1.35)	0.73 (0.34, 1.16)	0.77 (0.34, 1.21)	0.72 (0.43, 1.01)
ψ	45.0°	**36.45 (34.66, 38.14)**	42.09 (37.62, 45.80)	43.83 (40.93, 46.77)	44.47 (43.18 45.74)
λ_1	300.0	**390.4 (330.1, 399.6)**	279.0 (258.6, 311.0)	323.1 (289.9, 349.0)	302.6 (275.5, 330.2)
λ_2	50.0	**62.42 (52.7, 71.99)**	**79.41 (59.77, 103.40)**	61.47 (41.50, 84.30)	47.45 (40.03, 55.13)
σ^2	1.0	1.14 (0.87, 1.49)	1.02 (0.80, 1.54)	1.31 (0.83, 1.52)	0.95 (0.87, 1.05)
τ^2	0.2	**0.56 (0.53, 0.59)**	**0.45 (0.42, 0.49)**	**0.26 (0.21, 0.29)**	0.16 (0.13 0.22)
Prediction	95%	91%	92%	93%	95%

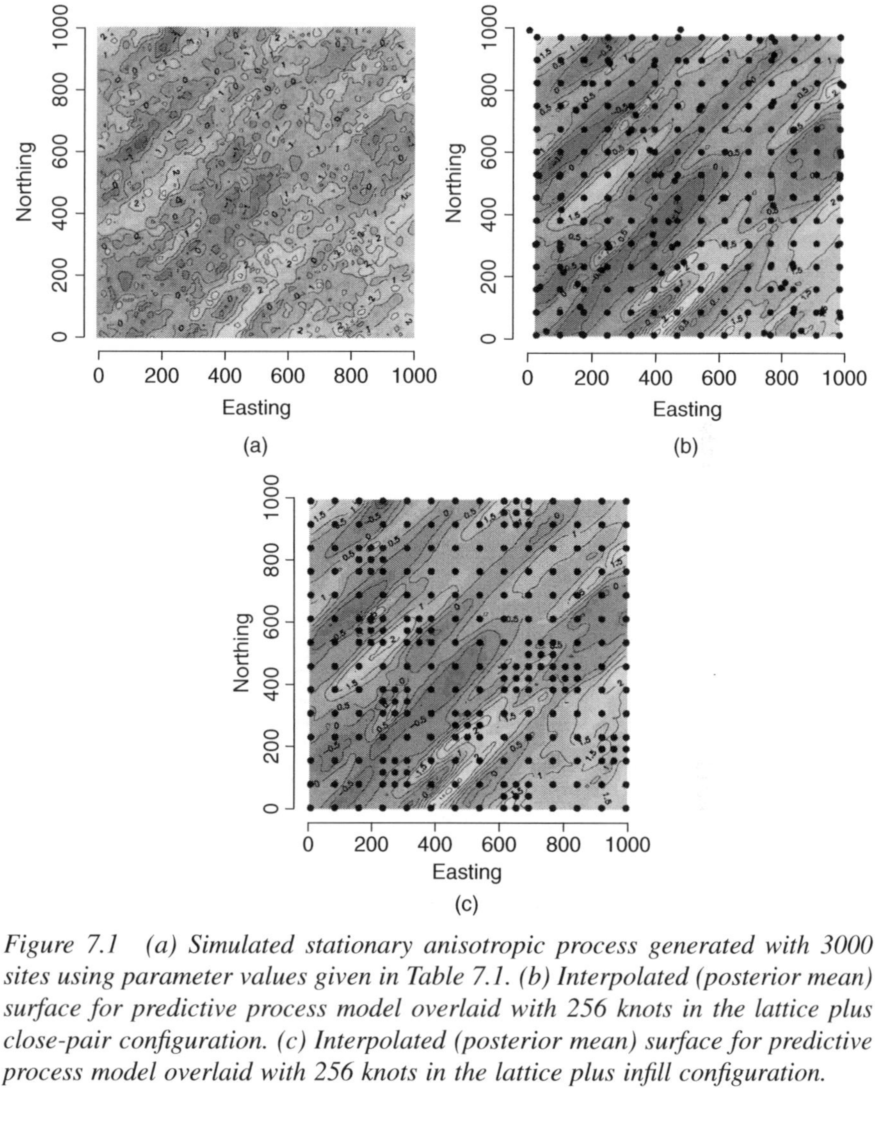

Figure 7.1 (a) Simulated stationary anisotropic process generated with 3000 sites using parameter values given in Table 7.1. (b) Interpolated (posterior mean) surface for predictive process model overlaid with 256 knots in the lattice plus close-pair configuration. (c) Interpolated (posterior mean) surface for predictive process model overlaid with 256 knots in the lattice plus infill configuration.

45.0° orientation of the process. We assign a flat prior to the intercept β, a $U(0, \pi/2)$ prior for the rotation parameter ψ, and $U(10, 400)$ prior for the λ's. This support for the λ's corresponds to about 30 to 1200 distance units for the effective spatial ranges along those axes (i.e., approximately 3λ is the distance at which the correlation drops to 0.05). The remaining process parameters σ^2 and τ^2 followed $IG(2, 1)$ and $IG(2, 0.2)$ priors, respectively. We kept $\nu = 0.5$ as fixed for the analysis in this subsection.

We carried out several simulation experiments with varying knot sizes and configurations. In addition to regular lattices or grids, we also explored two different knot configurations described by Diggle and Lophaven (2006) in spatial design contexts. The first, which is called the *lattice plus close pairs* configuration, considers a regular $k \times k$ lattice of knots but then intensifies this grid by randomly choosing m' of these lattice points and then placing an additional knot close to each of them, say, within a circle having the lattice point as center and a radius that is some fraction of the spacing on the lattice. The second configuration, called the *lattice plus infill* design, also starts with knots on a regular $k \times k$ lattice but now intensifies the grid by placing a more finely spaced lattice within m' randomly chosen cells of the original lattice.

Here we present illustrations using the above designs with knot sizes of 144, 256 and 529. For the uniform grid these were arranged on a square lattice with knots spaced at 91.0, 66.8 and 45.5 units, respectively. For the close-pair and infill designs we held the number of knots at 144, 256 and 529 (for a fair comparison with the uniform lattice) and adjusted the lattice accordingly. For instance, Figure 7.1(b) shows the close-pair design with 256 knots by randomly selecting 60 knots from a 14×14 lattice and then adding a knot to each of them to form close pairs. Similarly, Figure 7.1(c) shows the infill design with 256 knots formed by randomly selecting 12 cells of the original 14×14 lattice and adding a finely spaced sublattice in each of these cells. This results in five additional knots in each of those cells making the total number of knots: $14^2 + 12 \times 5 = 256$.

For each knot configuration, three parallel MCMC chains were run for 5000 iterations each. Convergence diagnostics revealed 1000 iterations to be sufficient for initial burn-in so the remaining 12000 (4000×3) samples were used for posterior inference. Table 7.1 provides parameter estimates for the three knot intensities on a uniform grid along with those from the parent model, i.e., with each of the 3000 locations as a knot, while Tables 7.2 and 7.3 provide those from the close-pair and infill designs, respectively. CPU times with the machine specifications described earlier were approximately 0.75, 1.5 and 4.25 h for the 144, 256 and 529 knot models, respectively, while for the parent model it was approximately 18 h.

All the tables reveal the improvement in estimation with increasing number of knots, irrespective of the design. In all three tables we find substantial overlaps in the credible intervals of the predictive process models with those from the original model. While 144 knots is adequate for capturing the regression term (β), higher knot densities are required for capturing the anisotropic field parameters and the nugget variance (τ^2). The latter, in particular, is a difficult parameter to estimate here with much of the variability being dominated by σ^2, yet we see a substantial improvement in moving from 256 to 529 knots. These tables suggest that estimation is more sensitive to the number of knots than to the underlying design, although the close-pair designs appear to improve estimation of the shorter ranges as seen for λ_2 with 256 knots. Predictions, on the other hand, are much more robust as is seen from the last row of Tables 7.1, 7.2 and 7.3. These show the empirical coverage of 95% prediction intervals based upon a hold-out set of 100 locations. The coverage is expected to be lower since there is

Table 7.2 Parameter credible intervals, 50% (2.5% 97.5%) for the predictive process models with 3000 locations using the lattice plus close-pair design with different knot intensities. Bold entries indicate where the true value is missed. The last row provides the empirical coverage of 95% prediction intervals for a set of 100 hold-out locations.

Parameter	True	144	256	529
β	1.0	1.07 (0.77, 1.40)	0.63 (0.26, 1.01)	0.72 (0.35, 1.10)
ψ	45.0°	**40.58 (38.65, 42.59)**	44.77 (42.68, 46.74)	43.76 (42.35, 45.98)
λ_1	300.0	**386.62 (344.0,1 399.69)**	291.29 (267.57, 386.78)	330.00 (295.33, 358.85)
λ_2	50.0	49.24 (43.86, 54.58)	53.40 (46.20, 60.72)	51.08 (43.98, 60.07)
σ^2	1.0	1.34 (1.0, 1.70)	1.42 (0.89, 1.65)	1.39 (0.91, 1.66)
τ^2	0.2	**0.55(0.52, 0.58)**	**0.45 (0.43, 0.48)**	**0.24 (0.22, 0.29)**
Prediction	95%	91%	92%	92%

Table 7.3 Parameter credible intervals, 50% (2.5% 97.5%) for the predictive process models with 3000 locations using the lattice plus infill design. Bold entries indicate where the true value is missed. The last row provides the empirical coverage of 95% prediction intervals for a set of 100 hold-out locations.

Parameter	True	144	256	529
β	1.0	1.18 (0.76, 1.66)	0.77 (0.39, 1.17)	0.71 (0.39, 1.02)
ψ	45.0°	**42.12 (40.70, 43.21)**	45.55 (44.46, 46.75)	43.45 (41.89, 45.13)
λ_1	300.0	**392.34 (343.23, 399.74)**	316.73 (270.11, 368.57)	345.85 (286.32, 369.79)
λ_2	50.0	58.26 (45.79, 66.46)	56.05 (47.97, 64.27)	47.31 (39.64, 55.38)
σ^2	1.0	1.72 (0.98, 2.66)	1.35 (0.92, 1.76)	1.14 (0.94, 1.37)
τ^2	0.2	**0.57 (0.54, 0.60)**	**0.48 (0.45, 0.50)**	**0.25 (0.22, 0.29)**
Prediction	95%	92%	92%	93%

less uncertainty in the predictive process than in the parent process (Section 3.2); however, it is only slightly so.

7.5.2 A simulation example using the two-step analysis

We generated 1000 data points in a [0, 100] × [0, 100] square and then generated the dependent variable from model (7.9), but using an exponential covariance function with range parameter $\phi = 0.06$ (i.e., an effective range of ~50 units), scale $\sigma = 1$ for the spatial process $w(\boldsymbol{s})$, and nugget variance $\tau^2 = 1$. We illustrate a comparison among three design strategies: regular grids, sequential search over all the observed locations, and sequential search over a fine regular lattice. In Figure 7.2, we plot the averaged predictive variances under each strategy. The sequential search algorithm is clearly better than choosing a regular grid as knots. For instance, with 180 sites selected, the sequential search over the observed locations yielded an averaged predictive variance of approximately 0.15. For the regular grids, roughly 150 additional sites are needed to achieve the same level of performance.

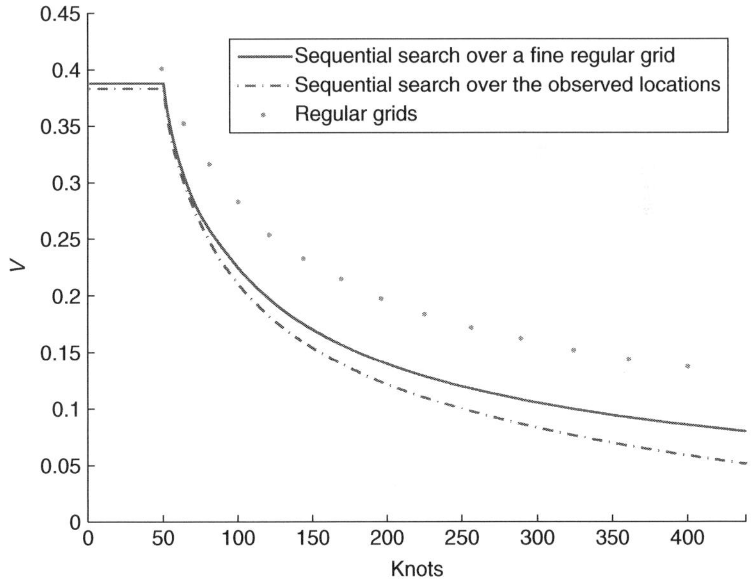

Figure 7.2 Averaged prediction variance (V) versus number of knots. Solid dots denote results for regular grids; the dash-dot line denotes results for the sequential search over the observed data locations (starting with 49 randomly chosen sites from the observed locations); and the solid line denotes results for the sequential search over a 60 × 60 *regular grid (starting with a* 7 × 7 *regular grid).*

7.5.3 Tree height and diameter analysis

Tree diameter at breast height (DBH; 1.37 m above the forest floor) and total height (TH) are used, in combination with other information, to assess the economic and ecological value of a tree. Measuring a tree's TH is very time consuming compared with measuring its DBH. Therefore TH is often measured on a small subset of those trees that receive DBH measurements. Then a statistical model that relates TH to DBH is used to predict TH for the complement of this subset. Here, we illustrate how the proposed methods can help improve the predictive performance of such models for a moderately large dataset.

The relationship between DBH and TH is influenced by many individual and environmental factors. Individual factors include age, species, and genetics, whereas, environmental factors include quality of soil, quantity of water and light, and competition for these resources. These factors often vary spatially across the domain. For example, a cohort of trees of the same species, age, and parentage will likely depict a similar DBH and TH relationship. An example of an environmental factor influencing tree growth characteristics is soil productivity which can result in parent material or disturbance history. Given these unobserved covariates, we can expect some level of spatial dependence in TH, even after accounting for DBH.

The analysis was based on 2391 mature trees measured on a 200×200 m portion of an uneven-aged softwood stand located near Sault Ste. Marie, Ontario (Ek 1969), illustrated in Figure 7.3(a). The major tree species are balsam fir (*Abies balsamea*) and black spruce (*Picea mariana*). Location, DBH, and TH were recorded for each tree. Here, the candidate models use an intercept and single covariate $\log(DBH)$ to explain the variability in the outcome variable $\log(TH)$. Candidate models include a simple nonspatial regression and predictive process models.

Knot locations were chosen following: (1) a geometric space-filling design, referred to as a 'coverage design' (Johnson *et al.*, 1990; Nychka and Saltzman, 1998), computed using the *spatial.design* function in the *fields* R package; (2) the approximately optimal design computed using the algorithm detailed in Section 7.4.2. In the approximately optimal design algorithm, the initial knot set, m_0, consisted of a single point in the middle of the domain and the 2391 observed locations served as the *possible sampling locations* set. Similar to the simulation example, we considered a range of knot intensities.

Based on results from an initial variogram analysis of the residuals from a nonspatial model, Figure 7.3(b), the adopted priors for τ^2 and σ^2 followed $IG(2\ 0.01)$. Assuming an exponential spatial correlation function the prior for the spatial decay parameter ϕ followed a $U(0.012\ 1)$, which corresponds to support from 1 to 275 m. Again, this is a broad range of support given the maximum distance between any two trees is 276 m. Note, Figure 7.3(b) suggests an effective spatial range (i.e., the range at which the spatial correlation drops

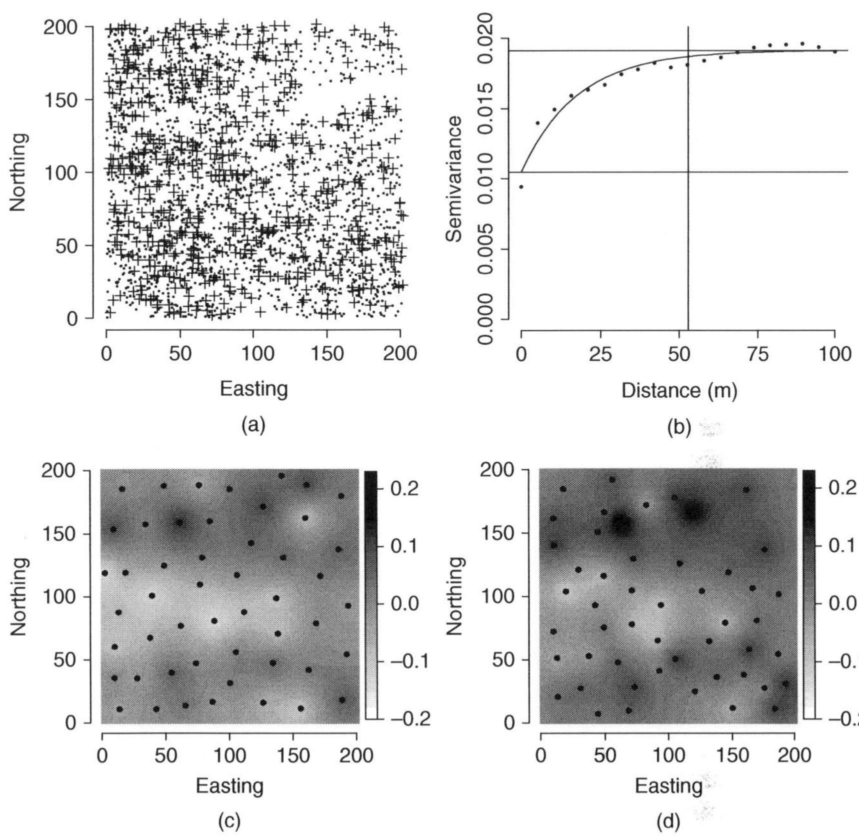

Figure 7.3 Tree height data set and associated candidate model output: (a) location of trees used to fit the candidate models (•) and those used to assess predictive performance (+); (b) empirical semivariogram of residuals from regressing log(*HT*) *on* log(*DBH*) *along with maximum likelihood estimates of nugget* τ^2 *(lower horizontal line), sill* $\tau^2 + \sigma^2$ *(upper horizontal line), and effective spatial range (vertical line); (c) 50 knot locations based on a coverage design and estimated spatial random effects surface; and (d) 50 knot locations based on the proposed approximately optimal design and estimated spatial random effects surface.*

to 0.05) of ~50 m. For all models the regression coefficients, β_0 and β_{DBH}, associated with the intercept and covariate log(*DBH*) each received *flat* priors.

Performance of candidate models was compared based on their fit to observed data as well as predictive performance. Model fit to the observed data was assessed using *independent* replicates for each observed outcome: for each $\boldsymbol{s}_i \in \mathcal{S}$, we draw $y_{rep}(\boldsymbol{s}_i)^{(l)}$ from $N\left(\boldsymbol{x}(\boldsymbol{s}_i)^T \boldsymbol{\beta}^{(l)} + \tilde{w}(\boldsymbol{s}_i)^{(l)}, \tau^{2(l)}\right)$, one for one with the posterior samples. Letting $\mu_{rep,i}$ and $\sigma^2_{rep,i}$ be the posterior predictive mean and variance for each $y_{rep}(\mathbf{s}_i)$, we compute $G = \sum_{i=1}^{n}(y(\mathbf{s}_i) - \mu_{rep,i})^2$ and $P = \sum_{i=1}^{n} \sigma^2_{rep,i}$. We use $D = G + P$ (Gelfand and Ghosh 1998) as a model selection criterion; smaller values of D indicate better models. Further, a hold-out set was used to assess the predictive performance of each model by computing the mean squared prediction error (MSPE), $\frac{1}{q}\sum_{i=1}^{q}(y(\mathbf{s}_i) - \tilde{y}(\mathbf{s}_i))^2$, where $\tilde{y}(\mathbf{s}_i)$ is the predicted outcome at the ith hold-out location and q is the number of locations in the hold-out set. The hold-out set consisted of $q = 598$ observations (i.e., 25%) selected randomly from the dataset. The remaining 1793 observations were used to fit the candidate models. The location of the observed and hold-out observations are depicted in Figure 7.3(a), with solid dot and plus symbols, respectively.

The median and 95% CI parameter estimates for the nonspatial model are 0.825 (0.769, 0.880), 0.611 (0.591, 0.631), and 0.018 (0.016, 0.019) for β_0, β_{DBH}, and τ^2, respectively. Posterior predictive loss and associated values were G = 31.33, P = 31.42 and D = 62.75. The model's hold-out set MSPE was 6.52. Again, as suggested by the variogram estimated from this model's residuals, Figure 7.3(b), there is relatively strong spatial dependence even after accounting for DBH. This residual dependence encourages the addition of spatial random effects.

Table 7.4 offers parameter estimates for the predictive process models fit using the two knot designs and at the 25 and 50 knot intensities. All models produce nearly identical estimates of β_0 and β_{DBH}. The approximately optimal design produces consistently smaller τ^2 and larger σ^2 compared with estimates based on the coverage design. Also, as suggested by the simulation example, the approximately optimal design results in marginally improved model fit and predictive performance. For instance, the 50 knot models produce D of 53.56 versus 52.79 and MSPE of 5.63 versus 5.43 for the coverage and approximately optimal designs, respectively.

Figure 7.3(c) and 7.3(d) was generated by interpolating over the median of each location's spatial random effect posterior distribution for the 50 knot coverage and approximately optimal designs, respectively. In these figures the estimated knot locations are indicated with black circles. For both knot intensities, the approximately optimal design permits spatial random effects to more closely approximate the residual surface and hence provide improved model fit and prediction over the space-filling design.

7.5.4 Austria precipitation analysis

In practice, we employ the predictive process when the number of observations makes fitting the desired model computationally prohibitive. The Austria monitoring network dataset does not warrant dimension reduction in space or time; however, to connect with running analyses in the text, we consider a few models for these data using predictive processes. The dataset comprises monthly weather records from January 1994 through December 2009 at $n = 37$ monitoring stations

Table 7.4 Parameter estimates and credible intervals (in parentheses) for candidate predictive process models along with mean squared prediction error (MSPE) for the tree height dataset.

	Coverage design		Approximately optimal design	
	25	50	25	50
β_0	0.874 (0.805, 0.944)	0.883 (0.818, 0.951)	0.892 (0.828, 0.957)	0.862 (0.798, 0.925)
β_{DBH}	0.596 (0.577, 0.616)	0.598 (0.579, 0.618)	0.593 (0.574, 0.612)	0.593 (0.574, 0.612)
σ^2	0.007 (0.004, 0.012)	0.005 (0.004, 0.009)	0.010 (0.006, 0.015)	0.009 (0.006, 0.013)
τ^2	0.012 (0.008, 0.014)	0.012 (0.011, 0.014)	0.009 (0.006, 0.012)	0.010 (0.007, 0.012)
ϕ	0.033 (0.019, 0.053)	0.041 (0.023, 0.063)	0.041 (0.025, 0.058)	0.056 (0.037, 0.077)
G	27.62	26.02	27.10	25.86
P	28.21	27.53	27.74	26.92
D	55.84	53.56	54.84	52.79
MSPE	5.75	5.63	5.73	5.43

in Austria. For the analysis presented here, we consider only the January 1994 precipitation data, which due to missing records consists of $n = 35$ observations.

Our analysis follows the same set-up as the forestry data analysis presented in Section 7.5.3. Candidate models include the *full* model (7.4) and its predictive process specification (7.5). For the predictive process model, we consider two configurations consisting of 18 knots: (1) a regular grid bounded by the domain; and (2) the approximately optimal design detailed in Section 7.4.2. The resulting designs are illustrated in Figure 7.4. In addition to an intercept the candidate models included station elevation as a covariate. The three candidate models are assessed using predictive performance. Specifically, we consider MSPE calculated using a leave-one-out (single site deletion) approach–prediction for the ith location is made using the remaining $n - 1$ observations. The full model results in the smallest MSPE of 0.076 with the predictive process model based on the approximately optimal knot design at 0.08 and the knot grid at 0.087. These results are consistent with those of the simulated and forestry data analyses.

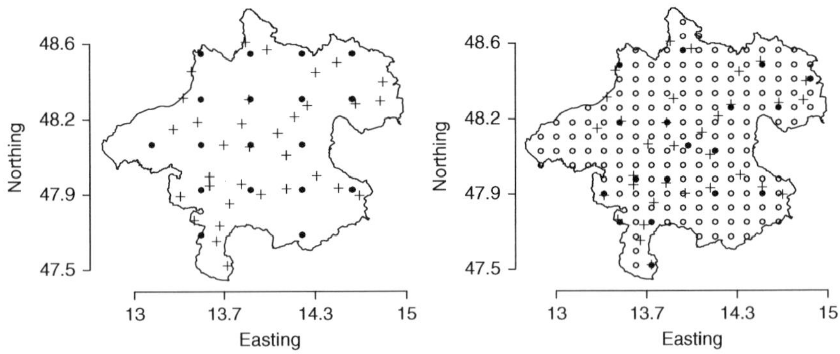

Figure 7.4 Upper Austria monitoring network analysis: (a) weather station locations (+) and predictive performance knots on a sparse grid (•); (b) weather station locations (+), candidate knot locations used in the approximately optimal design algorithm (○), and knot locations chosen using the approximately optimal design algorithm (•).

A spatio-temporal analysis of these data could view them as a time series (at $T = 12 \times 16 = 184$ time points) of spatial process realizations and work in the setting of dynamic models that accommodate residual spatial and temporal dependence, see, e.g., Tonellato (1997), West and Harrison (1997), Stroud *et al.* (2001), and Gelfand *et al.* (2005). Again, if this dataset was large in the spatial and/or temporal dimension, we would encounter a computational challenge when fitting a hierarchical Bayesian specification of such models. Recently, Finley *et al.* (2012) extended the space-time modeling framework detailed in Gelfand *et al.* (2005) to reduce the dimensionality of a similar weather monitoring dataset and ease the computational burden of estimating the spatio-temporal process of interest. Although, they only considered a simple space-filling design, a more sophisticated approach such as the one described in Section 7.4.2 could have been used.

7.6 Discussion and future work

Our overall objective has been to investigate the issue of knot selection in fitting hierarchical spatial models incorporating Gaussian processes to large datasets. To do so, we propose simply replacing the parent spatial process specification by its induced predictive process specification. As in existing low-rank kriging approaches, knot selection is required and as we demonstrated in Section 7.5 some sensitivity to the number of knots is expected. While for most applications a reasonable grid of knots should lead to robust inference, with fewer knots the separation between them increases and estimating random fields with fine-scale spatial dependence becomes difficult. Indeed, learning about fine scale spatial dependence is always a challenge (Cressie, 1993, p.114).

A further challenge noted in our simulated examples was the situation where the variance components ratio ($\sigma^2/\tau^2 = 5.0$) is large (attractive since the spatial story dominates) so that estimation of τ^2 becomes difficult. One possible remedy is reparametrizing (σ^2, τ^2) in terms of their ratio and the larger variance component (Diggle and Ribeiro Jr 2007). Another option to explore is to modify the predictive process as $\dot{w}(\boldsymbol{s}) = \tilde{w}(\boldsymbol{s}) + \tilde{\epsilon}(\boldsymbol{s})$, where $\tilde{\epsilon}(\boldsymbol{s})$ is an independent Gaussian process with variance $C(\boldsymbol{s}, \boldsymbol{s}) - \boldsymbol{c}(\boldsymbol{s}; \boldsymbol{\theta})^T \boldsymbol{C}^{*-1}(\boldsymbol{\theta})\boldsymbol{c}(\boldsymbol{s}; \boldsymbol{\theta})$ (Finley *et al.* 2009).

Faster model fitting alternatives that avoid MCMC can also be employed (Rue *et al.* 2009). In fact, due to the enormous amount of matrix multiplication, the predictive process approach within a full MCMC implementation is perhaps limited to order 10^4 observations on modest single processor machines (see Section 7.5); strategies that are empirical Bayesian in flavor, combining deterministic and simulation aspects, are likely the future for fitting very large space time datasets, a setting which is becoming more and more common. For instance, Cressie and Johannesson (2008) work with data on the order of hundreds of thousands. It is also not uncommon to find space-time datasets with a very large number of distinct time points, possibly with different time points observed at different locations (e.g., real estate transactions). With multiple processors, substantial gains in computing efficiency can be realized through parallel processing of matrix operations (Heroux *et al.* 2006); this path merits further investigation.

In summary, there is much possible future work; here, we have only just opened the door to possible problems to explore. Indeed, we have raised more questions than we have answered. For instance, we could certainly return to the design approaches discussed in Section 7.4.1 with regard to knot selection adopting alternative criteria in place of the one adopted in Section 7.4.2. We could make further comparison with geometric and simple random sampling designs. We could look at other dimension reduction approaches besides the predictive process including the modified predictive process (Finley *et al.* 2009), kernel convolution, and the newly proposed approach in Sang and Huang (2012).

We note that knot selection carries over directly to the spatio-temporal setting. Conceptually, knot selection in time is more straightforward; we might simply use equally spaced knots along the observed timescale. However, we could argue that such choice ignores possible space-time interaction. Also, we do face the potential for the so-called 'curse of dimensionality,' here, for a fixed number of knots, taking the form of sparser coverage of the domain of interest, as we move from two to three dimensions.

Finally, in very recent work (Guhaniyogi *et al.*, 2011), the authors experimented with random knot selection, allowing the number of knots and their locations to vary with iteration of the MCMC model fitting. This seems like a promising path to consequentially reduce the effective number of knots. However, investigation to date has yielded only modest gains in exchange for substantial increase in computing time.

References

Banerjee, S., Carlin, B. P. and Gelfand, A. E. (2004). *Hierarchical Modeling and Analysis for Spatial Data*. Boca Raton, FL: Chapman and Hall/CRC Press.

Banerjee, S., Gelfand, A. E., Finley, A. O. and Sang, H. (2008). Gaussian predictive process models for large spatial datasets. *Journal of the Royal Statistical Society, Series B*, **70**, 825–848.

Barry, R. P. and Ver Hoef, J. M. (1996). Blackbox Kriging: Kriging without specifying variogram models. *Journal of Agricultural, Biological and Environmental Statistics*, **1**, 297–322.

Berk, R. A. (2008). *Statistical Learning from a Regression Persepective*. New York: Springer.

Breiman, L., Friedman, J. H., Olshen, R. A. and Stone, C. J. (1984). *Classification and Regression Trees*. Monterey, CA: Wadsworth & Brooks/Cole Advanced Books & Software.

Brimkulov, U., Krug, G. and Savanov, V. (1986). *Design of Experiments for Random Fields*. Nauka: Moscow.

Chipman, H. A., George, E. and McCulloch, R. (1998) Bayesian CART model search (with discussion). *Journal of the American Statistical Association*, **93**, 935–960.

Chipman, H. A., George, E. I. and McCulloch, R. E. (2010). BART: Bayesian Additive Regression Trees. *Annals of Applied Statistics*, **4**, 266–298.

Crainiceanu, C. M., Diggle, P. J. and Rowlingson, B. (2008). Bivariate binomial spatial modeling of Loa loa Prevalence in tropical Africa (with discussion). *Journal of the American Statistical Association*. **103**, 21–37.

Cressie, N. A. C. (1993). *Statistics for Spatial Data*, 2nd edition. New York: John Wiley & Sons, Ltd.

Cressie, N. A. C. and Johannesson, G. (2008). Spatial prediction for massive datasets. *Journal of the Royal Statistical Society, Series B*, **70**, 209–226.

Cressie, N. A. C. and Wikle, C. K. (2011). *Statistics for Spatio-Temporal Data*. New York: John Wiley & Sons, Ltd.

Denison, D. G. T., Holmes C. C., Mallick B. K. and Smith A. F. M. (2004). *Bayesian Methods for Nonlinear Classification and Regression*. New York: John Wiley & Sons, Ltd.

Diggle, P. and Lophaven S. (2006). Bayesian geostatistical design *Scandinavian Journal of Statistics*, **33**, 53–64

Diggle, P. and Ribeiro Jr, P. J. (2007). *Model Based Geostatistics*. New York: Springer.

Du, J., Zhang, H. and Mandrekar, V. S. (2009). Fixed-domain asymptotic properties of tapered maximum likelihood estimators. *Annals of Statistics*, **37**, 3330–3361

Eidsvik, J., Finley, A. O., Banerjee, S. and Rue, H. (2011). Approximate Bayesian inference for large spatial datasets using predictive process models. *Computational Statistics and Data Analysis* http://dx.doi.org/10.1016/j.bbr.2011.03.031.

Ek, A. R. (1969). Stem map data for three forest stands in northern Ontario. For. Res. Lab., Sault Ste. Marie, Ontario. Inf. Rep. O-X-113.

Fedorov, V. V. (1996). *Design of Spatial Experiments, Model Fitting and Prediction*. Oak Ridge, TN: Oak Ridge National Laboratory.

Finley, A. O., Banerjee, S. and Gelfand, A. E. (2012) Bayesian dynamic modeling for large space-time datasets using Gaussian predictive processes. *Journal of Geographical Systems*, **14**, 29–47.

Finley, A. O., Sang, H., Banerjee, S. and Gelfand, A. E. (2009). Improving the performance of predictive process modeling for large datasets. *Computational Statistics and Data Analysis*, **53**, 2873–2884.

Friedman, J. H. (1991). Multivariate adaptive regression splines. *Annals of Statistics*, **19**, 1–67.

Fuentes, M. (2007). Approximate likelihood for large irregularly spaced spatial data. *Journal of the American Statistical Association*, **102**, 321–331.

Furrer, R., Genton, M. G. and Nychka, D. (2006). Covariance tapering for interpolation of large spatial datasets. *Journal of Computational and Graphical Statistics*, **15**, 502–523.

Gelfand, A. E., Banerjee, S. and Gamerman, D. (2005) Univariate and multivariate dynamic spatial modelling. *Environmetrics*, **16**, 465–479.

Gelfand, A. E., Diggle, P., Guttorp, P. and Fuentes, M. (eds) (2010). *Handbook of Spatial Statistics*. Boca Raton, FL: Taylor and Francis.

Gelfand, A. E. and Ghosh, S. K. (1998). Model Choice: a minimum posterior predictive loss approach. *Biometrika*, **85**, 1–11.

Gelfand, A. E., Kim, H., Sirmans, C. F. and Banerjee, S. (2003). Spatial modelling with spatially varying coefficient processes. *Journal of the American Statistical Association*, **98**, 387–396.

Gelfand A. E., Schmidt, A., Banerjee S. and Sirmans, C. F. (2004). Nonstationary multivariate process modelling through spatially varying coregionalization. *Test*, **13**, 263–312.

Gelman, A., Carlin, J. B., Stern, H. S. and Rubin, D. B. (2004). *Bayesian Data Analysis*, 2nd edition. Boca Raton, FL: Chapman and Hall/CRC Press.

Gu, C. (2002). *Smoothing Spline ANOVA Models*. New York: Springer.

Guhaniyogi, R., Finley, A. O., Banerjee, S. and Gelfand, A. E. (2011). Adaptive Gaussian predictive process models for large spatial datasets. *Environmetrics* in press.

Hastie, T., Tibshirani, R. and Friedman J. H. (2009) *The Elements of Statistical Learning*, 2nd edition. New York: Springer.

Heroux, M. A. Padma, R. and Simon, H. D. (eds) (2006) *Parallel Processing for Scientific Computing*. Philadelphia, PA: Society for Industrial and Applied Mathematics.

Higdon, D. (2002). Space and space-time modeling using process convolutions. In *Quantitative Methods for Current Environmental Issues*, eds C. Anderson, V. Barnett, P. C. Chatwin and A. H. El-Shaarawi, New York: Springer, 37–56.

Higdon, D., Swall, J. and Kern, J. (1998). Non-stationary spatial modeling. In *Bayesian Statistics 6*. Oxford University Press, 761–768.

Johnson, M. E., Moore, L. M. and Ylvisaker, D. (1990). Minimax and maximin distance designs. *Journal of Statistical Planning and Inference*, **26**, 131–148.

Kamman, E. E. and Wand, M. P. (2003). Geoadditive models. *Journal of the Royal Statistical Society, Series C*, **52**, 1–18.

Kaufman, C. G., Schervish, M. J. and Nychka, D. W. (2009). Covariance tapering for likelihood-based estimation in large spatial data sets. *Journal of the American Statistical Association*, **103**, 1545–1555.

Lang, S. and Brezger, A. (2004). Bayesian P-splines. *Journal of Computational and Graphical Statistics*, **13**, 183–212.

Le, N. and Zidek, J. V. (1992). Interpolation with uncertain spatial covariances: a Bayesian alternative to kriging. *Journal of Multivariate Analysis*, **43**, 351–374.

Lin, X., Wahba, G., Xiang, D., Gao, F., Klein, R. and Klein, B. (2000). Smoothing spline ANOVA models for large data sets with Bernoulli observations and the randomized GACV. *The Annals of Statistics*, **28**, 1570–1600.

Müller, W. G. (2001). *Collecting Data: Optimum Design of Experiments for Random Fields*. New York: Springer.

Nychka, D. and Saltzman, N. (1998). Design of air-quality monitoring networks. In *Case Studies in Environmental Statistics, Lecture Notes in Statistics*, eds D. Nychka, L. Cox and W. Piegorsch. New York: Springer, 51–76.

Paciorek, C. J. and Schervish, M. J. (2006). Spatial modelling using a new class of nonstationary covariance functions. *Environmetrics*, **17**, 483–506.

Pukelsheim F. (1993). *Optimum Design of Experiments*. New York: John Wiley & Sons, Ltd.

Ramsay, J. O. and Silverman, B. W. (2005). *Functional Data Analysis*. New York: Springer.

Rasmussen, C. E. and Williams, C. K. I. (2006). *Gaussian Processes for Machine Learning*. Cambridge, MA: MIT Press.

Robert, C. P. and Casella, G. (2004). *Monte Carlo Statistical Methods*. New York: Springer.

Rue, H. and Held, L. (2005). *Gaussian Markov Random Fields: Theory and Applications*, Volume 104 of Monographs on Statistics and Applied Probability. London: Chapman & Hall.

Rue, H., Martino, S. and Chopin, N. (2009). Approximate Bayesian inference for latent Gaussian models by using integrated nested Laplace approximations (with discussion). *Journal of the Royal Statistical Society, Series B*, **71**, 1–35.

Ruppert, D., Wand, M. P. and Caroll, R. J. (2003). *Semiparametric Regression*. Cambridge: Cambridge University Press.

Sang, H. and Huang, J.Z. (2012). A full-scale approximation of covariance functions for large spatial data sets. *Journal of the Royal Statistical Society*: Series B (Statistical Methodology), **74**(1), 111–132.

Schabenberger, O. and Gotway, C. A. (2004). *Statistical Methods for Spatial Data Analysis (Texts in Statistical Science Series)*. Boca Raton, FL: Chapman and Hall/CRC.

Stein, M. L. (1999). *Interpolation of Spatial Data: Some Theory for Kriging*. New York: Springer.

Stein, M. L. (2007). Spatial variation of total column ozone on a global scale. *Annals of Applied Statistics*, **1**, 191–210.

Stein, M. L. (2008). A modeling approach for large spatial datasets. *Journal of the Korean Statistical Society*, **37**, 3–10.

Stein, M. L., Chi, Z. and Welty, L. J. (2004). Approximating likelihoods for large spatial data sets. *Journal of the Royal Statistical Society, Series B*, **66**, 275–296.

Stroud, J. R., Muller, P. and Sansó, B. (2001) Dynamic models for spatiotemporal data. *Journal of the Royal Statistical Society, Series B*, **63**, 673–689.

Tonellato, S. (1997). Bayesian dynamic linear models for spatial time series. *Technical Report*, Rapporto di riceria 5/1997, Dipartimento di Statistica, Universita CaFoscari di Venezia, Venice.

Vecchia, A. (1988). Estimation and model identification for continuous spatial processes. *Journal of the Royal Statistical Society, Series B*, **50**, 297–312.

Ver Hoef, J. M., Cressie, N. A. C. and Barry, R. P. (2004). Flexible spatial models based on the fast Fourier transform (FFT) for cokriging. *Journal of Computational and Graphical Statistics*, **13**, 265–282.

Wackernagel, H. (2006). *Multivariate Geostatistics: An Introduction with Applications*, 3rd edition. New York: Springer.

Wahba, G. (1990) *Spline Models for Observational Data*. Philadelphia, PA: SIAM.

West, M. and Harrison, P. (1997) *Bayesian Forecasting and Dynamic Models*, 2nd edition. New York: Springer.

Wikle, C. K. and Cressie, N. (1999). A dimension reduced approach to space-time Kalman filtering. *Biometrika*, **86**, 815–829.

Xia, G. and Gelfand, A. E. (2006). Stationary process approximation for the analysis of large spatial datasets. Technical Report, ISDS, Duke University, Durham, NC.

Xia, G., Miranda, M. L. and Gelfand, A. E. (2006). Approximately optimal spatial design approaches for environmental health data. *Environmetrics*, **17**, 363–385.

Zhang, H. (2004). Inconsistent estimation and asymptotically equal interpolations in model-based geostatistics. *Journal of the American Statistical Association*, **99**, 250–261.

Zhu, Z. and Stein, M. (2005). Spatial sampling design for parameter estimation of the covariance function. *Journal of Statistical Planning and Inference*, **134**, 583–603.

Zidek, J. V., Sun, W. and Le, N. D. (2000). Designing and integrating composite networks for monitoring multivariate Gaussian pollution fields. *Applied Statistics and Computing*, **49**, 63–79.

8

Exploratory designs for assessing spatial dependence

Agnes Fussl[1], Werner G. Müller[1] and Juan Rodríguez-Díaz[2]

[1]*Department of Applied Statistics, Johannes Kepler University Linz, Austria*

[2]*Faculty of Science, University of Salamanca, Spain*

8.1 Introduction

Efficient data acquisition as introduced in Chapter 1 requires some prior understanding of the process to be observed, ideally in the form of a spatio-temporal model or at least a narrow enough class of such models. If this is not the case as it is frequently at the beginning of a study, the employed design must guarantee that an appropriate model can be identified as data collection progresses. These designs, usually called exploratory designs (cf. Chapter 4 in Müller, 2007), typically employ minimum assumptions which make random sampling a reasonable choice.

However, random sampling can be very inefficient even for simple tasks that arise at the beginning of a study. One of the simplest and most initial (albeit quite important) tasks in a spatio-temporal context is the assessment of whether spatial or temporal dependence is present or not and if present at what intensity

Spatio-temporal design: Advances in efficient data acquisition, First Edition.
Edited by Jorge Mateu and Werner G. Müller.

(and form). In this chapter we would like to adress these questions and possible improvements over random sampling in coping with them. As the treatment of the temporal dimension usually involves straightforward regular observations or simple extensions from the spatial case, we will in the following mainly concentrate on the latter.

As a first stage one should then attempt to detect whether there is any spatial dependence in the data or not. Should they be spatially independent, the respective statistical design and analysis usually reduces to the application of a classical and well established toolbox. Thus, the decision of whether one can confine oneself to this well understood body of knowledge or whether one has to resort to the rather freshly developed methodologies of spatial design is a crucial element of any serious spatial investigation. Spending some effort on efficient designs for testing for spatial dependence could save considerable overall effort. If, for instance, one detects at an early stage of an investigation that effectively no spatial correlation is present, one can return to the classical, rather simple to construct optimal designs treated, e.g., in Chapter 3 of Müller (2007). Therefore, in the following sections we will concentrate on the question of how to optimally select coordinates/predictor values to detect the spatial structure, if it exists, and how to avoid spuriously detecting spatial dependence if there is no such structure. Later we will also consider evaluating the specific form of this dependence.

In particular, this chapter deals with statistical modeling of areal data which are observed on polygon entities with defined boundaries. Typical examples for such areal spatial objects are areal entities like states or counties. The aim of the chapter is to give a short overview of how to collect and analyze a dataset containing information on areal spatial objects with regard to the following questions:

- How can we define spatial neighbors?
- Which spatial weights should be assigned to the identified neighbor links?
- Are there any spatial dependencies in the data?
- Which statistical models are adequate for the data?
- How are the spatial modeling approaches related to each other, what are the differences between them and what are the consequences on design?

To answer these questions an exemplary spatial data analysis is performed on a dataset concerning the grassland usage in Upper Austria (see Section 8.1.1). The data are analyzed using the statistical software R which provides a wide range of packages and functions to work on spatial data; the used R packages and the R code are given in Appendix 8.1. For an extensive introduction to spatial data analysis in R see Bivand *et al.* (2008). We will concentrate our exposition on the lattice type of data for two reasons: the continuous regions can be covered by an arbitrarily fine grid; and the continous random field models can be well approximated by lattice based ones such as Gaussian Markov random fields (cf. Rue and Held 2005).

8.1.1 The dataset and its visualization

The dataset contains information on the grassland usage in the 445 municipalities of Upper Austria over the years 1995, 1999, 2003, 2005, 2007 and 2008. A municipality's grassland usage is measured by

$$\log(\text{area of arable land} + 1) - \log(\text{area of grassland} + 1),$$

i.e., the log ratio of these two areas, making it scale free (variables *R95* to *R08*). The value of this log ratio is positive if the area of arable land is larger than the area of grassland and negative if the proportion of arable land compared with the grassland is smaller. Other variables provided in the dataset are: *LBBGG* (identification number of municipality), *BEZNR* (identification number of the municipality's district), *Longitude* and *Latitude* (local coordinates of municipality), *FLKM2* (area of the municipality in square kilometers) and *ALTITUDE* (height above sea level in meters). For graphical display of the data a shape file of the boundaries of the municipalities is available.

Before a spatial analysis can be started, both data and shape file must be imported in R and combined to a `SpatialPolygonsDataFrame` object. To make a first check on spatial correlation for the important variables in the dataset it is reasonable to display them in a so-called choropleth map (Waller and Gotway 2004). For color filling of the maps the R package `RColorBrewer` is used which is based on the web tool *ColorBrewer* (for the latest version see Brewer and Harrower 2009).

One way to visualize the spatio-temporal development of the grassland usage in Upper Austria is to create a slide show consisting of a series of maps of log ratios *R95* to *R08*. Alternatively, the function `spplot` can be used to show all the maps at once (Figure 8.1). According to the maps it is evident to assume some spatial dependencies in the data. Nearly all municipalities in the south of Upper Austria seem to have more grassland than arable land, whereas municipalities in the central east of Upper Austria and along the river Inn tend to have a higher percentage of arable land in relation to grassland. Examining the development of the log ratios, nearly no change can be identified over the years. For a few municipalities the sign of the log ratio switches between the years 1995 and 2008.

However, it seems to be obvious that the altitude of the municipalities influences the log ratios and therefore also the observable structure in Figure 8.1. A look at the map of the altitudes (Figure 8.2) shows that the higher elevated regions are in the northeast and south of Upper Austria, whereas the less elevated and flatter areas can be found in the center of Upper Austria and along the two largest rivers (Danube and Inn). These are also the regions where a concentration of more arable land in relation to grassland is present. Possible reasons for this dependency are that it might be easier to cultivate on lowlands than on the slopes of the hillier municipalities and that there is a higher availability of water resources along the two rivers, as well. Thus, the question remains if there are spatial dependencies in the data anyway or if they are only resulting from the different elevations of the municipalities above sea level.

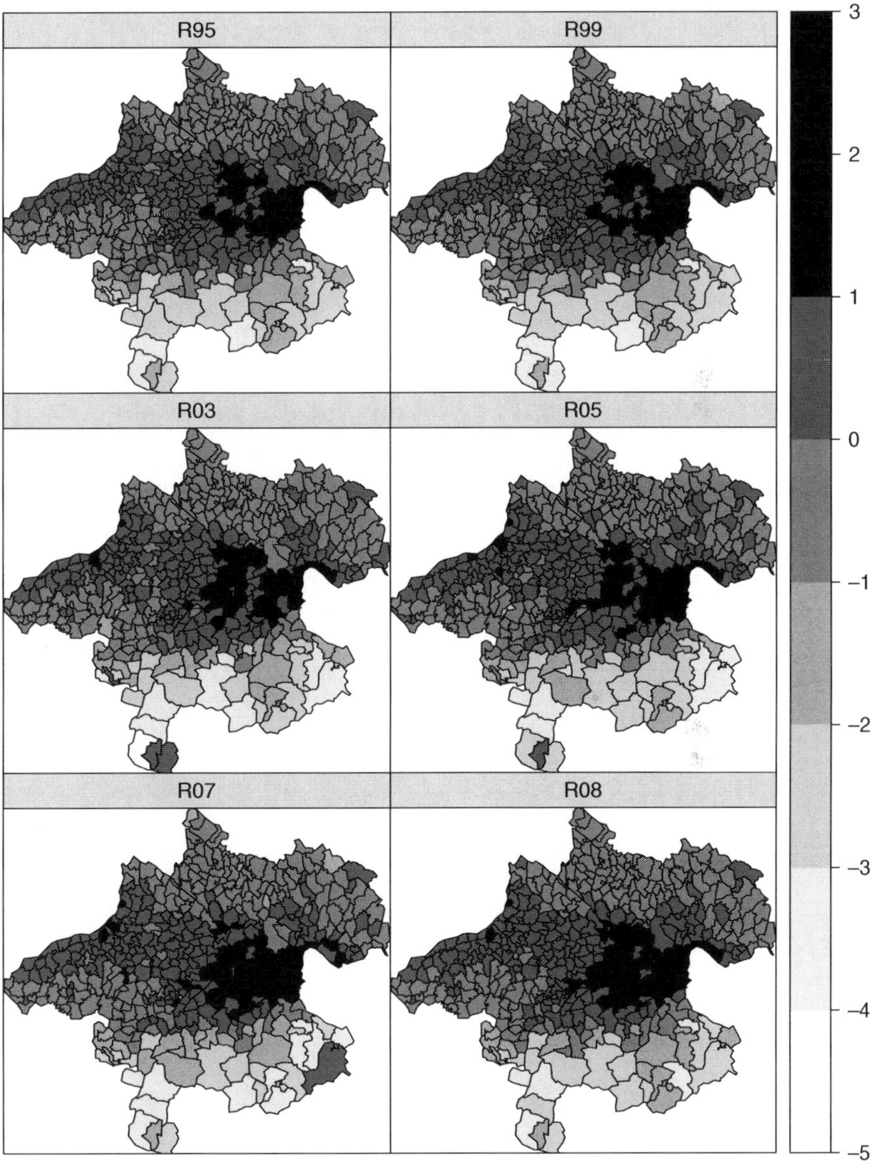

Figure 8.1 Maps of the log ratios R95 *to* R08. *(Please see plate section for color version of the figure.)*

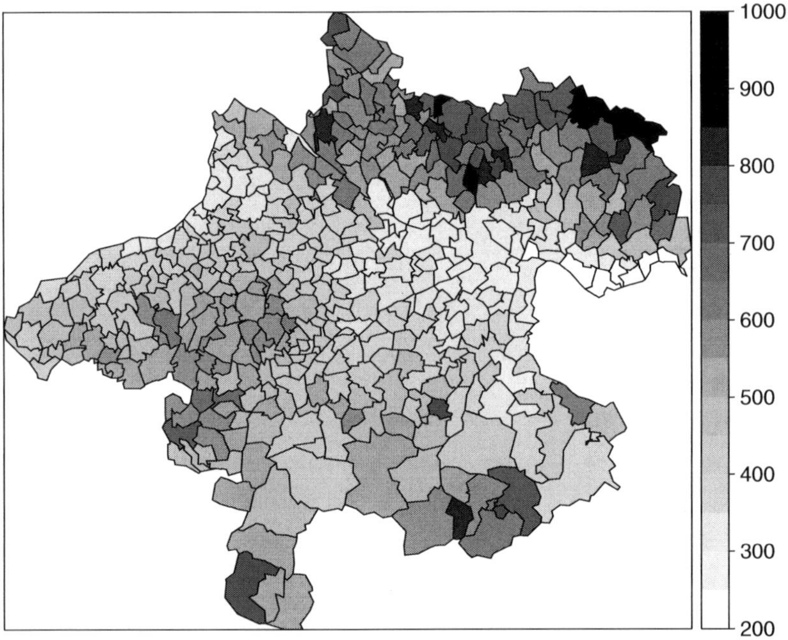

Figure 8.2 Map of the variable ALTITUDE.

8.2 Spatial links

The sampling design primarily affects the so-called spatial link matrices (or spatial weighting matrices), which represent the spatial relationships between observations and are frequently employed in spatial econometrics (for a characterization of this branch of statistics see Anselin, 1988 or more recently Arbia, 2006). In general, spatial link matrices represent similarities, e.g., connectivity, neighborhoods or inverse distances. A spatial link matrix **G** is an $n \times n$ matrix (n is the number of observations) with the following properties:

(i) $g_{ij} = 0$ for $i = j$;

(ii) $g_{ij} > 0$ if i and j are spatially connected.

Thus the key concept to analyze and model areal data is to define which sites in the dataset are connected and therefore so-called spatial neighbors. Subsequently, spatial weights may be assigned to each of the identified neighbor links. Both steps are essential issues for statistical modeling of areal data because the results of the spatial analysis are crucially dependent on the decisions made in constructing the spatial weights. Sections 8.2.1 and 8.2.2 give a short overview of the different approaches to define spatial neighbors and spatial weights. In the literature these topics are dealt with, e.g., by O'Sullivan and Unwin (2003),

Banerjee *et al.* (2004, pp. 70 and 71), Waller and Gotway (2004, pp. 223–225), Fortin and Dale (2005, pp. 113–118), and Schabenberger and Gotway (2005, pp. 18–19). Due to the practical relevance for programming in R, the two sections are structured following Bivand *et al.* (2008).

8.2.1 Spatial neighbors

Neighbor relationships between all objects are usually represented by a $n \times n$ binary connectivity matrix $\mathbf{C}$, where n is the number of observations (Fortin and Dale 200, pp. 113–118). The component c_{ij} of the connectivity matrix is defined as follows:

$$c_{ij} = \begin{cases} 1, & \text{if there exists a neighbor relationship between two objects } i \text{ and } j \\ 0, & \text{if two objects } i \text{ and } j \text{ are not in a neighbor relationship.} \end{cases}$$

Each component c_{ii} is set to zero since no region is a neighbor to itself. Generally, connectivity matrices can be symmetric or asymmetric, where asymmetry is present when i is a neighbor of j but j is not a neighbor of i or vice versa.

When working with the package `spdep` in R (Bivand *et al.* 2010), neighbor relationships are represented by a `nb` object (Bivand *et al.* 2008, p. 240). This object consists of a list of length n, where for each observation i an integer vector of index numbers of its neighbors is recorded. To objects without any neighbors an integer zero is assigned. Additionally, the `nb` object gives information about the symmetry, presence of no-neighbor observations, average number of links, link number distribution and least/most connected regions.

To complete the definition of a connectivity matrix the term neighborhood has to be specified more precisely: one of the most commonly used approaches in the literature to determine neighbor relationships is to create contiguity neighbors (R function `poly2nb`). Two polygon areas i and j are contiguity neighbors if they share

- at least one point on their boundaries (=Queen-style) or
- at least two points on their boundaries (=Rook-style).

To use other neighborhood criteria it is necessary to choose a point to represent each polygon entity, which is often the polygon centroid. Once representative points are available, neighbors can be determined, e.g., by means of graph measures like Delauney triangulation neighbors (R function `tri2nb`) or Gabriel graph neighbors (R function `graph2nb`). Another method to create neighbors is to choose the k nearest points as neighbors for each polygon (R function `knearneigh`). In many cases this method leads to an asymmetric connectivity matrix (Banerjee *et al.* 2004, p. 70), which can be made symmetrical in R using the function `make.sym.nb`. Alternatively, distance-based neighbors can be established by connecting points within an interpoint distance with fixed lower and upper distance bounds (R function `dnearneigh`).

For our example this last method is used as the basis for the connectivity matrix. The lower and upper distance bounds are set to 0 and 10.076 m, which is the minimum distance at which all areas have a distance-based neighbor. Thus, it can be guaranteed that all areas in the dataset are linked to at least one neighbor (Figure 8.3). For more details on creating spatial neighbors in R, see Bivand *et al.* (2008, pp. 242–251).

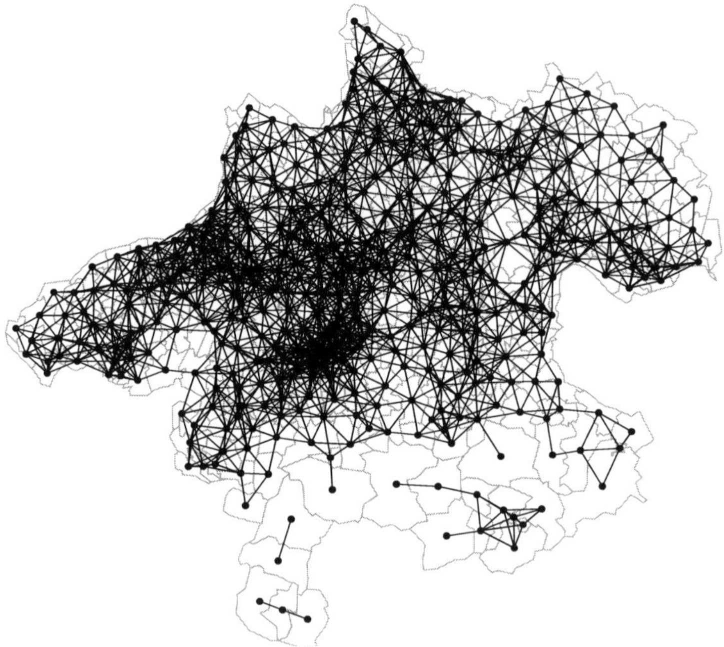

Figure 8.3 Distance-based neighbors within a radius of 10.076 m.

8.2.2 Spatial weights

After establishing the connectivity matrix **C**, spatial weights may be assigned to each neighbor relationship. For this purpose a spatial weights matrix **W** can be computed. The idea of spatial weights is to assign higher weights w_{ij} to connected areas i and j if area j is (in some sense) closer to area i than other connected areas to i. The definition of proximity may for example be based on:

- distance between two areas
- length of the shared border of two areas or
- relative sizes of the areas.

We could assume, for example, that the strength of neighbor relationships decreases with distance. Therefore, we use the inverse distance between two entities ($1/d_{ij}$) as weight to determine the component w_{ij} of matrix **W**. If area i is not a neighbor of area j (i.e., if $c_{ij} = 0$), w_{ij} is set to zero. If objects i and j are adjacent component w_{ij} in this example would then be defined as follows:

$$w_{ij} = \frac{c_{ij}}{d_{ij}}$$

Another variant is the exponential weighting

$$w_{ij} = e^{-\theta d_{ij}} - \delta_{ij}, \tag{8.1}$$

where θ is some decay parameter and δ_{ij} is the usual Kronecker δ. Note that this scheme corresponds to a spatial process with a specific version of the Matérn variogram $\gamma_M(h, \theta)$.

However, if there is no reason to assume more than the existence and absence of neighbor relationships, a spatial weights matrix **W** deviating noticeably from the binary connectivity matrix **C** should be avoided. In this case the simplest way to define the spatial weights matrix **W** would be to set **W** equal to the connectivity matrix **C** (i.e., $w_{ij} = c_{ij}$).

Spatial weights matrices are often converted by using coding schemes to cope with the heterogeneity which is induced by the different linkage degrees of the spatial object. A widely used coding scheme is the row-sum standardized coding scheme where the sum of each row of the standardized link matrix is equal to one (O'Sullivan and Unwin 2003, p. 42; Waller and Gotway 2004, p. 225). We simply obtain the components of this row standardized matrix $\mathbf{W}_{std}$ by dividing each w_{ij} by the sum of the neighbor weights for region i:

$$w_{std,ij} = \frac{w_{ij}}{\sum_{j=1}^{n} w_{ij}} \tag{8.2}$$

This row standardization is used to take the different numbers of neighbors per unit into account. Other coding schemes are for instance the globally standardized, and the variance stabilizing coding scheme (Tiefelsdorf 2000). For reasons of simplification, the spatial link matrix **W** is in the following always the row-standardized version.

In R we use the function `nb2listw` to convert a `nb` object into a spatial weights object `listw`. The argument `glist` of this function can be used to pass a list of vectors of weights corresponding to the neighbor relationships. Additionally, there are various different weight styles available to standardize the matrix. Conversion style `W` is the default value and creates a row standardized weights matrix. Style `B` retains a weight of unity for each link, in style `C` the complete set of weights sums to the number of observed entities n and in style `U` all weights together sum to 1 (Bivand *et al.* 2008, pp. 251–255).

In our example we set the spatial weights matrix **W** equal to the binary connectivity matrix **C** and use the row standardized matrix $\mathbf{W}_{std}$ as in (8.2) for further statistical analysis and modeling of the data. Extensions to the spatio-temporal setting are straightforward and can, e.g., be found in Dubé and Legros (2011).

8.3 Measures of spatial dependence

Several measures for quantifying spatial dependence have been proposed in the literature. Generally, they can be classified into two groups: the first ones are based on weighted covariance type expressions analogous to the Durbin–Watson statistic for time series (prototypically the Moran's $\mathfrak{I}$; Moran, 1950). The others are based on weighted averages of squared differences – often called semivariances (prototypically the Geary's $\mathfrak{c}$; Geary, 1954). Most of the available software tools applicable for spatial analysis provide a standard implementation of those measures, see, e.g., Rangel *et al.* (2010) or the many R-packages to be found on http://cran.r-project.org/web/views/Spatial.html. To get an overview concerning the relationships between those measures see Dale *et al.* (2002).

As a measure of the intensity of the spatial dependence and for detecting its potential existence probably the most popular statistic, the Moran's $\mathfrak{I}$, is used here. Therefore, to test for the presence of spatial dependence we employ in this chapter Moran's $\mathfrak{I}$ test as this is perhaps the most common global test for this kind of problem (Waller and Gotway 2004, p. 227; Schabenberger and Gotway 2005, p. 21; Bivand *et al.* p. 258). Moran's $\mathfrak{I}$ is calculated by dividing the product of the variable of interest and its spatial lag with the cross-product of the variable of interest and adjusting this term for the spatial weights:

$$\mathfrak{I} = \frac{n}{\sum_{i=1}^n \sum_{j=1}^n w_{ij}} \frac{\sum_{i=1}^n \sum_{j=1}^n w_{ij}(y_i - \bar{y})(y_j - \bar{y})}{\sum_{i=1}^n (y_i - \bar{y})^2},$$

where y_i is the ith observation of the variable of interest, $\bar{y}$ is the mean of the variable of interest and w_{ij} is the (standardized) spatial weight of the neighbor relationship between two areas i and j. Moran's $\mathfrak{I}$ will be positive, if neighboring areas tend to have similar values of the variable of interest, and negative, when they tend to have different values. For the testing procedure the observed value is standardized by subtracting its expected value and dividing this difference by the standard deviation under the null hypothesis of no spatial dependence. Usually, the test is one-sided, testing whether the observed statistic is significantly greater than its expected value. For details concerning the testing procedure see Schabenberger and Gotway (2005, p. 21). A different motivation and interpretation of $\mathfrak{I}$ is provided in Dray (2011).

In R the function `moran.test` implemented in the package `spdep` is used to perform the test for spatial autocorrelation. Examining the log ratio of the year 2008 *R08* in our example yields the following result:

```
        Moran's I test under normality

data:  data$R08
weights: ooe_W1

Moran I statistic standard deviate = 36.231, p-value < 2.2e-16
alternative hypothesis: greater
```

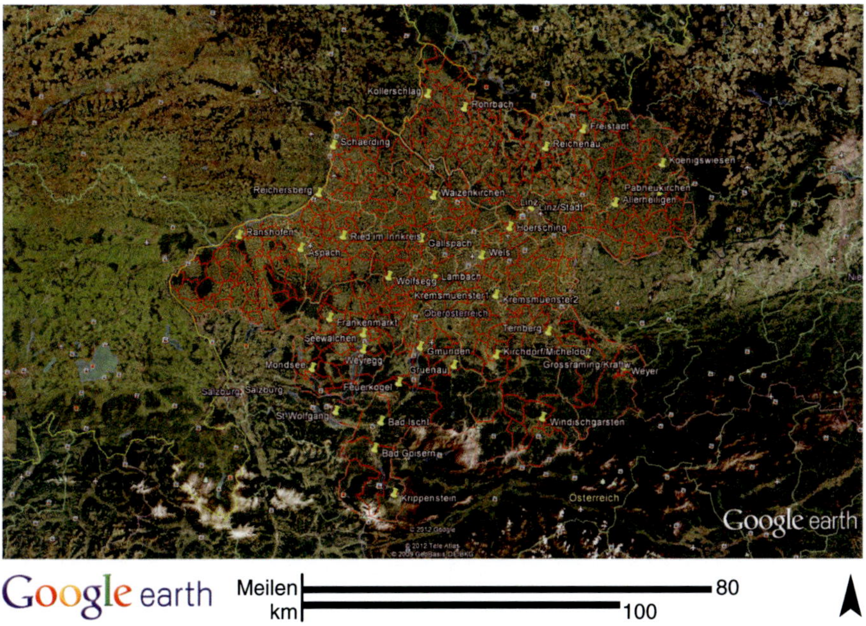

Figure 1.8 A Google-Earth view of observation regions and sites.

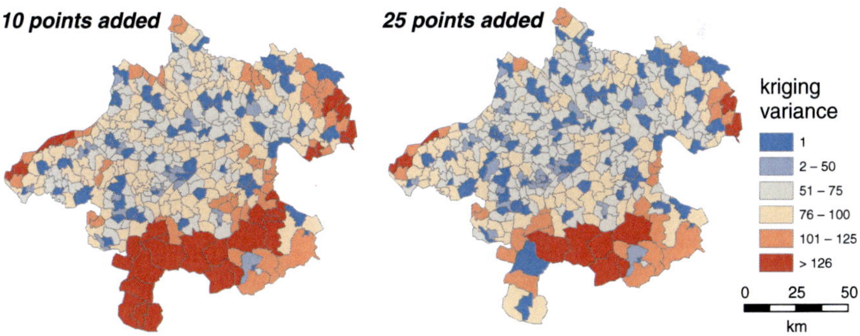

Figure 3.4 The kriging variance after the addition of new samples to the initial set (use in conjunction with Figure 3.3). Darker areas denote regions of high uncertainty since no points were located in those regions.

Spatio-temporal design: Advances in efficient data acquisition, First Edition.
Edited by Jorge Mateu and Werner G. Müller.

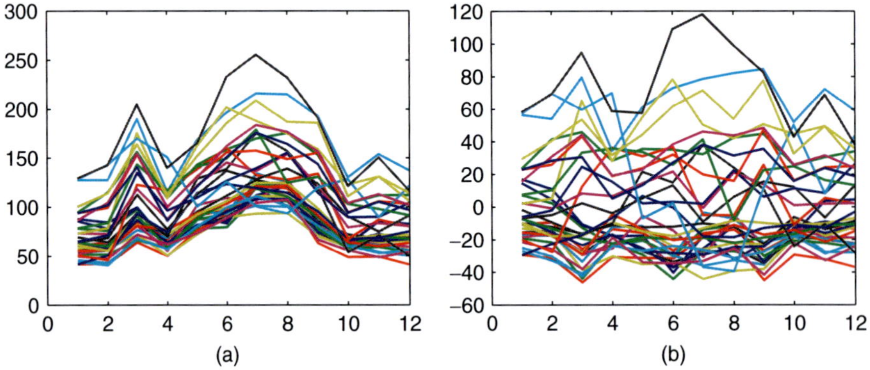

Figure 4.2 *(a) The average monthly rainfall over the years at each of the 36 stations, for each of the 12 months. (b) The residual rainfall at each of the 36 stations, for each of the 12 months.*

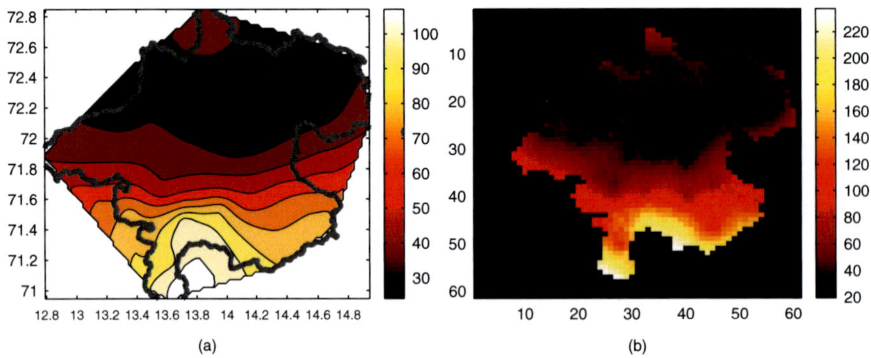

Figure 4.12 *(a) The median rainfall field +53 calculated from the predictive distributions of trans-Gaussian kriging applied to the 36 means of rainfall residuals. (b) Expected lengths of 95% predictive intervals corresponding to the trans-Gaussian kriging from (a).*

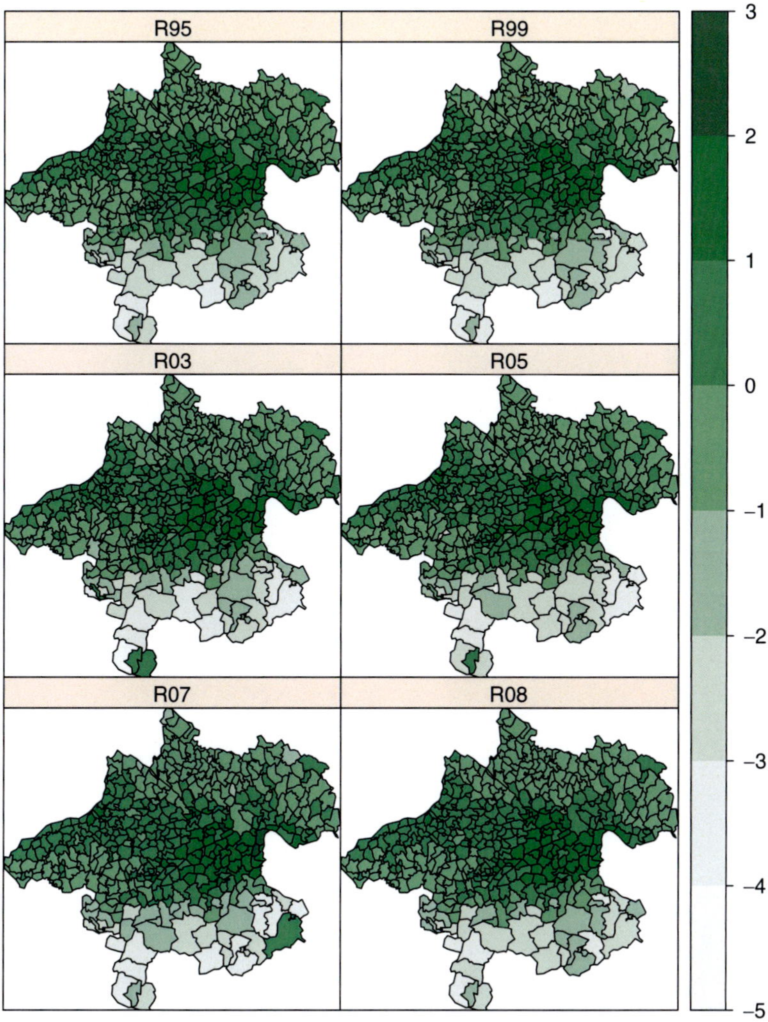

Figure 8.1 Maps of the log ratios R95 to R08.

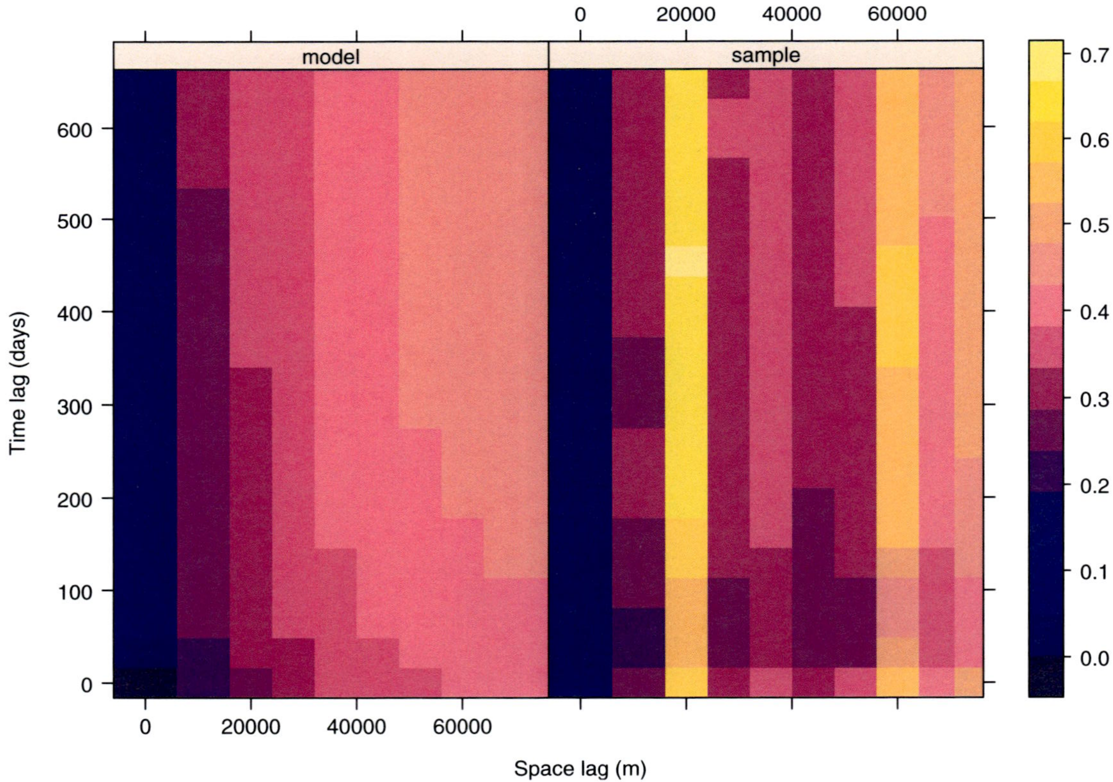

Figure 9.7 Sample variogram and fitted sum-metric model of residuals from the multiple linear regression of temperature data on MODIS and elevation.

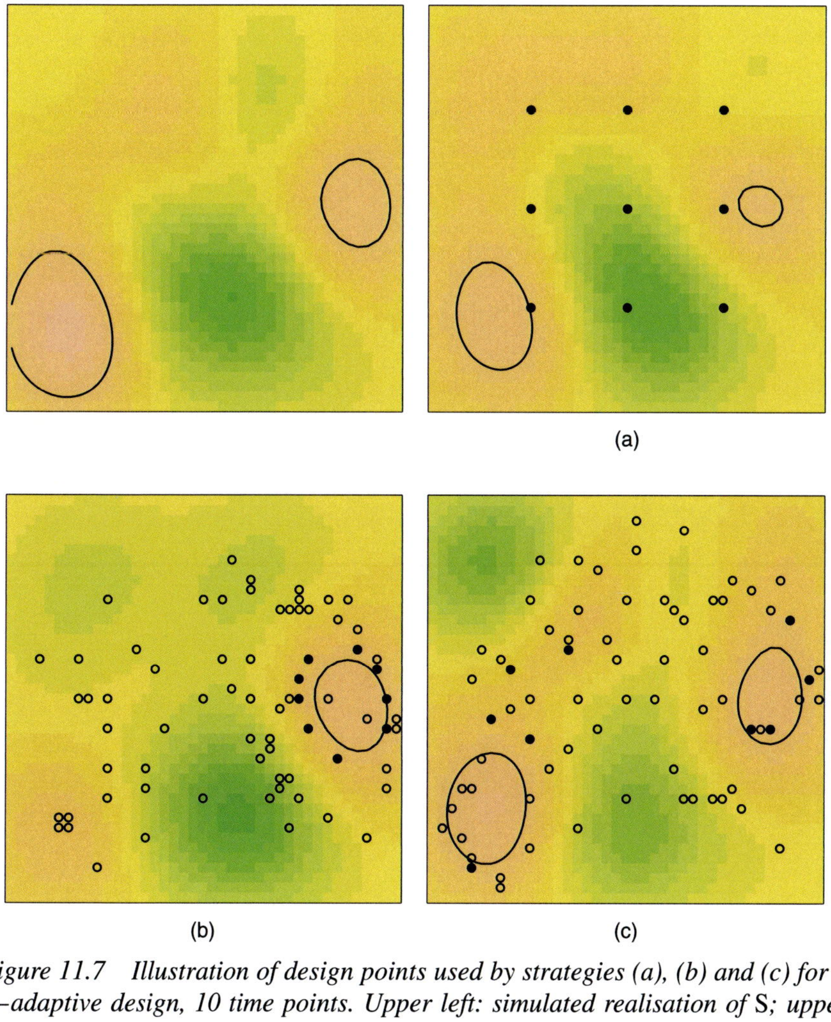

Figure 11.7 Illustration of design points used by strategies (a), (b) and (c) for a 9–adaptive design, 10 time points. Upper left: simulated realisation of S*; upper right: regular lattice design strategy (a); lower left: strategy (b); lower right: strategy (c). Predicted values of* S *are shown in each case. Solid circles indicate design points at the current time* ($t = 10$)*, and open time points design points for previous time points. Contour lines indicate regions where* $\hat{S} > 1$.

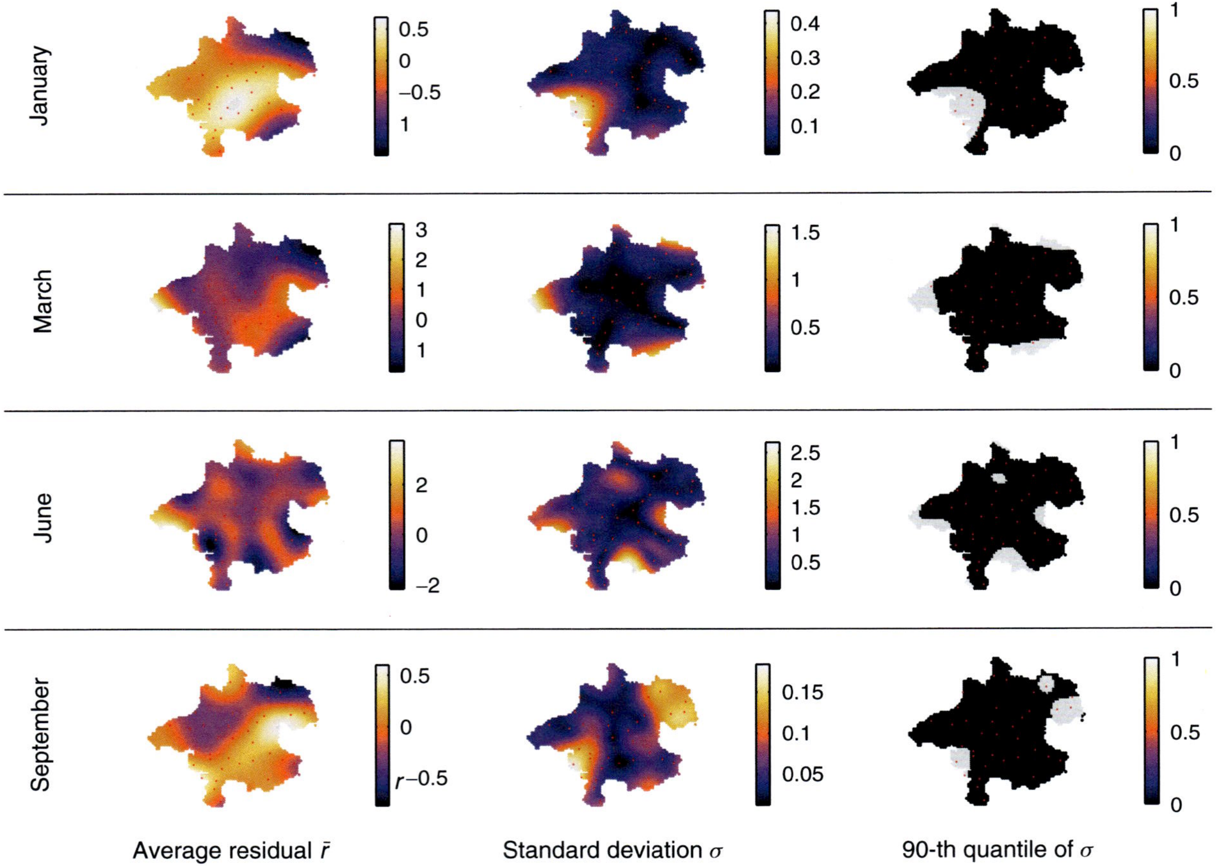

Figure 13.6 Predictions by the committee of SVR models for January, March, June and September residual average temperatures ($r = T - \hat{T}$). Left column: mean of the estimated average monthly temperature residuals $\bar{r}$; middle column: standard deviation of the 50 SVR predictions; right column: 90th quantile of the distribution of standard deviation, corresponding to uncertain areas for each month.

```
sample estimates:
Moran I statistic       Expectation          Variance
      0.795973350      -0.002252252       0.000485391
```

The test results seem to show a significant positive spatial autocorrelation in *R08*, but the interpretation has to be done in a more careful manner:

First, we have to consider the test assumptions of constant mean and variance of the y_i and should be aware of spurious autocorrelation or 'misspecification' (Schabenberger and Gotway 2005, p. 22). Any autocorrelation in our data may, e.g., simply be due to the altitudes of the municipalities and not to any spatial pattern in the log ratio of arable land versus grassland. To overcome this issue we could for example fit a mean model to the data including additional variables (see Section 8.4) and performing the test for spatial autocorrelation again for the residuals of this model using the function `lm.morantest`.

Secondly, we should keep in mind that the test outcome is also affected by the choice of the spatial weights and the standardizing scheme used for the weights (see the examples in Bivand *et al.* 2008 p. 262).

In Figure 8.4 two approaches to plot the spatial autocorrelation are shown (Bivand *et al.* 2008, p. 267). We can find the values of Moran's $\mathfrak{I}$ for sixteen successive lag orders of contiguous neighbors in Figure 8.4(a) (function `sp.correlogram`) and for a sequence of distance band neighbors in Figure 8.4(b) (function `correlog` in package `pgirmess`). In our example, the first four bands of 0–10, 10–20, 20–30 and 30–40 km have significant values of spatial autocorrelation (on the other hand the following significantly negative values may indicate some kind of nonstationarity). Another graphical tool to examine the data for spatial autocorrelation is the Moran scatterplot in Figure 8.5. It shows the relationship between the variable of interest (x-axis) and the spatially weighted average of neighboring values, also called the spatially lagged values (y-axis).

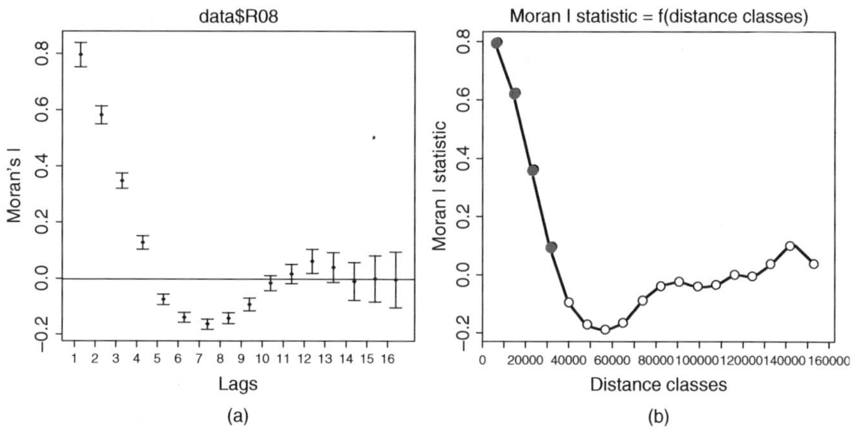

Figure 8.4 Correlograms: values of Moran's $\mathfrak{I}$ for sixteen successive lag orders (a); values of Moran's $\mathfrak{I}$ for a sequence of distance band neighbors (b).

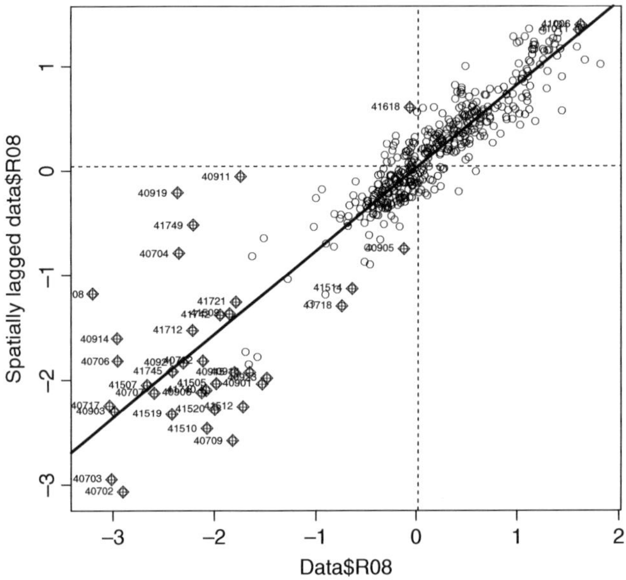

Figure 8.5 Moran scatterplot.

Most of the points in our example appear in the low–low and high–high quadrants representing a positive spatial dependency. Only a few locations can be found in the low–high and high–low quadrants referring to locations surrounded by dissimilar valued neighbors. The slope of the regression line corresponds to Moran's $\mathfrak{I}$ and represents the linear association between the observed values and the spatially lagged values (Anselin, 1993, 1995). In R the function `moran.plot` is used to visualize the data in a Moran scatterplot (Bivand *et al.* 2008, p. 268).

As already noted several alternative measures to $\mathfrak{I}$ exist that could equivalently be employed for our purposes, most notably the so-called contiguity ratio proposed by Geary (1954). Recently, López *et al.* (2011) provided extensive simulations on four candidate measures including $\mathfrak{I}$ and demonstrated its comparative value in an economic application.

8.4 Models for areal data

For simplification of exposition, we will in the rest of the section again assume that the considered models are linear and the error processes Gaussian, with the obvious extensions to locally linearized models as in previous chapters. The regression residuals from estimation of the model $\mathbf{y} = \mathbf{X}\boldsymbol{\beta} + \boldsymbol{\varepsilon}$ under the assumption $\boldsymbol{\varepsilon}$ i.i.d. will be used for the test of spatial dependence. The real data generating process, the true but unknown status of the world, is assumed to be one of the following:

H_0: spaceless;

H_A: spatial.

The distinction between these hypotheses will be made clear in the following subsections. Depending on the two examined cases, one either wants to reject or not reject the null hypothesis of spatial independence of Moran's $\mathfrak{I}$ test to make a correct decision. For more on this issue see, e.g., Anselin (1988). Further, one wants to find an optimal or nearly optimal design for a test strategy to receive either nonrejection or rejection of the null hypothesis for derivation of a model that matches the real status of the world.

8.4.1 H_0: A spaceless regression model

We start with the basic linear regression model, which reads:

$$\mathbf{y} = \mathbf{X}\boldsymbol{\beta} + \boldsymbol{\varepsilon}, \tag{8.3}$$

where $\mathbf{y}$ is the outcome variable of interest, $\mathbf{X}$ a matrix of explanatory variables, $\boldsymbol{\beta}$ the vector of parameters and $\boldsymbol{\varepsilon}$ an error term with errors assumed to be independently distributed.

For a standard regression model it is crucial to know whether the residuals are spatially dependent or not. If there is no spatial dependence in the residuals, one can use standard estimation methods, like OLS, but if the residuals show spatial dependence, one has to use special methods (cf. Section 8.5.2). When the OLS estimation method is applied instead, spatial autocorrelation in the error term leads to biased estimates of the residual variance and inefficient estimates of the regression coefficients. For regression residuals Moran's $\mathfrak{I}$ is defined as a scale invariant ratio of quadratic forms in the normally distributed regression residuals $\hat{\boldsymbol{\varepsilon}} = (\hat{\varepsilon}_1, \ldots, \hat{\varepsilon}_n)'$, i.e.,

$$\mathfrak{I} = \frac{\hat{\boldsymbol{\varepsilon}}' \frac{1}{2}(\mathbf{W} + \mathbf{W}')\hat{\boldsymbol{\varepsilon}}}{\hat{\boldsymbol{\varepsilon}}'\hat{\boldsymbol{\varepsilon}}} \tag{8.4}$$

where $\sum_{i=1}^{n} \sum_{j=1}^{n} w_{ij} = n$ (Tiefelsdorf 2000).

The classical Moran's $\mathfrak{I}$ as the two-dimensional analog of a test for univariate time series correlation is given in, e.g., Cliff and Ord (1981). For a random variable Y, measured in each of the n nonoverlapping subareas of the whole study area, Moran's $\mathfrak{I}$ is defined from the residuals of an intercept only regression, i.e., $\hat{\boldsymbol{\varepsilon}} = \mathbf{My}$ where $\mathbf{y} = (y_1, \ldots, y_n)'$, $\mathbf{M} = \mathbf{I}_n - \frac{1}{n}\mathbf{1}_n\mathbf{1}_n'$, where $\mathbf{I}_n$ is an $n \times n$ identity matrix and $\mathbf{1}_n$ is an $n \times 1$ vector of ones. In this case, and if the spatial link matrix $\mathbf{W}$ has full rank (i.e., there is no observation completely separated from all others), the expected value of the test statistic $\mathfrak{I}$ under independence is given by $E[\mathfrak{I}|H_0] = -\frac{1}{n-1}$, and the variance of $\mathfrak{I}$ can be given in terms of the eigenvalues γ_i of matrix $\mathbf{K} = \mathbf{M}'\frac{1}{2}(\mathbf{W} + \mathbf{W}')\mathbf{M}$ as $\mathrm{Var}[\mathfrak{I}|H_0] = \frac{2n}{n^2-1}\sum_{i=1}^{n}(\gamma_i - \bar{\gamma})^2 = \frac{2n}{n^2-1}\sigma_\gamma^2$. The test statistic $\mathfrak{I}$ is then asymptotically normally distributed.

The Moran's $\mathfrak{I}$ test is used for parametric hypotheses about the spatial autocorrelation level ρ, i.e., $H_0 : \rho = 0$ against $H_A : \rho > 0$ for positive spatial autocorrelation; or $H_0 : \rho = 0$ against $H_A : \rho < 0$ for negative spatial autocorrelation. Tests for positive correlation are much more relevant in practice, because negative spatial autocorrelation very rarely appears in the real world. Thus, from now on $\rho \geq 0$ will be assumed. The z-transformed Moran's $\mathfrak{I}$ is, for normally distributed regression residuals and well-behaved spatial link matrices under certain regularity conditions (Tiefelsdorf, 2000), asymptotically standard normally distributed, i.e., $z(\mathfrak{I})$ is defined as

$$z(\mathfrak{I}) = \frac{\mathfrak{I} - E[\mathfrak{I}|H_0]}{\sqrt{\text{Var}[\mathfrak{I}|H_0]}} \sim N(0, 1). \tag{8.5}$$

The exact small sample distribution of Moran's $\mathfrak{I}$ was seemingly independently obtained by Hepple (1998) and Tiefelsdorf and Boots (1995), but does not offer advantages with respect to the design task as shown in Müller *et al.* (2012). The behavior under deviations from normality is investigated in Griffith (2010).

Note that it turns out that a special class of spatial objects is relevant especially for design purposes. These are observations that belong to a design but are far apart from all other objects, in the sense that they have no spatial links to other observations. They have been termed far-off objects by Gumprecht (2007) and a discussion of their role in Moran's $\mathfrak{I}$ tests and corresponding designs are given therein.

Conveniently, we also start in our example with estimating the basic linear regression model to see which covariates contribute for explaining the variance in the response variable *R08*. As the spatial autocorrelation that we detected with the tests in Section 8.3 could actually be caused by model misspecification we also check if there still remains spatial autocorrelation in the residuals of the model. For this purpose a map of the residuals as in Section 8.1.1 can be plotted and a Moran's $\mathfrak{I}$ test as in Section 8.3 is performed for the residuals (in R: function `lm.morantest`).

In a first attempt we include all explanatory variables in the linear regression model except the variable *FLKM2* (area of the municipality in square kilometers), which would not make any sense in explaining the dependent variable *R08*. We call the R function `lm` and obtain the following result:

```
Call:
lm(formula = data$R08 ~ data$ALTITUDE + data$R07 + data$R05 +
    data$R03 + data$R99 + data$R95)

Residuals:
     Min        1Q    Median        3Q       Max
-0.75505 -0.04410 -0.00830  0.04535  0.68314

Coefficients:
              Estimate Std. Error t value Pr(>|t|)
(Intercept)  1.602e-01  2.196e-02   7.292 1.44e-12 ***
```

```
data$ALTITUDE -1.931e-04  4.556e-05  -4.238 2.75e-05 ***
data$R07       8.595e-02  1.463e-02   5.874 8.40e-09 ***
data$R05       4.727e-02  1.839e-02   2.570   0.0105 *
data$R03       8.107e-02  1.744e-02   4.650 4.41e-06 ***
data$R99       4.237e-01  5.116e-02   8.281 1.50e-15 ***
data$R95       3.669e-01  4.825e-02   7.604 1.77e-13 ***
---
Signif. codes:  0 '***' 0.001 '**' 0.01 '*' 0.05 '.' 0.1 ' ' 1

Residual standard error: 0.1135 on 438 degrees of freedom
Multiple R-squared: 0.9834, Adjusted R-squared: 0.9832
F-statistic:  4324 on 6 and 438 DF,  p-value: < 2.2e-16
```

All covariates seem to have a significant influence on the dependent variable and the Moran's $\mathfrak{I}$ test decides in favor of the hypothesis that there is no more spatial autocorrelation in the residuals (in fact it decides for negative spatial autocorrelation, which indicates some kind of overcorrection):

```
        Global Moran's I for regression residuals

data:
model: lm(formula = data$R08 ~ data$ALTITUDE + data$R07 + data$R05
+ data$R03 + data$R99 + data$R95)
weights: ooe_W1

Moran I statistic standard deviate = -6.4293, p-value = 1
alternative hypothesis: greater
sample estimates:
Observed Moran's I        Expectation          Variance
     -0.1445027577      -0.0061184919      0.0004632761
```

However, if we analyze the explanatory variables in detail, we can find that the ratios *R07* to *R95* are highly correlated (i.e., multicollinearity). According to Fahrmeir *et al.* (2009, p. 70), a multicollinearity problem can be identified by computing the variance inflation factor (VIF) for each explanatory variable and a multicollinearity problem is present for VIF > 10. This case arises for two of our covariates, i.e., *R95* (VIF = 56.82) and *R99* (VIF = 62.75). Therefore, we drop these two explanatory variables *R95* and *R99* in a second attempt and obtain the following output:

```
Call:
lm(formula = data$R08 ~ data$ALTITUDE + data$R07 + data$R05
 + data$R03)

Residuals:
     Min       1Q   Median       3Q      Max
-1.21197 -0.06322  0.00175  0.06983  1.29267
```

```
Coefficients:
                 Estimate Std. Error t value Pr(>|t|)
(Intercept)     1.774e-01  4.060e-02   4.370 1.55e-05 ***
data$ALTITUDE -2.892e-04  8.417e-05  -3.436 0.000646 ***
data$R07        3.292e-01  2.345e-02  14.036  < 2e-16 ***
data$R05        2.980e-01  3.050e-02   9.770  < 2e-16 ***
data$R03        2.633e-01  3.060e-02   8.604  < 2e-16 ***
---
Signif. codes:  0 '***' 0.001 '**' 0.01 '*' 0.05 '.' 0.1 ' ' 1

Residual standard error: 0.2109 on 440 degrees of freedom
Multiple R-squared: 0.9424,     Adjusted R-squared: 0.9419
F-statistic:  1800 on 4 and 440 DF,  p-value: < 2.2e-16
```

The regression coefficients of the remaining covariates are all significant again, but the Moran's $\mathfrak{I}$ test for the residuals yields a significant result now, an effect which also occurs for other choices of the spatial weights matrix $\mathbf{W}$:

```
        Global Moran's I for regression residuals

data:
model: lm(formula = data$R08 ~ data$ALTITUDE + data$R07 + data$R05
+ data$R03)
weights: ooe_W1

Moran I statistic standard deviate = 3.7525, p-value = 8.752e-05
alternative hypothesis: greater
sample estimates:
Observed Moran's I        Expectation           Variance
     0.0755249376      -0.0057114017       0.0004686499
```

Usually in this case, we fit a spatial regression model instead of the linear regression model to estimate the spatial effect as well. However, we should deal with another possible problem first of all: if the dependent variable $\mathbf{y}$ and one or more of the explanatory variables $\mathbf{X}$ are generated according to a spatial autoregressive process with a positive autoregression parameter ρ (for details see Section 8.4.2) and $\mathbf{y}$ is regressed on $\mathbf{X}$, there is a risk of spurious spatial regression (Fingleton, 1999; Lauridsen and Kosfeld 2006; Beenstock *et al.* 2012). To check whether we have the case of spurious spatial regression, it is necessary to test for spatial nonstationarity. Lauridsen and Kosfeld (2006) and Beenstock *et al.* 2012 propose two contrary procedures to test for spatial nonstationarity and it is not obvious to us which is the 'correct' approach. The fact is, if we observe spatial nonstationarity, this involves the risk of spurious spatial regression. Moreover, one should also test for spatial cointegration in a second step. Spatial cointegration concerns the case where two or more variables in the regression are nonstationary, while the errors are stationary (Lauridsen and Kosfeld 2006, p. 9). However, in case all variables are stationary, we can estimate a spatial regression model as in the next section.

8.4.2 H_0: Spatial regression models

As seen in the previous section, we can obviously not assume that the observations (and model errors) are independent of each other when we work with geographically referenced data and suppose correlations between neighboring areas (Bivand *et al.* 2008, p. 273; Gibbons and Overman 2012). Therefore in the literature one can find several approaches to include spatial dependencies in the regression equation (Kissling and Carl 2008; Gibbons and Overman 2012). These models are called *spatial simultaneous autoregressive models* and differ from each other in the assumption where the spatial autoregressive process occurs. The following paragraphs present the ideas of the different spatial regression models and try to point out the connections between them.

In a first step the linear regression model (8.3) is extended by the term $\lambda\mathbf{Wu}$ so that we get the regression equation in (8.6). It is assumed that the errors are no longer independent, but involve the spatial autoregressive process. This model is denoted as the *simultaneous autoregressive* (SAR) model (Bivand *et al.* 2008, p. 277)) or–in the spatial econometrics context–the *spatial error* (SE) model (Gibbons and Overman 2012):

$$\mathbf{y} = \mathbf{X}\boldsymbol{\beta} + \mathbf{u} \quad \text{with} \quad \mathbf{u} = \lambda\mathbf{Wu} + \boldsymbol{\varepsilon}, \tag{8.6}$$

where $\mathbf{X}$ is again the matrix of explanatory variables and $\boldsymbol{\beta}$ the corresponding parameter vector. $\mathbf{u}$ denotes the spatially dependent error term, $\mathbf{W}$ is the spatial weights matrix as in Section 8.2.2 and λ is the spatial autoregression parameter. $\boldsymbol{\varepsilon}$ represents the vector of (spatially) independent residual errors which are normally distributed with zero mean and diagonal covariance matrix $\boldsymbol{\Sigma}_\varepsilon$ with elements $\sigma^2_{\varepsilon_i}$, $i = 1, \ldots, n$ or often a joint variance $\sigma^2_{\varepsilon_i} = \sigma^2_\varepsilon$ (Schabenberger and Gotway 2005, p. 335; Bivand *et al.* 2008, p. 277). The model equation in (8.6) can be easily rewritten in the following way:

$$\begin{aligned}
\mathbf{y} &= \mathbf{X}\boldsymbol{\beta} + \lambda\mathbf{Wu} + \boldsymbol{\varepsilon} \\
&= \mathbf{X}\boldsymbol{\beta} + \lambda\mathbf{W}(\mathbf{y} - \mathbf{X}\boldsymbol{\beta}) + \boldsymbol{\varepsilon} \\
&= \mathbf{X}\boldsymbol{\beta} - \lambda\mathbf{WX}\boldsymbol{\beta} + \lambda\mathbf{Wy} + \boldsymbol{\varepsilon}
\end{aligned} \tag{8.7}$$

Some further manipulation of Equation (8.7) yields Equation (8.8):

$$\begin{aligned}
\mathbf{y} - \lambda\mathbf{Wy} &= \mathbf{X}\boldsymbol{\beta} - \lambda\mathbf{WX}\boldsymbol{\beta} + \boldsymbol{\varepsilon} \\
(\mathbf{I}_n - \lambda\mathbf{W})\mathbf{y} &= (\mathbf{I}_n - \lambda\mathbf{W})\mathbf{X}\boldsymbol{\beta} + \boldsymbol{\varepsilon} \\
\mathbf{y} &= \mathbf{X}\boldsymbol{\beta} + (\mathbf{I}_n - \lambda\mathbf{W})^{-1}\boldsymbol{\varepsilon},
\end{aligned} \tag{8.8}$$

assuming the invertibility of $\mathbf{I}_n - \lambda\mathbf{W}$. Equations (8.7) and (8.8) clearly show the similarity of the SE model to the linear regression model in Equation (8.3) (Schabenberger and Gotway 2005, p. 336). Instead of the uncorrelated errors $\boldsymbol{\varepsilon}$ in (8.3), spatial autocorrelation is induced by introducing the error term

$(\mathbf{I}_n - \lambda\mathbf{W})^{-1}\boldsymbol{\varepsilon}$ in (8.8). From the representation in Equation (8.7) one can see the two additional terms $\lambda\mathbf{WX}\boldsymbol{\beta}$ and $\lambda\mathbf{Wy}$ in the regression model compared with model (8.3). These terms are called the spatially lagged explanatory variables ($\lambda\mathbf{WX}\boldsymbol{\beta}$) and the spatially lagged values of the response variable ($\lambda\mathbf{Wy}$). According to Kissling and Carl (2008), the SE model is used if the analyst assumes that the covariates $\mathbf{X}$ do not completely explain the spatial autocorrelation in the data, denoting this case as 'induced spatial dependence'. This case occurs, e.g., if important spatially influenced covariates are not included in the analysis. Another motivating reason for this kind of model arises if spatial autocorrelation is an inherent characteristic of the response variable $\mathbf{y}$ itself, denominating this case as 'inherent spatial autocorrelation'.

Omitting the term $-\lambda\mathbf{WX}\boldsymbol{\beta}$ in Equation (8.7) yields the so-called *spatial lag* (SL) model (Bivand *et al.* 2008, p. 291; Kissling and Carl 2008) or – equivalently in some other references – the *spatial autoregressive* model (LeSage and Pace 2009, p. 8; Gibbons and Overman 2012). Apart from the influence of the explanatory variables $\mathbf{X}$, the response variable $\mathbf{y}$ also depends on its spatially lagged values $\rho\mathbf{Wy}$ so that we obtain the regression equation:

$$\mathbf{y} = \rho\mathbf{Wy} + \mathbf{X}\boldsymbol{\beta} + \boldsymbol{\varepsilon}, \tag{8.9}$$

where ρ is the spatial autoregression parameter, $\boldsymbol{\varepsilon}$ again represents the vector of (spatially) independent residual errors and the other terms are as above. In contrast to the SE model, the SL model assumes that there is only 'inherent spatial autocorrelation' present in the data and therefore the spatial autoregressive process is included in the response variable itself (Kissling and Carl 2008). The term $\rho\mathbf{Wy}$ describes the relation between the values of the dependent variable $\mathbf{y}$ and the neighboring values to each observation of $\mathbf{y}$ (LeSage and Pace 2009, p. 9). Rewriting the regression equation in (8.9) leads to the following alternative representation of the SL model:

$$\begin{aligned}
\mathbf{y} - \rho\mathbf{Wy} &= \mathbf{X}\boldsymbol{\beta} + \boldsymbol{\varepsilon} \\
(\mathbf{I}_n - \rho\mathbf{W})\mathbf{y} &= \mathbf{X}\boldsymbol{\beta} + \boldsymbol{\varepsilon} \\
\mathbf{y} &= (\mathbf{I}_n - \rho\mathbf{W})^{-1}\mathbf{X}\boldsymbol{\beta} + (\mathbf{I}_n - \rho\mathbf{W})^{-1}\boldsymbol{\varepsilon}
\end{aligned} \tag{8.10}$$

The special case of the SL model (8.9), where the response variable $\mathbf{y}$ only depends on its own spatial lag ($\rho\mathbf{Wy}$) and where no covariates are included in the regression equation, is called the *pure spatial lag* (pure SL) model:

$$\mathbf{y} = \rho\mathbf{Wy} + \boldsymbol{\varepsilon} \tag{8.11}$$

Assuming that the response variable $\mathbf{y}$ does not depend on its own spatial lag $\mathbf{Wy}$, but on the spatial lags of the explanatory variables $\mathbf{WX}$, yields the *spatial lag of X* (SLX) model (LeSage and Pace 2009, p. 30; Gibbons and Overman 2012):

$$\mathbf{y} = \mathbf{X}\boldsymbol{\beta} + \mathbf{WX}\boldsymbol{\gamma} + \boldsymbol{\varepsilon} \tag{8.12}$$

Finally, the combination of the SL (8.9) and SLX (8.12) models leads to the *spatial Durbin* (SD) model (Kissling and Carl 2008):

$$\mathbf{y} = \rho\mathbf{W}\mathbf{y} + \mathbf{X}\boldsymbol{\beta} + \mathbf{W}\mathbf{X}\boldsymbol{\gamma} + \boldsymbol{\varepsilon} \tag{8.13}$$

The SD model in (8.13) presumes that spatial autocorrelation affects both response and explanatory variables, but drops the assumption of spatial dependence in the error process. Apart from nesting the SL and the SLX models one can see that the model equation in (8.13) is also linked to the model equation (8.7) of the SE model.

As the SD model in (8.13) incorporates some other spatial regression models, this model is often estimated first and then tested against the more specific models. Moreover, it is common to compare models with different specifications of the spatial weights matrix and, of course, with different combinations of covariates (Gibbons and Overman 2012). A commonly used instrument for comparing models is the Akaike Information Criterion (AIC), accounting both for model fit and model complexity (Kissling and Carl 2008).

Some other well-known spatial regression models such as the *spatial conditional autoregressive* (CAR) model or the *simultaneous/spatial moving average* (SMA) model are not presented in this chapter. For details of these models see Haining (1993), Waller and Gotway (2004) and Schabenberger and Gotway (2005). For a recent review of what is known as spatial econometrics see Anselin (2007).

To fit spatial regression models in R the `spdep` package provides various functions, some of them using different methods to estimate the parameters (Bivand *et al.* 2008, pp. 277–296, 2010):

- `spautolm`: maximum likelihood estimation for the SE model (8.6), the CAR model and the SMA model; model type can be chosen by the option `family = ("SAR","CAR","SMA")`; is based on the function `errorsarlm`
- `lagsarlm`: maximum likelihood estimation for the SL model (8.9) and the SD model (8.13); the default setting of the option `type="lag"` is used for the SL model, and for the SD model set `type="mixed"`
- `stsls`: fits also a SL model (8.9) to the data using a two-stage least squares procedure in a simultaneous system of equations by using the spatial lags of the covariates as instruments for the spatially lagged dependent variable
- `errorsarlm`: maximum likelihood estimation of the SE model (8.6)
- `GMerrorsar`: a generalized moments estimator for the autoregressive parameter in a SE model (8.6)

For more details on the options of the different R functions please see the individual help files of the routines.

Finally, we present the estimation outputs for some of these spatial regression models. The first output shows the result of the ML estimation for the SE model using the function `spautolm` or, equivalently, the function `errorsarlm`. The regression coefficients estimated in this model are very similar to those estimated in the linear regression model (similar values, same signs, all significant). The spatial dependency in the data is estimated via the spatial autoregression parameter λ:

```
Call: spautolm(formula = data$R08 ~ data$ALTITUDE + data$R07 +
    data$R05 + data$R03, data = data, listw = ooe_W1)

Residuals:
       Min         1Q     Median         3Q        Max
-1.1343847 -0.0596381  0.0042328  0.0731940  1.2263243

Coefficients:
                Estimate  Std. Error z value  Pr(>|z|)
(Intercept)     1.9833e-01  4.6558e-02  4.2599 2.045e-05
data$ALTITUDE -3.3582e-04  9.5809e-05 -3.5051 0.0004565
data$R07        3.1967e-01  2.3904e-02 13.3733 < 2.2e-16
data$R05        2.7082e-01  2.9918e-02  9.0521 < 2.2e-16
data$R03        2.7501e-01  3.0169e-02  9.1155 < 2.2e-16

Lambda: 0.23115 LR test value: 7.5206 p-value: 0.0060997

Log likelihood: 67.4818
ML residual variance (sigma squared): 0.042974, (sigma: 0.2073)
Number of observations: 445
Number of parameters estimated: 7
AIC: -120.96
```

If we fit a SL model using the function `lagsarlm`, we once again obtain similar estimation results. However some parameter estimates of the SL model are only half as large as their SE model counterparts, and standard errors are consistently 10–20% smaller. Also, it is worth noting that the SL model fits much better than the SE model, as determined by the AIC.

```
Call:lagsarlm(formula = data$R08 ~ data$ALTITUDE + data$R07 +
    data$R05 + data$R03, data = data, listw = ooe_W1)

Residuals:
       Min         1Q     Median         3Q        Max
-0.9296895 -0.0713094  0.0025575  0.0701690  1.1748631

Type: lag
Coefficients: (asymptotic standard errors)
                Estimate  Std. Error z value  Pr(>|z|)
(Intercept)     1.0063e-01  3.6887e-02  2.7281   0.00637
data$ALTITUDE -1.6103e-04  7.5936e-05 -2.1206   0.03395
data$R07        2.3497e-01  2.2523e-02 10.4324 < 2.2e-16
data$R05        2.2526e-01  2.7750e-02  8.1174 4.441e-16
data$R03        2.3171e-01  2.7484e-02  8.4307 < 2.2e-16
```

```
Rho: 0.28677, LR test value: 96.53, p-value: < 2.22e-16
Asymptotic standard error: 0.02701
    z-value: 10.617, p-value: < 2.22e-16
Wald statistic: 112.72, p-value: < 2.22e-16

Log likelihood: 111.9866 for lag model
ML residual variance (sigma squared): 0.035063, (sigma: 0.18725)
Number of observations: 445
Number of parameters estimated: 7
AIC: -209.97, (AIC for lm: -115.44)
LM test for residual autocorrelation
test value: 1.1163, p-value: 0.29072
```

The SD model includes not only the covariates **X** and the spatial lag of the dependent variable **y**, but also the spatial lags of the explanatory variables. If we estimate such a model using the function `lagsarlm` (option `type="mixed"`), we obtain the following output:

```
Call:lagsarlm(formula = data$R08 ~ data$ALTITUDE + data$R07 +
   data$R05 + data$R03, data = data, listw = ooe_W1, type = "mixed")


Residuals:
       Min         1Q     Median          3Q        Max
-0.9252634 -0.0698749  0.0014853  0.0638801  1.1744096

Type: mixed
Coefficients: (asymptotic standard errors)
                    Estimate  Std. Error z value  Pr(>|z|)
(Intercept)           0.07277795  0.04885996  1.4895  0.136350
data$ALTITUDE        -0.00035993  0.00013149 -2.7372  0.006196
data$R07              0.23172127  0.02345894  9.8777 < 2.2e-16
data$R05              0.22601075  0.02818171  8.0198 1.110e-15
data$R03              0.22843739  0.02740654  8.3351 < 2.2e-16
lag.data$ALTITUDE     0.00025899  0.00017339  1.4937  0.135243
lag.data$R07          0.00923654  0.05165748  0.1788  0.858092
lag.data$R05          0.22267410  0.08011565  2.7794  0.005446
lag.data$R03         -0.23495689  0.08636231 -2.7206  0.006516

Rho: 0.29162, LR test value: 19.153, p-value: 1.2063e-05
Asymptotic standard error: 0.07763
z-value: 3.7565, p-value: 0.00017229
Wald statistic: 14.112, p-value: 0.00017229

Log likelihood: 117.9214 for mixed model
ML residual variance (sigma squared): 0.034128, (sigma: 0.18474)
Number of observations: 445
Number of parameters estimated: 11
AIC: -213.84, (AIC for lm: -196.69)
LM test for residual autocorrelation
test value: 0.013289, p-value: 0.90822
```

The regression coefficients of the covariates and the spatial autoregressive parameter are still very similar to the estimates of the previous models, while the intercept and the spatial lags of the altitude and the ratio *R07* do not have a significant influence on the dependent variable. However, it is somewhat suspicious that the spatially lagged ratio *R03* has a negative effect on the dependent variable, which should be analyzed more carefully.

Note that the assumed spatial model often not only determines the behavior under the alternative, but can also govern the choice of the spatial dependence measure. Recently Li *et al.* (2007) have suggested the APLE statistics, given by

$$\mathfrak{I}_{APLE} = \frac{\hat{\boldsymbol{\varepsilon}}'\frac{1}{2}(\mathbf{W}+\mathbf{W}')\hat{\boldsymbol{\varepsilon}}}{\hat{\boldsymbol{\varepsilon}}'[\mathbf{W}'\mathbf{W}+\mathrm{tr}(\mathbf{W}^2)\mathbf{I}_n/n]\hat{\boldsymbol{\varepsilon}}}, \tag{8.14}$$

for a better reflection of dependence under a SL alternative. Asymptotic and exact distributions for the APLE (and its potential consequent use in the next chapter) were provided by Reder and Müller (2009) and Li *et al.* (2010). Also for SL models a theoretical comparison of $\mathfrak{I}$ and Lagrange multiplier tests is given in Baltagi and Yang (2010). A simple regression based formulation can be found in Born and Breitung (2011).

8.5 Design considerations

Let us consider the first case (see Section 8.4.1), where we estimate a model under the assumption of spatial independence, and the true model is of the same form. The aim is then not to reject the null hypothesis (spatial independence). For the approximate test we require the moments of Moran's $\mathfrak{I}$, which can be expressed in terms of the eigenvalues of the matrix $\mathbf{K}$ (Tiefelsdorf 2000), with $\mathbf{M}=\mathbf{I}_n-\mathbf{X}(\mathbf{X}'\mathbf{X})^{-1}\mathbf{X}'$ denoting the general projection matrix. Since only the moments are of interest, the evaluation of eigenvalues can be passed by making use of the trace operator tr. In this case under the assumption of spatial independence, expected value and variance of $\mathfrak{I}$ are then given by

$$E[\mathfrak{I}\mid H_0] = \frac{\mathrm{tr}(\mathbf{K})}{n-k} = \frac{\mathrm{tr}\{\mathbf{M}\frac{1}{2}(\mathbf{W}+\mathbf{W}')\mathbf{M}\}}{n-k} = \frac{\mathrm{tr}(\mathbf{MW})}{n-k} \tag{8.15}$$

and

$$\begin{aligned}\mathrm{Var}[\mathfrak{I}\mid H_0] &= \frac{\mathrm{tr}(\mathbf{MWMW}')+\mathrm{tr}(\mathbf{MW})^2+\{\mathrm{tr}(\mathbf{MW})\}^2}{(n-k)(n-k+2)} - \{E[\mathfrak{I}\mid H_0]\}^2\\ &= \frac{2\{(n-k)\mathrm{tr}(\mathbf{K}^2)-\mathrm{tr}(\mathbf{K})^2\}}{(n-k)^2(n-k+2)}\end{aligned} \tag{8.16}$$

respectively (Henshaw 1966).

An application of the theoretical moments of Moran's $\mathfrak{I}$ is the approximation of the exact distribution of Moran's $\mathfrak{I}$ by well-known simple distributions, that allow fast assessment of the significance of an observed Moran's $\mathfrak{I}$ without numerical evaluation of its exact probability. If the skewness and the kurtosis of Moran's $\mathfrak{I}$ (Tiefelsdorf, 2000) do not differ substantially from their counterparts of the normal distribution, the z-transformation of Moran's $\mathfrak{I}$ can be used to obtain the significance of an observed Moran's $\mathfrak{I}$. However, if there is a marked difference between the skewness and the kurtosis of Moran's $\mathfrak{I}$ to those of the normal distribution, alternative approximation strategies need to be employed.

The null case is the simpler one, there is no spatial effect in the data, data follow an ordinary linear model, the correct model is estimated and the null hypothesis of no spatial dependence should be retained. The intention is to find an optimal design which gives the best locations for the observations in the sense that the rejection of the null hypothesis is minimized.

Under the alternative, the (wrongly) estimated model is still: $\mathbf{y} = \mathbf{X}\boldsymbol{\beta} + \boldsymbol{\varepsilon}$ and $\boldsymbol{\varepsilon}$ i.i.d., but now the true assumed (but unknown) data generating process is, e.g., a SAR error process (8.6). The variance–covariance matrix $\boldsymbol{\Omega}(\rho)$ of the error terms is

$$\boldsymbol{\Omega}(\rho) = E[\mathbf{u}\mathbf{u}'] = \sigma^2[(\mathbf{I}_n - \rho\mathbf{W})'(\mathbf{I}_n - \rho\mathbf{W})]^{-1} \tag{8.17}$$

To ensure that $\boldsymbol{\Omega}(\rho)$ is positive definite, ρ is restricted to the interval $]\frac{1}{\lambda_{min}}; \frac{1}{\lambda_{max}}[$, where λ_{min} and λ_{max} denote the smallest and largest eigenvalues of $\mathbf{W}$. Note that we are using SAR without being restricted to it, being well aware of its peculiar problems that might effect design considerations as described in Martellosio (2011). In fact, it is only necessary to be able to compute $\boldsymbol{\Omega}$ in the following. For the use of a CAR alternative, see Müller and Waldl (2011).

The model is estimated via OLS and the residuals $\hat{\boldsymbol{\varepsilon}} = \mathbf{y} - \mathbf{X}\hat{\boldsymbol{\beta}}$ are used for the calculation of Moran's $\mathfrak{I}$. If the real data generating process follows a SAR error process, the aim is to reject the null hypothesis of no spatial dependence. The task is to maximize the power of the test, i.e., the probability to reject the null hypothesis given the alternative (spatial dependence). For the normal approximation again only the conditional moments are needed. The conditional expectation of Moran's $\mathfrak{I}$ (cf. Tiefelsdorf, 2000) can be evaluated by the improper integral

$$E[\mathfrak{I}|H_A] = \int_0^\infty \prod_{i=1}^{n-k}(1 + 2\lambda_i t)^{-\frac{1}{2}} \cdot \sum_{i=1}^{n-k} \frac{h_{ii}^*}{1 + 2\lambda_i t}\, dt \tag{8.18}$$

where h_{ii}^* are the diagonal elements of matrix $\mathbf{H} = \mathbf{P}'\mathbf{A}\mathbf{P}$ with $\mathbf{A} = \boldsymbol{\Omega}'^{\frac{1}{2}}\mathbf{M}\frac{1}{2}(\mathbf{W} + \mathbf{W}')\mathbf{M}\boldsymbol{\Omega}^{\frac{1}{2}}$ and $\mathbf{P}$ is the matrix of the normalized eigenvectors of matrix $\mathbf{B} = \boldsymbol{\Omega}'^{\frac{1}{2}}\mathbf{M}\boldsymbol{\Omega}^{\frac{1}{2}}$. The eigenvalues and their associated eigenvectors are resequenced so that $0 < \lambda_1 \leq \lambda_2 \leq \ldots \leq \lambda_{n-k}$. The variance of $\mathfrak{I}$ under the alternative is given by

$$\mathrm{Var}[\mathfrak{I}|H_A] = E[\mathfrak{I}^2|H_A] - E[\mathfrak{I}|H_A]^2 \tag{8.19}$$

where

$$E[\mathfrak{I}^2|H_A] = \int_0^\infty \left[\prod_{i=1}^{n-k}(1+2\lambda_i t)^{-\frac{1}{2}}\right] \cdot \left[\sum_{i=1}^{n-k}\sum_{j=1}^{n-k} \frac{h_{ii}^* h_{jj}^* + 2(h_{ij}^*)^2}{(1+2\lambda_i t)(1+2\lambda_j t)}\right] t\, dt$$

and $E[\mathfrak{I}|H_A]$ is given in Equation (8.18). The upper truncation points for the integrals can be approximated by a formula from De Gooijer (1980). Following this, we obtain an approximation of the upper bound for the expected value (8.18) of

$$\left[\frac{(n-k)h_{max}}{2\lambda_1^{\frac{n-k}{2}}}\left(\frac{n-k}{2}-1\right)\frac{1}{\epsilon}\right]^{\frac{1}{\frac{n-k}{2}-1}} = \tau_1 \tag{8.20}$$

where h_{max} is the biggest absolute value of the elements of the diagonal of matrix $\mathbf{H}$ and ϵ is a given positive small number less than 1. An approximation of the upper bound for $E[\mathfrak{I}^2|H_A]$ is

$$\left[\frac{3(n-k)^2 h_{max}^{(2)}}{(2\lambda_1)^{\frac{n-k}{2}}}\left(\frac{n-k}{2}-2\right)\frac{1}{\epsilon}\right]^{\frac{1}{\frac{n-k}{2}-2}} = \tau_2 \tag{8.21}$$

with $h_{max}^{(2)}$ denoting the biggest absolute value of the elements of matrix $\mathbf{H}$. Tiefelsdorf (2000) suggests to use $\frac{1}{n-k}\sum_{i=1}^{n-k}\lambda_i$ instead of λ_1. For more details and an implementation of the above in R, see Bivand *et al.* (2009).

8.5.1 A design criterion

In both cases, where a linear regression model is estimated and the corresponding residuals are used to calculate Moran's $\mathfrak{I}$ test, the test result, whether to reject or not reject the null hypothesis of no spatial autocorrelation in the error term, depends on the true data generating process. As the true process is unknown, a general design criterion $\mathcal{J}$ (which does not depend on the knowledge of the true data generating process) is needed. The aim is to minimize the probability that, given the alternative, the Moran's $\mathfrak{I}$ test does not reject the null hypothesis of no spatial autocorrelation. The test statistic $z(\mathfrak{I}) = \frac{\mathfrak{I}-E(\mathfrak{I}|H_0)}{\sqrt{\mathrm{Var}(\mathfrak{I}|H_0)}}$ is asymptotically normally distributed and therefore we minimize:

$$\min_{H_A} P\left(\frac{\mathfrak{I}-E(\mathfrak{I}|H_0)}{\sqrt{\mathrm{Var}(\mathfrak{I}|H_0)}} \le \Phi^{-1}(1-\alpha)\right),$$

where Φ denotes the cumulative distribution function of the standard normal distribution. This leads to

$$\min_{H_A} P\left(\mathfrak{I} \le \Phi^{-1}(1-\alpha)\sqrt{\mathrm{Var}(\mathfrak{I}|H_0)} + E(\mathfrak{I}|H_0)\right)$$

Using the z-transformation for $\mathfrak{I}$ under the alternative gives $\frac{\mathfrak{I}-E[\mathfrak{I}|H_A]}{\sqrt{\text{Var}[\mathfrak{I}|H_A]}}$, which is also asymptotically standard normally distributed. The final criterion to be maximized is therefore given by

$$\mathcal{J}_{\mathfrak{I}}(\xi) = 1 - \Phi\left(\frac{\Phi^{-1}(1-\alpha)\sqrt{\text{Var}[\mathfrak{I}|H_0]} + E[\mathfrak{I}|H_0] - E[\mathfrak{I}|H_A]}{\sqrt{\text{Var}[\mathfrak{I}|H_A]}}\right). \tag{8.22}$$

The maximization of $\mathcal{J}_{\mathfrak{I}}$ over $S_\xi \in \mathcal{X}$ gives the final optimal locations for the observation sites and thus maximizes the power of the Moran's $\mathfrak{I}$ test. To calculate $\mathcal{J}_{\mathfrak{I}}$, the expected value (8.15) and the variance (8.16) of $\mathfrak{I}$ under the null hypothesis, and the expected value (8.18) and the variance (8.19) of $\mathfrak{I}$ under the alternative hypothesis are needed. Unfortunately, the given criterion is not convex and thus we can not employ the sort of equivalence theorems from the well developed optimum design theory (cf. Atkinson *et al.* 2007) but must resort to alternative algorithmic approaches, as given below. This criterion was first suggested by Gumprecht *et al.* (2009) and later extended by Müller *et al.* (2012) for the exact distribution. They show that optimizing the design serves as a regulatory device for the validity of the normal approximation, so it is sufficient to consider the approximation in what follows.

Evidently, the global optimal design can be found by evaluating all possible designs, i.e., in an m-point grid there are $\binom{m}{r}$ possible r-point designs, r goes from $4 + k + 1$ to m, where k is the number of the regressors in the model. This minimum number of points in a design follows from the approximation of the upper truncation points for the integrals (8.20) and (8.21). The number of possible designs increases very fast with the size of the grid. This leads to a high runtime, as the numerical integration needs a considerable amount of time. From this point of view it is worth noticing that not all possible designs are different in the sense that they have different criterion values. Some of the r-point designs are only rotations, reflections or translations of other r-point designs, and therefore give the same value of the criterion $\mathcal{J}_{\mathfrak{I}}$; let us call the respective designs 'symmetric'. To avoid calculating $\mathcal{J}_{\mathfrak{I}}$ for those designs which are known to be symmetric to others, an appropriate symmetry check can be performed before the computation of $\mathcal{J}_{\mathfrak{I}}$ (Gumprecht *et al.* 2009).

For our asymmetric set-up, however, there is no hope for that because on our 445-point lattice there are (440-k)! different potential designs and only very few, if any, of them can be considered symmetric. Thus, the complete evaluation of all truly different designs is mostly impractical and can only be performed for very small designs.

Gumprecht *et al.* (2009) suggest a simple search algorithm for finding a 'nearly' optimal design. This algorithm is much faster than the full enumeration algorithm as for the r-point design the number of evaluated $(r-1)$-point designs is r. This algorithm can also be performed in an acceptable amount of time for relatively large grids. The procedure is quite simple:

1. Start with an initial design ξ_0 with $S_{\xi_0} = \mathcal{X}$, called the 'base' design. Thus in the first iteration the number of points r in ξ_0 is m.

2. Delete each point, one at a time, to get $(r - 1)$ designs ξ_e, and compute $\mathcal{J}_{\mathfrak{J}e}$. The symmetries can be checked before the criterion is calculated.

3. Take the best $(r - 1)$ design ξ_e, i.e., the design with the largest $\mathcal{J}_{\mathfrak{J}e}$, and put it as new base design.
   Go to Step 2.

The algorithm stops if $r = (4 + k + 1)$. The r-point design that gives the largest $\mathcal{J}_{\mathfrak{J}}$ is the 'nearly' optimal one. Note the similarities to the 'coffee-house' procedure given in Müller (2007): the disadvantage of these algorithms is, that once a r-point design is chosen, all smaller $r - i$ point designs are restricted to this set of points. As a result it can happen quite easily that the algorithm is trapped in a local maximum. To avoid this one could alternatively employ methods of stochastic optimization such as in Haines (1987) or more recently Ver Hoef (2012).

Computation of all designs in this section is again what is known as a NP-hard problem and a rigorous search would be prohibitive. Besides, the criteria are not convex and typically exhibit multiple local optima and therefore most of the cited papers employ simulated annealing algorithms (cf. van Groenigen and Stein 1998). However, by using simple exchange type procedures (cf. Royle 2002) considerable gains in the criteria can be achieved within a few steps, as can be seen from the respective examples in the next section.

8.5.2 Example

In the example of the Upper Austrian municipalities it would clearly be too demanding to search through the whole 445-item grid of available locations. We employ the spatial link matrix **G** implied by Figure 8.3. In the search- and exchange algorithms the corresponding row-standardized spatial weight matrices **W** are used. The regression model of the optimal design procedure is an intercept only model. As an exemplary value of the spatial autoregressive parameter $\rho = 0.28677$ (as estimated from the SL model) was used, which is needed for evaluating expressions under the alternative hypothesis (other values give qualitatively similar results though differing designs). Using this parameter and sequential elimination (simple search) leads to the optimal criteria values (maximum power of the test) for the respective n-point designs displayed in Figure 8.6. From this graph it can be seen that the best is the 122-point design with $\mathcal{J}_{\mathfrak{J}} = 0.996311$. The strong decay for larger numbers of observations is another instance of the power-trap described in Krämer (2005) and Martellosio (2010). One should note the following aspects about the optimization:

- In the first part of the optimization procedure there seem to be a lot of problems in the computations, obtaining several NaNs and oscillations in

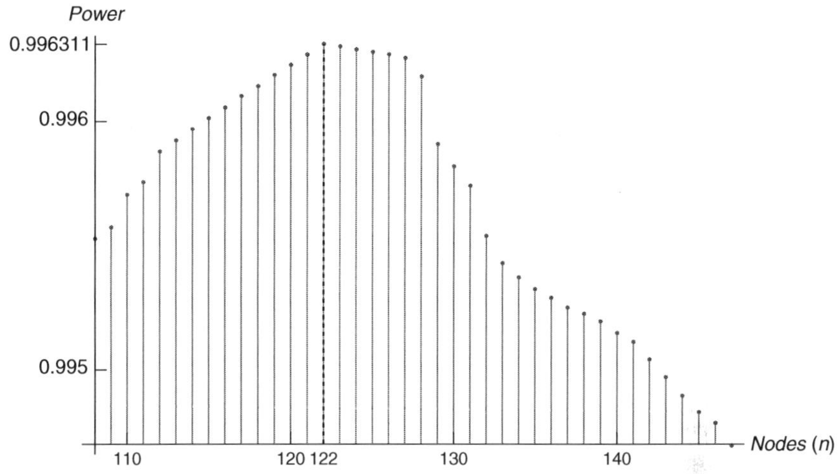

Figure 8.6 Maximum powers of Moran's $\mathfrak{I}$ test for various design sizes n.

powers. Nobody can be sure that the selection of the nodes to remove is the optimal one (another reason to say that the final design obtained is quasi-optimal).

- Only after reaching the 381-point design the power begins to increase steadily, getting the maximum value for a 122-point design.
- The 108-point design has all the nodes connected in pairs. After reaching this design, zero powers begin to appear when one of the nodes of a pair is removed. From that point on, the results of selecting the best node to remove are no longer very reliable (in most cases the first node of the remaining ones is removed).

Note that we assume here the number of sampling sites to be freely chosen; some considerations on the effectiveness can be found in Griffith (2005, 2008). Thus a corresponding 122-point design can be selected exclusively on the basis of the criterion, which is displayed in Figure 8.7. We can observe that this design only consists of connected couples, triples and at maximum quadruples.

8.6 Discussion

To conclude we have to admit that there are several different methods to fit a regression model to areal data, but it is not obvious up to now which one of the presented methods is optimal for the given dataset, nor are the implications on a proper design procedure straightforward. Some questions–which go beyond the scope of this chapter–still remain unsettled:

- Should we prefer the linear regression model, where we face the problem of collinearity, but eliminate the spatial effect in the data?

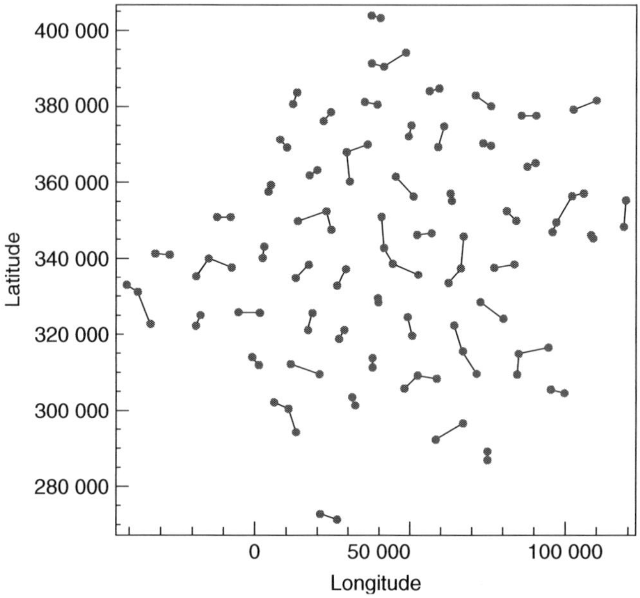

Figure 8.7 A quasi-optimum design for detecting spatial dependence.

- Do we have the case of spurious spatial regression in the given data? How can we test for spatial nonstationarity?
- If it is feasible to fit a spatial regression model, which one should we use in this context?

To answer these questions further work has to be done and the mentioned topics must be examined in detail.

Note that the techniques given in this section have also some impact on methods for spatial filtering, where the main idea is to separate the regional interdependencies by partitioning the original variable into a filtered nonspatial (so-called 'spaceless') variable and a residual spatial variable. Afterwards conventional statistical techniques that are based on the assumption of spatially uncorrelated errors can be used for the filtered part. One of the most common filters is based on an eigenfunction decomposition related to Moran's $\mathfrak{I}$ (cf. Getis and Griffith 2002), and thus may be improved by appropriately selecting supporting sites.

Once spatial dependence in the data is detected the so-called variogram plays a central role in the analysis of spatial data. A valid variogram model is selected and the parameters of this model are estimated before kriging (spatial prediction) is performed. These inference procedures are generally based on the examination of the empirical variogram (Figure 8.8), which consists of average squared differences of data taken at sites lagged the same distance apart in the same direction. The ability of an investigator to estimate the parameters of a variogram model

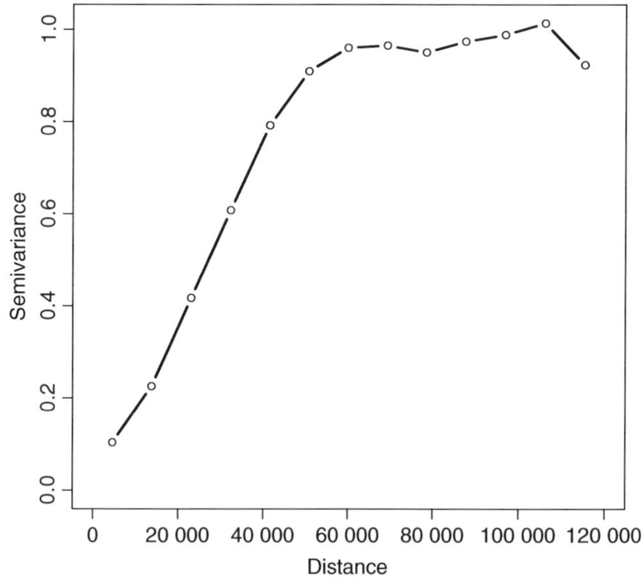

Figure 8.8 Variogram for the grassland dataset.

efficiently is again significantly affected by the sampling design, i.e., the locations of the sites $x_1, \ldots, x_n \in \mathcal{X}$ where data $\mathbf{y}$ are observed. For the relationship of variograms to g-ratios and its implications see Bellehumeur and Legendre (1998).

In Müller and Zimmerman (1999) several practical approaches for constructing sampling designs for estimation of the variogram were compared by Monte Carlo simulations. Those designs could be adopted at the early stages of a sampling program until the variogram is sufficiently well-estimated, after which one could shift to an existing approach that emphasizes prediction. Alternatively, the two objectives could be combined in a compound design, as suggested in Müller and Stehlík (2010).

Instead of directly going after the variograms a number of alternative methods were suggested which allow correlations to be ignored. A two-stage strategy for instance was suggested by Müller and Zimmerman (1995):

(a) Find the optimal configuration of distances $\xi^*_{\mathcal{L}}$ in the space spanned by all possible point pair distances (the so-called lag space $\mathcal{L}$);

(b) and map this configuration into the original site space $\mathcal{X}$ (find a site space design).

Finding the solution of (b) was also the purpose of Warrick and Myers (1987). The usefulness of such a distance algorithm can only be assessed in two ways: via simulation, or by comparing it with a technique that directly employs ideas for optimum designs for correlated observations. First results in this direction can be found in Müller and Zimmerman (1999). They conclude that algorithms based on ignoring correlations are much quicker but marginally less efficient with respect to a design criterion than the augmentation procedures from the previous section.

It is traditional practice that the covariance/variogram parameters are estimated in a separate stage. However, if one is willing to make distributional assumptions, it is natural to employ likelihood based estimation techniques. In the following we will thus assume that the errors in our spatial model follow a stationary Gaussian process.

In particular one could now assume that the trend is known and fixed and maximize the log likelihood

$$2L(\boldsymbol{\theta}) = -n\log(2\pi) - \log\det \mathbf{C}(\boldsymbol{\theta}) - \mathbf{y}'\mathbf{C}^{-1}(\boldsymbol{\theta})\mathbf{y}. \tag{8.23}$$

It is now natural to base a design criterion on the information matrix associated with the corresponding estimate of the parameter $\boldsymbol{\theta}$, which is given by (note that it depends upon a design ξ via $\mathbf{C}$)

$$\mathbf{M}''(\xi, \boldsymbol{\theta})_{jj'} = \frac{1}{2}\mathrm{tr}\left\{\mathbf{C}^{-1}(\boldsymbol{\theta})\frac{\partial \mathbf{C}(\boldsymbol{\theta})}{\partial \theta_j}\mathbf{C}^{-1}(\boldsymbol{\theta})\frac{\partial \mathbf{C}(\boldsymbol{\theta})}{\partial \theta_{j'}}\right\}, \tag{8.24}$$

where the $\frac{\partial \mathbf{C}(\boldsymbol{\theta})}{\partial \theta_j}$ are $n \times n$ matrices with entries $\frac{\partial c(\mathbf{x},\mathbf{x}';\boldsymbol{\theta})}{\partial \theta_j}$, $\mathbf{x}, \mathbf{x}' \in \xi$.

Designs maximizing the determinant of $\mathbf{M}''$ have been suggested by Zhu and Stein (2006) (they also employ a minimax and Bayesian criterion to avoid undesirable effects due to the linearizations) and Zimmerman (2006), who calls them CP (covariance parameter) optimal. Both demonstrate the behavior of the criteria for numerous artificial and real examples. A related discussion is found in Chapter 2; for a Bayesian adaptive approach, see Marchant and Lark (2006).

Appendix 8.1: R code

```
# Packages
require(maptools)
require(maps)
require(spdep)
require(RColorBrewer)
require(pgirmess)
require(HH)
require(lmtest)
require(sandwich)

# Map of Upper Austria
data=data.frame(read.csv("grasslandmunratio_cs.csv",header=TRUE,
  sep=";",dec=".")[,1:12])
years=names(data[,7:12])
row.names(data)=data[,3]
```

```
getinfo.shape("ooegemeinden.shp")
ooe=readShapePoly("ooegemeinden.shp", IDvar="LBBGG")
plot(ooe, border="blue", axes=TRUE, las=1)
grassland=SpatialPolygonsDataFrame(ooe,data=data)

# Video
dir.create("moviegrassland")
max(data[,7:12])
min(data[,7:12])
at_green=pretty(as.vector(unlist(data[,7:12])),n=10)
cols_green=colorRampPalette(rev(brewer.pal(10,"RdYlBu")))
  (length(at_green)-1)   # rev() re-sorts
for(i in years){
    png(file=paste("moviegrassland/pic",i,".png",sep=""),
      width=960,height=600)
    print(spplot(grassland,i,col.regions=cols_green,
      at=at_green, main=paste("log Ratio: area of arable
      land and area of grassland",i,sep=" ")))
    dev.off()
}
# Plot of log ratios R95 to R08
pdf("Plots.pdf")
at_green=pretty(as.vector(unlist(data[,7:12])),n=9)
cols_green=colorRampPalette(brewer.pal(9,"Greens"))(length(at_green)-1)
spplot(grassland,years,col.regions=cols_green,at=at_green,as.table=T)
dev.off()

# Influence of Altitude
pdf("Altitude.pdf")
at_altitude=pretty(as.vector(unlist(data[,6])),n=11)
cols_altitude=colorRampPalette(brewer.pal(9,"YlOrBr"))(length(at_altitude)-1)
altitude=names(data[6])
spplot(grassland,altitude,col.regions=cols_altitude,at=at_altitude,as.table=T)
dev.off()

# Creating Neighbours

pdf("Neighbours.pdf")
par(mfrow=c(1,2))
ooe_nb1=poly2nb(grassland)          # Queen-style contiguities
plot(grassland,border="grey60")
plot(ooe_nb1,coordinates(grassland),pch=19,cex=0.6,add=TRUE)
title(main="Queen-style contiguities")

ooe_nb2=poly2nb(grassland, queen=FALSE)      # Rook-style contiguities
plot(grassland,border="grey60")
plot(ooe_nb2,coordinates(grassland),pch=19,cex=0.6,add=TRUE)
title(main="Rook-style contiguities")

coords=coordinates(grassland)
IDs=row.names(as(grassland,"data.frame"))
ooe_nb3=tri2nb(coords,row.names=IDs)         # Delauney triangulation
plot(grassland,border="grey60")
```

```
plot(ooe_nb3,coordinates(grassland),pch=19,cex=0.6,add=TRUE)
title(main="Delauney triangulation neighbours")

ooe_nb4=graph2nb(gabrielneigh(coords),row.names=IDs)          # Gabriel graph
plot(grassland,border="grey60")
plot(ooe_nb4,coordinates(grassland),pch=19,cex=0.6,add=TRUE)
title(main="Gabriel graph neighbours")

ooe_nb5=knn2nb(knearneigh(coords,k=1),row.names=IDs)          # k=1 neighbors
plot(grassland,border="grey60")
plot(ooe_nb5,coordinates(grassland),pch=19,cex=0.6,add=TRUE)
title(main="All areas k=1 neighbours")
dev.off()

pdf("Distbased.pdf")
dists=unlist(nbdists(ooe_nb5,coords))
summary(dists)
max_dist=max(dists)
# Distance based neighbors within 1*max_dist
ooe_nb6=dnearneigh(coords,d1=0,d2=1*max_dist,row.names=IDs)
is.symmetric.nb(ooe_nb6)
summary(ooe_nb6)
plot(grassland,border="grey60")
plot(ooe_nb6,coordinates(grassland),pch=19,cex=0.6,add=TRUE)
#title(main="Distance based neighbours within 1*max_dist")
dev.off()

# Spatial Weights

# available styles: W, B, C, U
# error when given an nb argument with areas with no neighbors = default

# row standardized weights matrix --> sums of weights in each row = 1
ooe_W1=nb2listw(ooe_nb6,style="W")

# weight of unity for each neighbor relationship
ooe_W2=nb2listw(ooe_nb6,style="B")

# equal weights for all links --> complete set of weights sums to number
  of areas
ooe_W3=nb2listw(ooe_nb6,style="C")

# equal weights for all links --> complete set of weights sums to 1
ooe_W4=nb2listw(ooe_nb6,style="U")

# Connectivity Matrix/Spatial Lag Matrix
W1=listw2mat(ooe_W1)
W2=listw2mat(ooe_W2)
W3=listw2mat(ooe_W3)
W4=listw2mat(ooe_W4)

# Spatial autocorrelation tests

# Moran's I
```

```
moran=moran.test(data$R08,listw=ooe_W1)     # randomisation=F
moran
moran1=moran.test(data$R08,listw=ooe_W1,randomisation=T)
moran1

# Spatial Correlogram
correlo=sp.correlogram(neighbours=ooe_nb6,var=data$R08,order=16,method="I",
          style="W",zero.policy=TRUE)
correlo
dist_correlo=correlog(coords,data$R08,method="Moran")
dist_correlo

pdf("Correlogram R08.pdf")
plot(correlo)
plot(dist_correlo)
# Moran Scatterplot (different for different spatial weight styles)
moran.plot(data$R08,listw=ooe_W1)
title(main="Moran Scatterplot")
dev.off()

# Linear Regression Models

# Scatterplot-Matrix (covariates)
pairs(data[,5:11])
cor(data[,5:11])

# Linear regression model
linreg=lm(data$R08~data$ALTITUDE+data$R07+data$R05+data$R03+data$R99+data$R95)
linreg=lm(data$R08~data$ALTITUDE+data$R07+data$R05+data$R03)
summary(linreg)

# Collinearity - Variance inflation factor
vif(linreg)
# Plot of the residuals
grassland$lmresid=residuals(linreg)
at_res=pretty(as.vector(unlist(grassland$lmresid)),n=9)
cols_res=colorRampPalette(brewer.pal(9,"Greens"))(length(at_res)-1)
spplot(grassland,"lmresid",col.regions=cols_res,at=at_res,as.table=T)

# Moran's I test for residuals
moran_res=moran.test(grassland$lmresid,listw=ooe_W1,randomisation=F)
# under randomisation/under normality --> randomisation=T/F
moran_res

lm.morantest(linreg,ooe_W1)

# Spatial regression models

# SAR
sar=spautolm(data$R08~data$ALTITUDE+data$R07+data$R05+data$R03,
     data=data,listw=ooe_W1)
summary(sar)

# CAR
```

```
car=spautolm(data$R08~data$ALTITUDE+data$R07+data$R05+data$R03,
     data=data,family="CAR",listw=ooe_W1)
summary(car)

# SMA
sma=spautolm(data$R08~data$ALTITUDE+data$R07+data$R05+data$R03,
     data=data,family="SMA",listw=ooe_W1)
summary(sma)

# Spatial lag model
lag=lagsarlm(data$R08~data$ALTITUDE+data$R07+data$R05+data$R03,
     data=data,listw=ooe_W1)
summary(lag)

# Plot of the residuals
grassland$lagresid=residuals(lag)
at_res1=pretty(as.vector(unlist(grassland$lagresid)),n=9)
cols_res1=colorRampPalette(brewer.pal(9,"Greens"))(length(at_res1)-1)
spplot(grassland,"lmresid",col.regions=cols_res1,at=at_res1,as.table=T)

# Spatial Durbin model (spatially lagged explanatory variables)
mix=lagsarlm(data$R08~data$ALTITUDE+data$R07+data$R05+data$R03,
     data=data,listw=ooe_W1,type="mixed")
summary(mix)
anova(lag,mix)      # AIC

# Spatial error model
er=errorsarlm(data$R08~data$ALTITUDE+data$R07+data$R05+data$R03,
     data=data,listw=ooe_W1)
summary(er)

# Alternatives
# two stage least squares procedure
stsls=stsls(data$R08~data$ALTITUDE+data$R07+data$R05+data$R03,data=data,
       listw=ooe_W1)
summary(stsls)
stslsR=stsls(data$R08~data$ALTITUDE+data$R07+data$R05+data$R03,data=data,
       listw=ooe_W1,robust=TRUE)
summary(stslsR)
#Generalized Moments estimator
GMerr=GMerrorsar(data$R08~data$ALTITUDE+data$R07+data$R05+data$R03,data=data,
       listw=ooe_W1)
summary(GMerr)
```

Acknowledgments

We are grateful to an attentive referee, whose comments led to an improvement of this contribution. Some work on this contribution was financed by Acciones Integradas 2008-2009 (Project No. ES 18/2008).

References

Anselin, L 1988 *Spatial Econometrics: Methods and Models*. Kluwer Academic Publishers, Dordrecht.

Anselin, L 1993 The Moran Scatterplot as an ESDA Tool to Assess Local Instability in Spatial Association. In *Spatial Analytical Perspectives on GIS* (eds Fischer MM, Scholten HJ and Unwin D). Taylor and Francis, London, pp. 111–125.

Anselin, L 1995 Local Indicators of Spatial Association – LISA. *Geographical Analysis*, **27**, 93–115.

Anselin L 2007 Spatial econometrics. In *Palgrave Handbook of Econometrics*: Volume 1: Econometric Theory (eds Patterson K and Mills TC). Palgrave Macmillan, New York, pp. 310–330.

Arbia, G 2006 *Spatial Econometrics: Statistical Foundations and Applications to Regional Convergence*. Springer Verlag, Berlin.

Atkinson A, Donev A and Tobias R 2007 *Optimum Experimental Designs, with SAS (Oxford Statistical Science Series)*. Oxford University Press, Oxford.

Baltagi BH and Yang Z 2010 Standardized LM Tests for Spatial Error Dependence in Linear or Panel Regressions. Technical Report 11-2010, Singapore Management University, School of Economics.

Banerjee S, Carlin BP and Gelfand AE 2004 *Hierarchical Modeling and Analysis for Spatial Data*. Chapman & Hall, London.

Beenstock M and Felsenstein D 2012 Testing for unit roots and cointegration in cross section data. *Spatial Economic Analysis* (in press).

Bellehumeur C and Legendre P 1998 Multiscale sources of variation in ecological variables: modeling spatial dispersion, elaborating sampling designs. *Landscape Ecology* **13**(1), 15–25.

Bivand RS, Müller WG and Reder M 2009 Power Calculations for Global and Local Moran's I. *Computational Statistics & Data Analysis* **53**, 2859–2872.

Bivand RS, Pebesma EJ and Gómez-Rubio V 2008 *Applied Spatial Data Analysis with R*. Springer, New York.

Bivand RS *et al.* 2010 *spdep: Spatial dependence: weighting schemes, statistics and models*. R package version 0.5-26. http://CRAN.R-project.org/package = spdep.

Born B and Breitung J 2011 Simple regression-based tests for spatial dependence. *The Econometrics Journal* **14**(2), 330–342.

Brewer CA and Harrower M 2009 *ColorBrewer 2.0*. http://www.colorbrewer2.org, [accessed: January 25, 2012].

Cliff AD and Ord JK 1981 *Spatial Processes. Models & Applications*. Pion Limited, London.

Dale MRT, Dixon P, Fortin MJ, Legendre P, Myers DE and Rosenberg MS 2002 Conceptual and mathematical relationships among methods for spatial analysis. *Ecography* **25**(5), 558–577.

De Gooijer JG 1980 Exact moments of the sample autocorrelations from series generated by general ARIMA processes of order (p,d,q), d = 0 or 1. *Journal of Econometrics*, **14**, 365–379.

Dray S 2011 A new perspective about Moran's coefficient: spatial autocorrelation as a linear regression problem. *Geographical Analysis* **43**(2), 127–141.

Dubé J and Legros D 2011 A spatio-temporal measure of spatial dependence: an example using real estate data. *Papers in Regional Science*, doi: 10.1111/ j.1435-5957. 2011.00402.x.

Fahrmeir L, Kneib T and Lang S 2009 *Regression. Modelle, Methoden und Anwendungen*. Springer, Berlin.

Fingleton B 1999 Spurious spatial regression: some Monte Carlo results with a spatial unit root and spatial cointegration. *Journal of Regional Science* **39**, 1–19.

Fortin M-J and Dale M 2005 *Spatial Analysis: A Guide for Ecologists*. Cambridge University Press, Cambridge.

Geary RC 1954 The Contiguity Ratio and Statistical Mapping. *The Incorporated Statistician*, **5**, 115–127, 129–146.

Getis A and Griffith D 2002 Comparative spatial filtering in regression analysis. *Geographical Analysis* **34**, 130–140.

Gibbons S and Overman HG 2012 Mostly pointless spatial econometrics? *Journal of Regional Science* **52**, 172–191.

Griffith DA 2005 Effective geographic sample size in the presence of spatial autocorrelation. *Annals of the Association of American Geographers* **95**(4), 740–760.

Griffith DA 2008 Geographic sampling of urban soils for contaminant mapping: how many samples and from where. *Environmental Geochemistry and Health* **30**(6), 495–509.

Griffith DA 2010 The Moran coefficient for non-normal data. *Journal of Statistical Planning and Inference* **140**(11), 2980–2990.

Gumprecht D 2007 Treatment of far-off objects in Moran's $\mathfrak{I}$ test. *Research Report Series, Vienna University of Economics and Business Administration* **46**.

Gumprecht D, Müller WG and Rodríguez-Díaz JM 2009 Designs for detecting spatial dependence. *Geographical Analysis* **41**(2), 127–143.

Haines LM 1987 The application of the annealing algorithm to the construction of exact optimal designs for linear regression models. *Technometrics* **29**, 439–447.

Haining RP 1993 *Spatial Data Analysis in the Social and Environmental Sciences*. Cambridge University Press, Cambridge.

Henshaw RC 1966 Testing single-equation least squares regression models for autocorrelated disturbances. *Econometrica* **34**, 646–660.

Hepple LW 1998 Exact testing for spatial autocorrelation among regression residuals. *Environment and Planning A* **30**(1), 85–108.

Kissling WD and Carl G 2008 Spatial autocorrelation and the selection of simultaneous autoregressive models. *Global Ecology and Biogeography* **17**(1), 59–71.

Krämer W 2005 Finite sample power of Cliff-Ord-type tests for spatial disturbance correlation in linear regression. *Journal of Statistical Planning and Inference* **128**(2), 489–496.

Lauridsen J and Kosfeld R 2006 Spurious Spatial Regression, Spatial Cointegration and Heteroscedasticity. *Working Paper, Southern University of Denmark, Odense*.

LeSage J and Pace RK 2009 *Introduction to Spatial Econometrics*. Chapman & Hall/CRC, Boca Raton, FL.

Li H, Calder CA and Cressie N 2007 Beyond Moran's I: testing for spatial dependence based on the spatial autoregressive model. *Geographical Analysis* **39**(4), 357–375.

Li H, Calder CA and Cressie N 2010 One-step estimation of spatial dependence parameters: Properties and extensions of the APLE statistic. Technical Report 846, Department of Statistics, The Ohio State University.

López FA, Matilla-García M, Mur J and Marín MR 2011 Four tests of independence in spatiotemporal data. *Papers in Regional Science* **90**(3), 663–685.

Marchant BP and Lark RM 2006 Adaptive sampling and reconnaissance surveys for geostatistical mapping of the soil. *European Journal of Soil Science* **57**(6), 831–845.

Martellosio F 2010 Power properties of invariant tests for spatial autocorrelation in linear regression. *Econometric Theory* **26**(01), 152–186.

Martellosio F 2011 Nontestability of equal weights spatial dependence. *Econometric Theory* **27**(06), 1369–1375.

Moran PAP 1950 A test for the serial dependence of residuals. *Biometrika* **37**, 178–181.

Müller WG 2007 *Collecting Spatial Data: Optimum Design of Experiments for Random Fields* 3rd rev. and extended edn. Springer, Heidelberg.

Müller WG and Stehlík M 2010 Compound optimal spatial designs. *Environmetrics* **21**(3–4), 354–364.

Müller WG and Waldl H 2011 Discussion of Lindgren, Rue and Lindström 'An explicit link between Gaussian fields and Gaussian Markov random fields: the stochastic partial differential equation approach'. *Journal of the Royal Statistical Society - Series B* **73**(4), 483–485.

Müller WG and Zimmerman DL 1995 An algorithm for sampling optimization for semivariogram estimation. In *Model-Oriented Data Analysis 4* (eds Kitsos CP and Müller WG). Physica, Heidelberg, pp. 173–178.

Müller WG and Zimmerman DL 1999 Optimal design for variogram estimation. *Environmetrics* **10**, 23–37.

Müller WG, Rodríguez-Díaz JM and Rivas López MJ 2012 Optimal design for detecting dependencies with an application in spatial ecology. *Environmetrics* **23**(1), 37–45.

O'Sullivan D and Unwin DJ 2003 *Geographic Information Analysis*. John Wiley & Sons, Ltd, Hoboken, NJ.

Rangel TF, Diniz-Filho JA and Bini LM 2010 SAM: a comprehensive application for Spatial Analysis in Macroecology. *Ecography* **33**(1), 46–50.

Reder M and Müller W 2009 More on the APLE statistic. *Mathematica Slovaca* **59**(5), 565–578.

Royle JA 2002 Exchange algorithms for constructing large spatial designs. *Journal of Statistical Planning and Inference* **100**(2), 121–134.

Rue H and Held L 2005 *Gaussian Markov Random Fields: Theory and Applications (Chapman & Hall/CRC Monographs on Statistics & Applied Probability)* Chapman and Hall/CRC, Boca Raton, FL.

Schabenberger O and Gotway CA 2005 *Statistical Methods for Spatial Data Analysis*. Chapman & Hall, London.

Tiefelsdorf M 2000 *Modelling Spatial Processes: the Identification and Analysis of Spatial Relationships in Regression Residuals by Means of Moran's I (Lecture Notes in Earth Sciences)*. Springer, Berlin.

Tiefelsdorf M and Boots B 1995 The exact distribution of Moran's I. *Environment and Planning A* **27**(6), 985–999.

van Groenigen JW and Stein A 1998 Constrained optimization of spatial sampling using continuous simulated annealing. *Journal of Environment Quality* **27**(5), 1078–1086.

Ver Hoef JM 2012 Practical considerations for experimental designs of spatially autocorrelated data using computer intensive methods. *Statistical Methodology* **9**, 172–184.

Waller LA and Gotway CA 2004 *Applied Spatial Statistics for Public Health Data*. John Wiley & Sons, Ltd, New York.

Warrick AW and Myers DE 1987 Optimization of sampling locations for variogram calculations. *Water Resources Research* **23**, 496–500.

Zhu Z and Stein ML 2006 Spatial sampling design for prediction with estimated parameters. *Journal of Agricultural, Biological, and Environmental Statistics* **11**(1), 24–44.

Zimmerman DL 2006 Optimal network design for spatial prediction, covariance parameter estimation, and empirical prediction. *Environmetrics* **17**(6), 635–652.

9

Sampling design optimization for space-time kriging

Gerard B.M. Heuvelink[1], Daniel A. Griffith[2], Tomislav Hengl[3] and Stephanie J. Melles[4]

[1]*Department of Environmental Sciences, Wageningen University, The Netherlands*
[2]*School of Economic, Political and Policy Sciences, University of Texas at Dallas, USA*
[3]*ISRIC – World Soil Information, Wageningen, The Netherlands*
[4]*Biology Department, Trent University, Ontario, Canada*

9.1 Introduction

Space-time geostatistics has received increasing attention over the past few decades, both on the theoretical side with publications that extend the set of valid space-time covariance structures (Ma 2003; Fuentes *et al.* 2008; Fonseca and Steel 2011), and on a practical side with publications of real-world applications (Gething *et al.* 2007; Heuvelink and Griffith 2010; see citations in Cressie and Wikle 2011). Public domain software solutions to space-time kriging are becoming more user-friendly (Yu *et al.* 2007; Spadavecchia 2008; Cameletti 2012; Pebesma 2012; Rue 2012), although not as much as is the case with spatial kriging. These developments, combined with almost all environmental

Spatio-temporal design: Advances in efficient data acquisition, First Edition.
Edited by Jorge Mateu and Werner G. Müller.

variables varying in both space and time, indicate that the use of space-time kriging will likely increase dramatically in the near future.

One topic that to our knowledge has not yet been addressed in space-time geostatistics concerns the choice of the number and configuration of observation points in space and in time. Sampling design optimization in space has been well covered in the geostatistical literature (Angulo *et al.* 2005; Xia *et al.* 2006; Brus and Heuvelink 2007; Müller 2007; Müller and Stehlík 2010). From a methodological point of view, no principal difficulty should prevent an extension of these approaches to the space-time domain, and existing techniques may be used. However, there are important differences too. In space-time kriging, often strong anisotropies must be taken into account because variation in time can be quite different from variation in space. With space-time applications, one may choose from a much richer set of sampling designs (de Gruijter *et al.* 2006; Chapter 14). Data collection costs may be different in space-time applications because it may be relatively cheap to collect a time series of data at fixed spatial locations (e.g., using permanently installed devices and data-loggers), or it may be relatively cheap to collect observations at multiple spatial locations at (nearly) the same time instant (e.g., during the same field campaign). Thus, there are interesting aspects to be explored when extending sampling design optimization from the spatial to the spatio-temporal domain.

Sampling design optimization requires that an optimality criterion be defined to evaluate and compare different designs. Common criteria are minimizing (1) the errors associated with estimation of the trend and covariance structure of the statistical model that is adopted, and (2) the average kriging prediction error variance. Of these two, minimization of the estimation error of the covariance structure is the most difficult to address. Solutions for spatial optimization have been derived (Zhu and Stein 2006; Diggle and Ribeiro 2007; Müller 2007), but because of the (numerical) difficulties involved, this issue is not our first priority in an extension to the space-time case. In this chapter, we restrict ourselves to a criterion that simultaneously minimizes the variance of the estimation error of a linear trend and that of the interpolation error of the kriging residual. This criterion is essentially equivalent to minimizing the average universal kriging variance. See Brus and Heuvelink (2007) for a spatial analog.

A solution algorithm is used to minimize the defined criterion and obtain an optimal sampling design. Both analytical and numerical techniques have been developed and used (Pilz and Spock 2006; Müller 2007; Melles *et al.* 2011), although analytical methods typically involve simplifications to make a solution tractable. Here we work with one particular numerical optimization technique, known as spatial simulated annealing (van Groenigen *et al.* 1999; Guedes *et al.* 2011), which is generic and easily implemented, although computationally demanding, and yields the global optimum only in theory. More attractive and efficient algorithms exist, such as genetic algorithms and greedy algorithms (ter Braak 2006; Vrugt *et al.* 2008; Baume *et al.* 2011). However, because the emphasis of this chapter is on statistical aspects, we restrict ourselves to the one technique for which we have much experience.

Minimization of the space-time average universal kriging variance is achieved in this chapter for monthly temperatures of the Upper Austria region (see Chapter 1). The geostatistical model that we use treats monthly temperature as the sum of a trend and a space-time correlated Gaussian stochastic residual. The trend is taken as a linear function of elevation and a remotely sensed dynamic estimate of temperature (MODIS data). The period that we consider spans three years (2001, 2002 and 2003). Time-critical sampling decisions impact the notion of optimality in terms of the number and configuration of observation points in space and time. Accordingly, we consider three optimization scenarios. The first considers optimization of a static design, in which 10 locations are to be selected from 35 current stations such that increases in the average space-time universal kriging variance caused by the thinning are as small as possible. The second scenario also considers a static design, but this time the 10 stations can be located anywhere within the Upper Austria region. Establishing a fixed sampling network suffers from duplication of information over time, and information redundancy can be reduced by allowing stations to be relocated. The third optimization scenario considers a dynamic design, in which the 10 station locations may be changed at the start of each year.

Before we present the case study, discuss results and draw conclusions, we briefly review space-time universal kriging and spatial simulated annealing.

9.2 Methodology

9.2.1 Space-time universal kriging

Consider a variable $z = \{z(\mathbf{s}, t) | s \in S, t \in T\}$ that varies within a spatial domain S and a time interval T. Let z be observed at n space-time points $(\mathbf{s}_i, t_i)$, $i = 1 \ldots n$. Although the number of observations n may be very large, a complete space-time coverage of z requires some form of interpolation. The objective thus is to obtain a prediction of $z(\mathbf{s}_0, t_0)$ at a point $(\mathbf{s}_0, t_0)$ at which z was not observed, where $(\mathbf{s}_0, t_0)$ frequently is associated with the nodes of a fine space-time grid. To do this, z is assumed to be a realization of a random function Z characterized by a full statistical model, including its space-time dependence structure. Next, $Z(\mathbf{s}_0, t_0)$ is predicted from the observations and possibly from auxiliary information.

The space-time variation of Z can be characterized by decomposing it into a deterministic trend m and a zero-mean stochastic residual V – with this decomposition being a subjective choice made by a modeler (Diggle and Ribeiro 2007)–as follows:

$$Z(\mathbf{s}, t) = m(\mathbf{s}, t) + V(\mathbf{s}, t). \tag{9.1}$$

The trend m is a deterministic, structural component representing large-scale variation (e.g., that part of Z that can be explained physically or empirically, using auxiliary information). The residual is a stochastic component representing small-scale, 'noisy' variation (i.e., the leftover part of Z, which still holds information when it is correlated in space and/or time). We assume that it is multivariate normal.

The trend component may be written as:

$$m(\mathbf{s}, t) = \sum_{i=0}^{p} \beta_i f_i(\mathbf{s}, t), \tag{9.2}$$

where the β_i are unknown regression coefficients, the f_i are covariates that must be exhaustively known over the space-time domain, and p is the number of covariates. Covariate f_0 is taken as unity, resulting in β_0 representing the intercept.

The residual V may be thought of as comprising three components: spatial, temporal, and space-time interaction. Assuming all three of these components stationary and mutually independent yields the 'sum-metric' space-time covariance structure (Snepvangers *et al.* 2003):

$$C(h, u) = C_S(h) + C_T(u) + C_{ST}\left(\sqrt{h^2 + (\alpha \cdot u)^2}\right), \tag{9.3}$$

where $C(h, u)$ denotes the covariance function of V with h units of distance in space and u units of distance in time, $C_S(h)$ denotes the purely spatial covariance function component, $C_T(u)$ the purely temporal covariance function component, and $C_{ST}(\sqrt{h^2 + (\alpha \cdot u)^2})$ the space-time interaction covariance function component, with α denoting a geometric anisotropy ratio that converts units of time distance, u, to space distance, h. It is needed because a unit of distance in space is not the same as a unit of distance in time. Note that we have taken h to be a scalar representing Euclidean distance. The first two terms on the right-hand side of Equation 9.3 allow for the presence of zonal anisotropies (i.e., variogram sills that are not the same in all directions), which occurs when the amount of variation in time is smaller or greater than that in space, and/or that in joint space-time.

Once the trend and covariance function of the residual term have been specified, (space-time) interpolation can be calculated in the usual way. Universal kriging yields the best linear unbiased predictor (with minimum expected mean squared error among all possible predictors under multivariate normality) of $Z(\mathbf{s}_0, t_0)$ as per (Bivand *et al.* 2008, Section 8.5):

$$\hat{z}(\mathbf{s}_0, t_0) = \mathbf{m}_0{}^T \hat{\boldsymbol{\beta}} + \mathbf{c}_0{}^T \mathbf{C}_n{}^{-1}(\overline{\mathbf{z}} - \mathbf{M}\hat{\boldsymbol{\beta}}), \tag{9.4}$$

where $\mathbf{M}$ is an $n \times p$ design matrix of predictor variables at the observation points, $\mathbf{m}_0$ is a vector of predictors at the prediction point, $\mathbf{C}_n$ is an $n \times n$ variance–covariance matrix for the n residuals at the observation points, $\mathbf{c}_0$ is a vector of covariances between the residuals at the observation and prediction points, T denotes matrix transpose, and $\overline{\mathbf{z}}$ is a vector of observations $z(\mathbf{s}_i, t_i)$. The regression coefficients are estimated in the usual way, using generalized least squares.

The variance of the universal kriging prediction error is given by:

$$\begin{aligned}\sigma^2(\mathbf{s}_0, t_0) &= Var(Z(\mathbf{s}_0, t_0) - \hat{Z}(\mathbf{s}_0, t_0)) \\ &= C(0,0) - \mathbf{c}_0^T \mathbf{C}_n^{-1} \mathbf{c}_0 + \\ &+ (\mathbf{m}_0 - \mathbf{M}^T \mathbf{C}_n^{-1} \mathbf{c}_0)^T (\mathbf{M}^T \mathbf{C}_n^{-1} \mathbf{M})^{-1} (\mathbf{m}_0 - \mathbf{M}^T \mathbf{C}_n^{-1} \mathbf{c}_0) \quad (9.5)\end{aligned}$$

These are familiar equations from a spatial universal kriging perspective, showing that space-time kriging and spatial kriging do not differ fundamentally either mathematically or statistically.

9.2.2 Sampling design optimization with spatial simulated annealing

Spatial simulated annealing (SSA; van Groenigen *et al.* 1999) is the spatial counterpart to simulated annealing (Kirkpatrick *et al.* 1983), and as such the algorithm starts from an initial random design, makes slight perturbations of previous designs to generate candidate new designs, which are either accepted or rejected depending on whether these improve or deteriorate a selected criterion. Candidate designs that improve the criterion are always accepted, while designs that deteriorate the criterion are accepted with some probability. The latter is done in order to escape from local optima. A 'cooling' schedule imposes a decreasing probability of accepting worsening designs as the number of iterations increases. The iterative procedure is stopped when a fixed number of iterations has been tried, or when no improvement occurs in the criterion over a set number of iterations.

Candidate designs can be derived from the current design in various ways. In this chapter, we randomly choose a station from the selected 10 stations and swap it with one of the remaining unused 25 stations (Scenario 1, optimal selection of 10 stations from the existing 35), or displace it geographically in a random direction (Scenario 2, optimal allocation of 10 stations in the entire Upper Austria region). The size of displacement is drawn from a uniform distribution between zero and a maximum displacement, while the maximum displacement is reduced as the number of iterations increases. Scenario 3 is a dynamic design that relocates the 10 stations at the start of each year first by randomly selecting a year and station number, and then by moving it, as in Scenario 2.

Each iteration step of SSA requires computing the space-time average universal kriging variance. This calculation was done by defining a space-time prediction grid with 1 km spatial resolution and a time resolution of 1 month. This scheme yields $11979 \times 36 = 431244$ prediction points. The universal kriging is based on $10 \times 36 = 360$ observations, which is sufficiently moderate to allow for global kriging (i.e., kriging is not restricted to observations in the local neighborhood of a prediction point). Computational load is substantial because SSA typically requires between 1000 and 5000 iterations, although this number is case dependent.

Recent examples of the application of SSA for optimization of spatial sampling designs are given in Brus and Heuvelink (2007) and Melles *et al.* (2011).

9.3 Upper Austria case study

9.3.1 Descriptive statistics

Figure 9.1 portrays the locations of the 35 stations in the Upper Austria region. The average monthly temperature at these stations for 2001, 2002 and 2003 yields 35 time series, with missing values for some stations. Figure 9.2 depicts the temperature over time at the 'Aspach' station, which is located in the western part of Upper-Austria at an altitude of 436 m asl (i.e., the site plotted in grey in Figure 9.1). This graph shows clear seasonal behaviour, with the lowest temperature in December-February and the highest in June-August. Figure 9.2 also shows a time series of temperature for Aspach as derived from Moderate Resolution Imaging Spectroradiometer (MODIS) remote sensing imagery. MODIS land surface temperature (LST) images were obtained directly from NASA's FTP data service.[1] Next these images were mosaicked together and resampled to the local projection system. The MODIS LST images are derived from MODIS thermal bands. Wan *et al.* (2004) show that the accuracy of MODIS LST images is about ±1°C. However, parts of a MODIS image can be missing pixel values,

Figure 9.1 Locations of 35 weather stations in the Upper Austria region. Station 'Aspach' is plotted in grey.

[1] https://lpdaac.usgs.gov/get_data/data_pool.

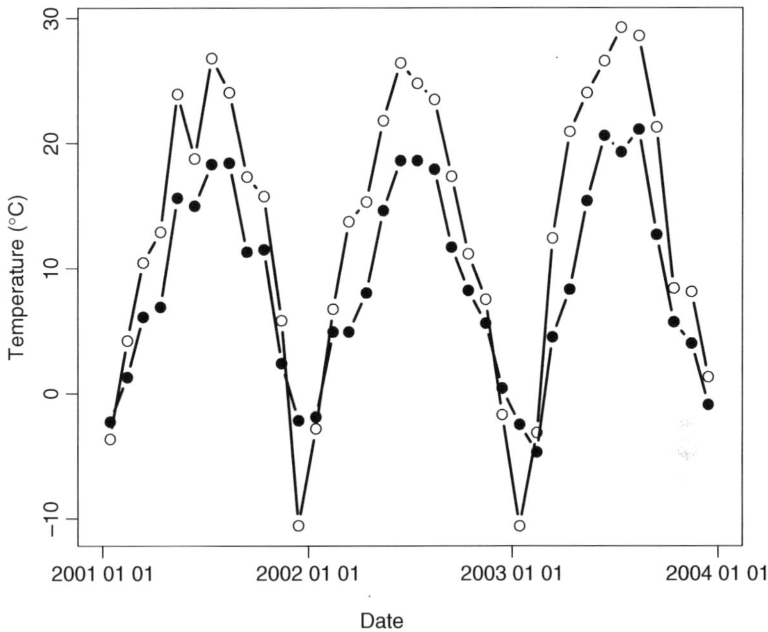

Figure 9.2 Time series of monthly temperature data during 2001–2003 at Station 'Aspach'. Solid circles are observations; open circles are MODIS derived temperatures.

which may be attributable to cloud cover and/or other unfavorable atmospheric conditions (Neteler 2010). For this case study, we imputed values for missing pixels by averaging the values of their neighboring dates.

Because MODIS data are exhaustively available for the region and time periods considered, and are correlated with observed monthly temperature, they furnish a sensible covariate in the geostatistical model specification to describe the trend. In addition, because temperature is known to be correlated with elevation, elevation also is included as a covariate, even though the MODIS data already contain part of the information furnished by elevation. Elevation was obtained from a 1 km resolution SRTM DEM of the area. All covariate layers (MODIS LST images, SRTM DEM) were prepared in the MGI Austria GK Central projection system.[2] All datasets and processing steps used in this chapter are available for download, together with detailed annotations, via the public repository.[3] Appendix 9.1 provides R code used for the space-time variography, space-time kriging and optimal removal of locations.

Figure 9.3 portrays the digital elevation model of Upper Austria. Figure 9.4 depicts a time series of MODIS maps for 2001. Figure 9.5 presents scatter plots of monthly temperature data versus elevation and MODIS data. Note that the

[2] http://spatialreference.org/ref/epsg/31255/.

[3] http://code.google.com/p/uppera/.

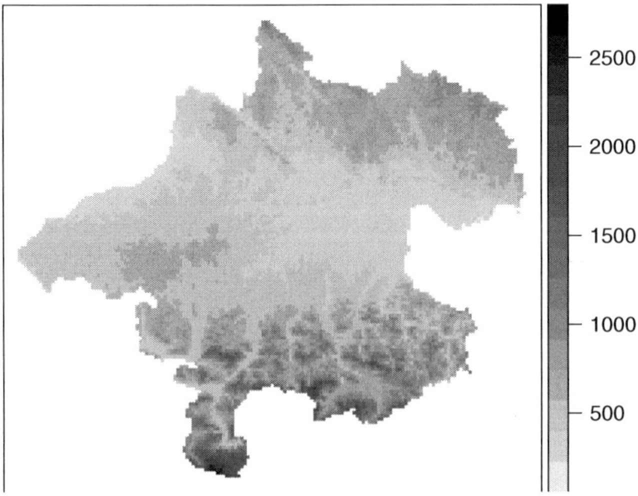

Figure 9.3 Digital elevation model of the Upper Austria region. Elevation in meters, grid cells are 1×1 *km.*

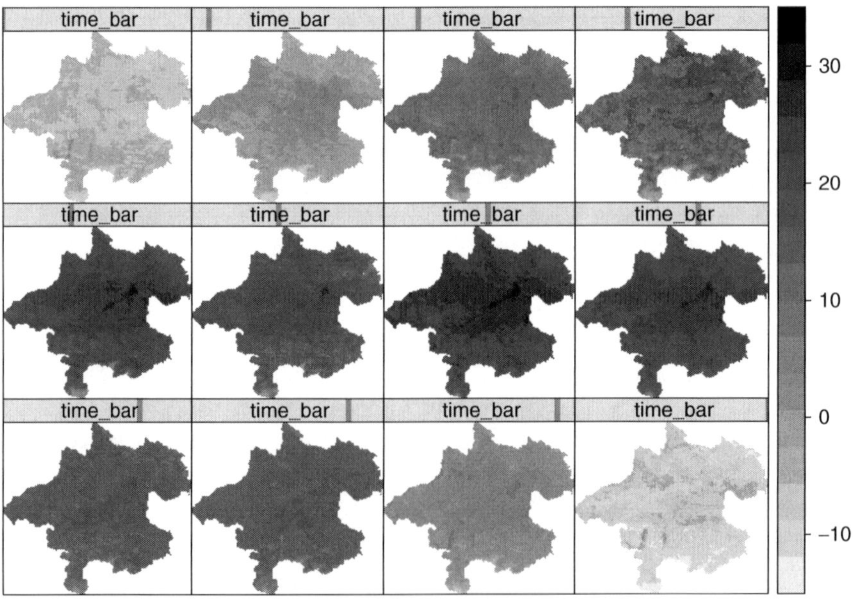

Figure 9.4 Time series of MODIS monthly temperature layers for year 2001. Vertical time bar indicates month of year (row-wise display with January in upper left and December in bottom right corner). Temperature in $^\circ C$*.*

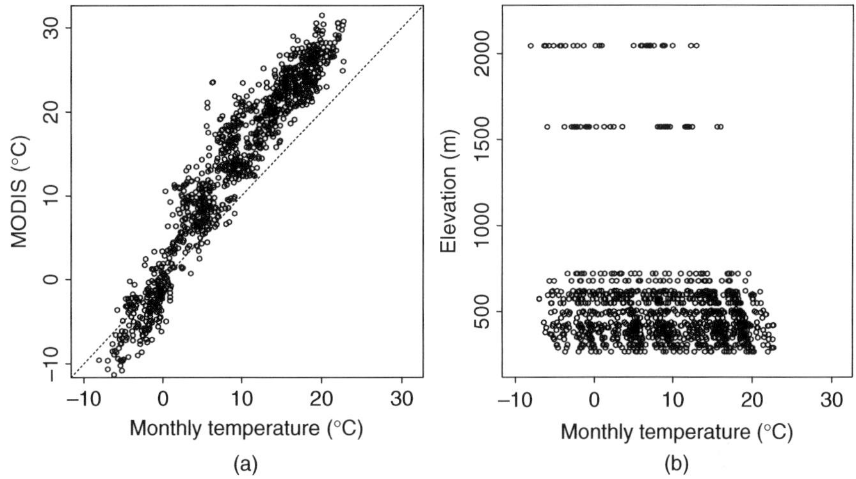

Figure 9.5 Scatter plot of monthly temperature against MODIS data (a) and elevation (b). Dashed line represents the 1:1 line. All observations from 35 stations during 2001–2003 were used (n = 1055).

covariation pattern of elevation and monthly temperature shows that two stations have high altitudes while the other 33 stations are in the range between 200 and 800 m asl. Figure 9.3 shows that MODIS data have a strong correlation with monthly temperature, but are more extreme. The relationship between monthly temperature and elevation is negligible, which is due to the stronger influence of seasonality. The relationship between elevation and the residual from a linear regression of temperature on MODIS data is stronger than that between monthly temperature and elevation, but still weak (data not shown).

A multiple linear regression (ignoring spatial and temporal dependencies)[4] of monthly temperature on elevation and MODIS data reveals that the MODIS data covariate is highly significant, whereas elevation is just significant (at the 5% level). Together these two covariates explain 92% of the variance in monthly temperature. The estimated regression coefficients are -0.38 °C/km for elevation, and 0.68 for MODIS. Figure 9.6 portrays the histogram of regression residuals. These residuals are fairly symmetrically distributed, with a few negative outliers and a moderate negative skewness (-0.64). Based on these results, both the MODIS data covariate and elevation were included in the trend component of the geostatistical model. No transformation of residuals was considered.

9.3.2 Estimation of the space-time model and universal kriging

The universal kriging model given in Equation (9.1) requires calculation of a sample variogram using the residuals rather than the monthly temperature data

[4] Accordingly, the significance tests are unreliable.

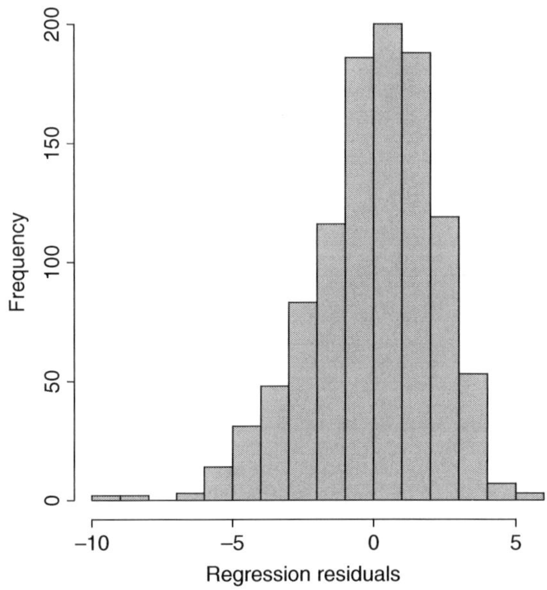

Figure 9.6 Histogram of residuals of a multiple linear regression of monthly temperature on MODIS and elevation.

themselves. This task was done using R *gstat* software (Bivand *et al.* 2008). The default approach in *gstat* is to assume independence when calculating the residuals, although a work-around exists that calculates the residuals for an imposed correlation structure (Bivand *et al.* (2008), section 8.4.6), allowing an iterative approach to be used to estimate the trend and residual variogram. We used the default approach and calculated the sample residual variogram for spatial lags of 8 km and temporal lags of 1 month. The right-hand side of Figure 9.7 summarizes these results. Next, we fitted the sum-metric model with the *optim* function in R, using a quasi-Newton method that allows box constraints (Byrd *et al.* 1995). The left-hand side of Figure 9.7 portrays this fitted model. Table 9.1 summarizes the parameter estimates of this fitted model.

Space-time universal kriging of monthly temperature for the 3-year period yields prediction and prediction error standard deviation maps for each of the 36 months. As an example, Figure 9.8 depicts the result for August 2001. The spatial pattern of the prediction map closely resembles that of the MODIS data (Figure 9.4), with low values in the southern part and higher values in the central and eastern parts of the region. The prediction error standard deviation map has the lowest values near station locations, and high values along the boundaries and in the southern part of the region, where the covariates are more extreme and cause larger errors in the trend estimates. Overall, the kriging standard deviation is small, indicating that prediction errors in many instances are smaller than 1°C and rarely greater than 2°C.

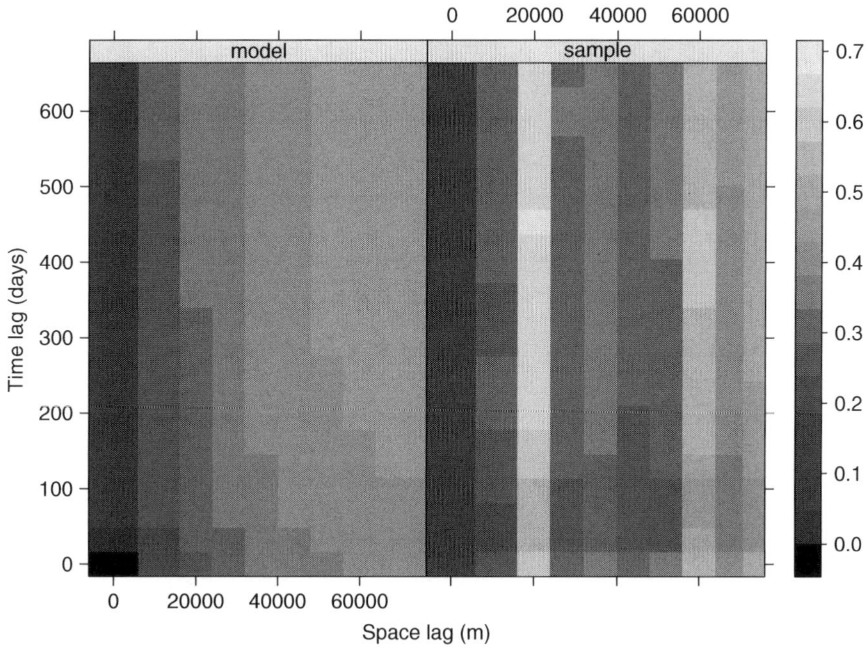

Figure 9.7 Sample variogram and fitted sum-metric model of residuals from the multiple linear regression of temperature data on MODIS and elevation. (Please see plate section for color version of the figure.)

Table 9.1 Parameters of the fitted sum-metric variogram model.

	Nugget	Sill	Range parameter	Anisotropy ratio
Spatial	0.075	0.381	25 km	
Temporal	0	0.062	150 days	
Space-time	0.035	0.035	25 km	75 m/day

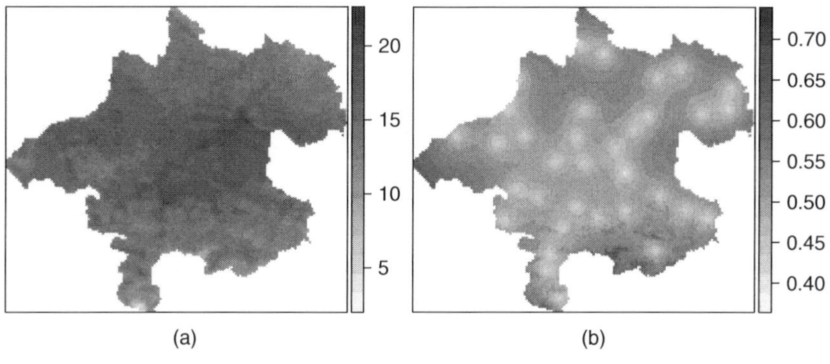

Figure 9.8 Space-time universal kriging prediction (a) and prediction error standard deviation (b) maps for the average temperature (°C) in August 2001.

9.3.3 Optimal design scenario 1

Scenario 1 selects 10 stations from the existing 35 such that the space-time average universal kriging variance is minimized. Six SSA runs were made, each producing somewhat different results, but all obtaining a very similar mean universal kriging variance (MUKV). A maximum of 2000 iterations was imposed, but the algorithm also was stopped if 200 successive iterations did not improve the design. Figure 9.9 shows the evolution of the SSA criterion for the run that achieved the smallest MUKV. As expected, initially worsening designs are accepted because the probability of accepting worsening designs is still relatively large. After 225 iterations, the criterion only improves and some station swaps still yield a noticeable decrease in the MUKV, although the MUKV reduction from the initial to the final design is less than 7%.

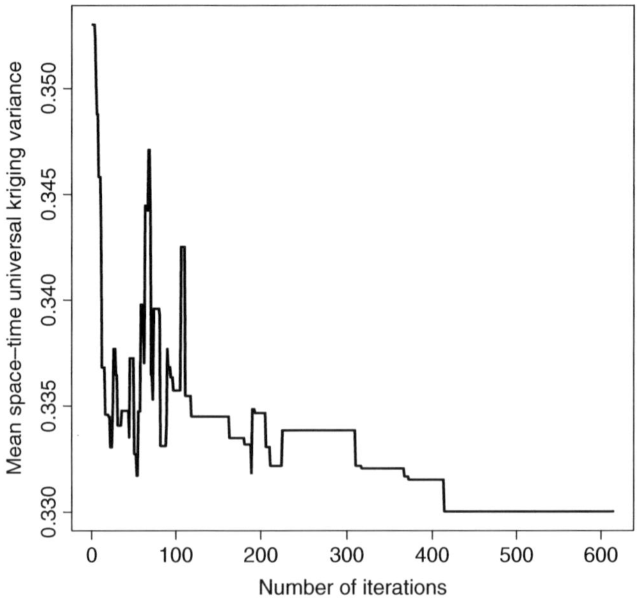

Figure 9.9 Evolution of the SSA criterion for Scenario 1. Execution was stopped after 622 iterations, when no improvement occurred for more than 200 successive iterations.

Figure 9.10 portrays the corresponding optimized design. The selected stations are fairly uniformly spread over the region, with a tendency to over-represent stations near the border of the study area. These results are as expected. All six runs selected the most southern station, which can be explained by this station being at a high altitude, and hence having extreme values for elevation as well as MODIS temperature. Measurements from these stations substantially reduce the trend parameter estimation error that is part of the MUKV.

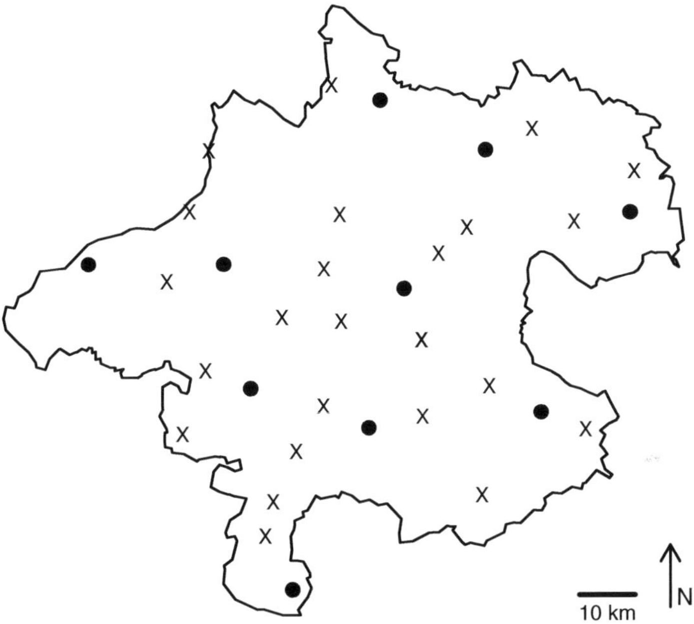

Figure 9.10 *An optimal static design for selecting 10 from 35 stations within the Upper Austria region. Filled circles represent selected stations, crosses those that were not selected.*

9.3.4 Optimal design scenario 2

Scenario 2 places 10 stations optimally within the region without restriction. This scenario also was executed six times. The design of the run with the smallest MUKV is shown in Figure 9.11. The MUKV value is 0.32134, which is smaller than that obtained under Scenario 1. The reduction is achieved by an even more uniform distribution of stations across the region, which is possible because station locations were not confined to those of the existing 35 stations. Again, a station is located in the southern 'appendage' where elevation is large and MODIS temperature small. This occurs in only two of the six runs, which were indeed the two runs with the smallest MUKV.

9.3.5 Optimal design scenario 3

Scenario 3 allows the repositioning of stations at the start of each calendar year. Results show that the added flexibility of this dynamic design pays off because the MUKV can be dramatically reduced to a value of 0.25301. Figure 9.12 shows the optimal design. Again, the locations are uniformly spread over the region, but this time the density can be increased because there are effectively 30 locations. However, note that for each single year the geographic spread is far from uniform. Apparently this outcome does not lead to a substantial increase of

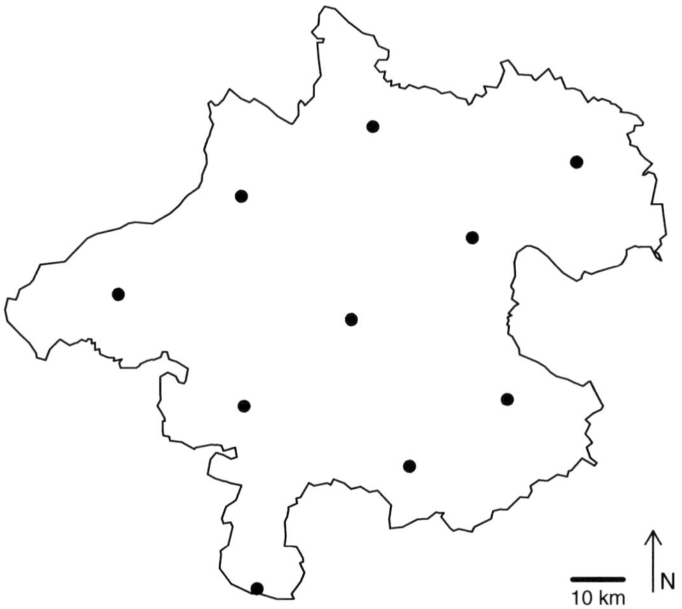

Figure 9.11 *An optimal static design for selecting 10 stations anywhere within the Upper Austria region.*

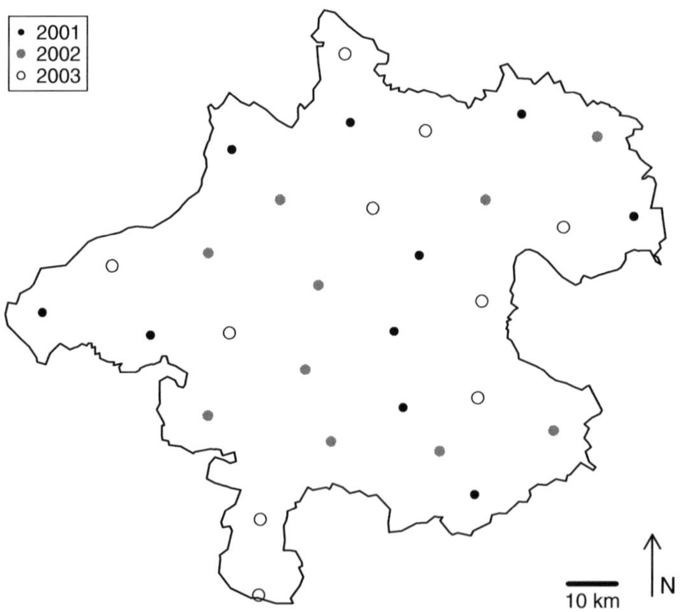

Figure 9.12 *An optimal dynamic design for selecting 10 stations anywhere within the Upper Austria region, allowing relocation of each station at the start of each year.*

the MUKV, which can be explained by the relatively small temporal and spatio-temporal variation of the stochastic residual of the random process (Table 9.1). This outcome also explains why restricting measurements to only one year does not impair the 'information content' of each single station. Results could have been quite different if no trend was assumed and ordinary kriging had been used, because in that case, temporal variability would likely not have been small compared with spatial variability.

Comparison of the MUKV of Scenario 3 with that obtained with the original data set (35 stations 'continuously' measured, MUKV $= 0.25548$) shows that the sampling design of Scenario 3 is (slightly) more accurate, while it uses far fewer observations. Even though several stations have missing values in the original dataset, which reduces the total number of observations from a potential of $3 \times 12 \times 35 = 1266$ to 1055, Scenario 3 has only 360 observations, which is a reduction of one-third.

9.4 Discussion and conclusions

In this chapter we extend the use of SSA for optimization of sampling designs in space to optimization of sampling designs in space-time. This requires use of a space-time geostatistical model for which we use a common model that treats the space-time variable of interest as the sum of a trend, which is linear in known covariates, and a stationary Gaussian stochastic residual. Recent advances in space-time statistical modeling enhance flexibility and allow comparisons of optimal designs for different models. One interesting exercise would be to analyze the sensitivity of optimal designs to the structure and parameters of the space-time geostatistical model. Another interesting exercise would be to analyze how the optimal design is influenced by including parameter estimation errors in the space-time covariance structure of the stochastic residual. From a numerical perspective, evaluation of alternative optimization techniques also would be worthwhile, particularly because SSA is known to be computer-intensive, and because space-time kriging typically deals with much larger datasets than spatial kriging.

The extension to space-time also allows detailed analysis of the efficiency of various types of designs in space and time. Here we focus on static designs and consider only one basic dynamic design. However, many more design options exist, such as synchronous, static-synchronous, and rotational (de Gruijter *et al.* 2006). The efficiency of these designs can be analyzed from a purely model-based perspective, which adds to existing studies that take a design-based or hybrid approach (Brus and de Gruijter 2011, 2012).

For the case study considered in this chapter, the dynamic design turns out to be much more attractive than the static designs in terms of the total number of observations that are required to achieve a given level of accuracy. However, the cost of a design is not necessarily proportional to the total number of observations. Relocating stations may be very costly, whereas maintaining a station in operation mode once it is installed may be relatively cheap. This situation may well be the case for the temperature monitoring considered here, but cost considerations may turn out differently when measurement instruments are expensive and relocating stations is cheap. Examples are the use of mobile stations that measure greenhouse gas emission, air pollution or nuclear radiation (Kukkonena *et al.* 2005; Heuvelink *et al.* 2010; Angelbratt *et al.* 2011). Accessibility of locations also becomes an issue, such as when mobile stations are mounted on cars or when the study area of interest is in remote or hostile terrain. These factors can be taken into account and the application of sampling design algorithms, such as those used here, may yield insightful solutions.

Appendix 9.1: R code

```
# Packages
library(rgdal)
library(gstat)
library(colorspace)
library(spacetime)
library(lattice)
library(raster)
library(sp)

# ----------------------------------
# Space-time variography and kriging
# ----------------------------------

# Convert space-time data frame to an STFDF
locs <- uppera.xyt.s[,c("ST_ID","X","Y")]
coordinates(locs) <- ~X+Y
proj4string(locs) <- uppera@proj4string
uppera.st <- STIDF(locs,time=uppera.xyt.s$Date,
  data=uppera.xyt.s[,c("TEMP","MODIS.LST","dem","res")])
uppera.std <- as(uppera.st,"STFDF")

# Compute and plot sample variogram
vv <- variogram TEMP~MODIS.LST+dem,data=uppera.std,
  width=8000,cutoff=75000,tlags=0:20)
plot(vv, ylab="Time lag (days)")

# Initialize parameters of sum-metric variogram model
pars.init <- c(nugget.s=0.05,sill.s=0.2,range.s=25000,
 nugget.t=0.01,sill.t=0.01,range.t=150,nugget.st=0.01,
 sill.st=0.1,range.st=25000,anis=75)
 pars.l <- c(nugget.s=0,sill.s=1E-2,range.s=1000,nugget.t=0,
  sill.t=1E-4,range.t=10,nugget.st=0,sill.st=1E-4,
  range.st=1000,anis=1)
pars.u <- c(nugget.s=0.5,sill.s=1,range.s=1E6,nugget.t=0.5,
  sill.t=1,range.t=1000,nugget.st=0.5,sill.st=1,
  range.st=1E6,anis=1000)

# Function to compute variogram model
ExpVgmSumMetric=function(pars,s,t) {
```

```
  vs=ifelse(s==0,0,pars[1]+pars[2]*(1-exp(-s/pars[3])))
  vt=ifelse(t==0,y0,pars[4]+pars[5]*(1-exp(-t/pars[6])))
  h=sqrt(s^2+(pars[10]*as.numeric(t))^2)
  vst=ifelse(h==0,0,pars[7]+pars[8]*(1-exp(-h/pars[9])))
  vs+vt+vst }

# Computes mean squared difference between sample
     # variogram and variogram model
ExpFitFn=function(pars,gfn,v,trace=FALSE) {
  mod=gfn(pars,v$spacelag,v$timelag)
  resid=v$gamma-mod
  if (trace)
    print(c(pars,MSE=mean(resid^2)))
  mean(resid^2) }

# Variogram fitting
pars.fit=optim(pars.init,ExpFitFn,gfn=ExpVgmSumMetric,
  v=vv,gr=NULL,method="L-BFGS-B",lower=pars.l,
  upper=pars.u,control=list(maxit=1000))

# Define space-time variogram of sum-metric model
vgm.st=vgm(pars.fit$par[1],"Exp",1E12,anis=c(0,90,0,1E-15,
     1E-15)) # spatial nugget
vgm.st=vgm(pars.fit$par[2],"Exp",pars.fit$par[3]*1E6,
  anis=c(0,90,0,1E-6,1E-6),add.to=vgm.st) # spatial partial
       # sill
vgm.st=vgm(pars.fit$par[4],"Exp",1E12,anis=c(0,0,0,1,1E-15),
  add.to=vgm.st) # temporal nugget
vgm.st=vgm(pars.fit$par[5],"Exp", ypars.fit$par[6]*1E6,
  anis=c(0,0,0,1,1E-6),add.to=vgm.st) # temporal partial sill
vgm.st=vgm(pars.fit$par[7],"Exp",1E-12,add.to=vgm.st)
     # space-time nugget
vgm.st=vgm(pars.fit$par[8],"Exp",pars.fit$par[9],
  anis=c(0,0,0,1,1/pars.fit$par[10]),add.to=vgm.st)
       # space-time partial sill

# Define space-time prediction grid (grids1km object in
     # workspace contains MODIS data)
n_begin <- 11 # January 2001
n_end <- 46 # December 2003
srtmdem$mask <- readGDAL("mask.sdat")$band1
range.X <- grids1km@coords[1]+grids1km@grid@cellsize[1]*
  (0:(grids1km@grid@cells.dim[1]-1))
range.Y <- rev(grids1km@coords[3]+grids1km@grid@cellsize[2]*
  (0:(grids1km@grid@cells.dim[2]-1)))
range.timedata <- n_begin:n_end
st.grid <- expand.grid(X=range.X,Y=range.Y,
  timedata=range.timedata,KEEP.OUT.ATTRS=FALSE)
st.grid[,"MODIS.LST"] <- unlist(unclass(grids1kmf@data
     [,n_begin:n_end]))
st.grid[,"dem"] <- as.data.frame(srtmdem)[,1]
st.grid[,"mask"] <- as.data.frame(srtmdem)[,"mask"]
st.grid.s <- subset(st.grid,!is.na(st.grid$mask))
gridded(st.grid.s)=~X+Y+timedata
```

```
# Define data set with observations
uppera.obs <- data.frame(X=uppera.std@sp@coords[,1],
  Y=uppera.std@sp@coords[,2],timedata=uppera.std@time,
  TEMP=uppera.std$TEMP,MODIS.LST=uppera.std$MODIS.LST,
  dem=uppera.std$dem,row.names=NULL)
coordinates(uppera.obs)=~X+Y+timedata

# Kriging
uppera.uk <- krige(formula=TEMP~MODIS.LST+dem,uppera.obs,
  st.grid.s, model=vgm.st,debug.level=-1)


# ----------------------------------------------------------
# Optimal removal of points by spatial simulated annealing
# ----------------------------------------------------------

# candidates is a SpatialPolygonsDataFrame of the study area
# uppera.obs is a df with X,Y,timedata,TEMP,MODIS.LST,dem
# grids1kmf is an sgdf with monthly LST MODIS data in cols
# st.grid.s is an spdf prediction grid in space-time

projxy <- "+proj=tmerc +lat_0=0+lon_0=13.33333333333333+k=1
  +x_0=0+y_0=-5000000 +ellps=bessel +units=m +no_defs"
proj4string(candidates) <- projxy
nspaceDiff <- 25 # number of points to delete
ntimeSteps <- 12
range.timedata <- n_begin:n_end
num_periods <- length(range.timedata)/ntimeSteps
xy <- c("X","Y")
nn_space <- unique(uppera.obs[xy])
nn_time <- unique(uppera.obs["timedata"])
nn <-dim(nn_space)[1]-nspaceDiff # pts to sample in space
delPts_vec <- as.vector(as.matrix(nn_space[sample(nrow
(nn_space),size=nspaceDiff),][1]))
uppera.obsNet <- subset(uppera.obs,!(uppera.obs$X %in%
     delPts_vec))
uppera.obsDel <- subset(uppera.obs, y(uppera.obs$X %in%
     delPts_vec))
coordinates(uppera.obsNet)= ~X+Y+timedata
uppera.uk.init <- krige(formula=TEMP~MODIS.LST+dem,
uppera.obsNet,st.grid.s,model=vgm.st,debug.level=-1)
avkrigvar.init <- mean(uppera.uk.init$var1.var,na.rm=TRUE)
oldpoints <- unique(as.data.frame(uppera.obsNet)[xy])
     # save initial design
# options
  nr_iterations <- 1000  # number of ssa iterations to run
  max_points_shift <- 1  # number of points to move
  start_p <- 0.2  # initial p of accepting a worsening design
  maxShiftFactorX <- 0.2  # multiplier of max possible shift
      # for point in X range
  minShiftFactorX <- 0  # multiplier of min possible shift for
      # point in X range
  maxShiftFactorY <- 0.2  # multiplier of max possible shift
      # for point in Y range
```

```
  minShiftFactorY <- 0  # multiplier of min possible shift
       # for point in Y range
  coolingFactor <- nr_iterations/10  # helps controls rate of
       # cooling
  countMax <- 200  # max number of ssa iterations to process
       # without change in criterion
  plotOptim <- TRUE  # logical plot or not
# settings for simulated annealing
  nr_designs <- 1  # counter for number of accepted designs
  old.uppera.obsNet <- uppera.obsNet  # save initial design
  criterionInitial <- avkrigvar.init  # initial criterion
  oldcriterion <- criterionInitial  # save current criterion
  oldDelPoints <- unique(uppera.obsDel[xy])  #delPts in space
  old.uppera.obsDel <- uppera.obsDel  # delPts with all times
  criterionIterf <- NULL  # keep track of improvement in
      # objective function
  bestCriterion <- Inf  # store best crit in case final design
      # is not best design
  count <- 0
  cnames <- dimnames(coordinates(uppera.obsNet))[[2]]
       # coordinate & time of data.frame
  vnames <- dimnames(uppera.obsNet@data)[[2]]
       # predictor variables of data.frame
  cvar <- which(names(as.data.frame(uppera.obsNet))
       %in% cnames)
  x_bounds <- bbox(candidates)[1,]
  y_bounds <- bbox(candidates)[2,]
  x_extent <- x_bounds[2]-x_bounds[1]
  y_extent <- y_bounds[2]-y_bounds[1]
  max_shift_x <- maxShiftFactorX*x_extent # max shift x-dir
  min_shift_x <- minShiftFactorX*x_extent # min shift x-dir
  max_shift_y <- maxShiftFactorY*y_extent # max shift y-dir
  min_shift_y <- minShiftFactorY*y_extent # min shift y-dir

for (k in 1:nr_iterations) {
  selected_shifts_del <- sample(nspaceDiff)
  selected_shifts_net <- sample(nn)
  oldDelPt <- oldDelPoints[which(selected_shifts_del<=max_
       points_shift),]
  newDelPt <- oldpoints[which(selected_shifts_net<=max_points_
       shift),]
  delPts <- rbind(oldDelPoints[which(selected_shifts_del>max_
       points_shift),],newDelPt)
  netPts <- rbind(oldpoints[which(selected_shifts_net>max_
       points_shift),],oldDelPt)
  overdelPts <- over(SpatialPoints(delPts,proj4string=CRS
  (as.character(projxy))),ygrids1kmf)
  overdelPts.t <- subset(overdelPts,select=range.timedata)
  overdelPts.t <- as.vector(t(overdelPts.t))
  overst.grid <- over(SpatialPoints(delPts),st.grid.s)
  uppera.obsDel <- cbind(X=rep(delPts["X"][1:nspaceDiff,],
   each=ntimeSteps),Y=rep(delPts["Y"][1:nspaceDiff,],
   each=ntimeSteps),timedata=rep(range.timedata,nspaceDiff),
   TEMP=rep(1,ntimeSteps*nspaceDiff),MODIS.LST=overdelPts.t,
   dem=rep(overst.grid$dem,each=ntimeSteps))
```

```
uppera.obsDel <- as.data.frame(uppera.obsDel)
overnetPts <- over(SpatialPoints(netPts,proj4string=CRS
(as.character(projxy))),grids1kmf)
overnetPts.t <- subset(overnetPts,select=range.timedata)
overnetPts.t <- as.vector(t(overnetPts.t))
overst.grid <- over(SpatialPoints(netPts),st.grid.s)
uppera.obsNet <- cbind(X=rep(netPts["X"][1:nn,],each=
 ntimeSteps),Y=rep(netPts["Y"][1:nn,],each=ntimeSteps),
 timedata=rep(range.timedata,nn),TEMP=rep(1,ntimeSteps*nn),
 MODIS.LST=overnetPts.t,dem=rep(overst.grid$dem,
 each=ntimeSteps))
uppera.obsNet <- as.data.frame(uppera.obsNet)
coordinates(uppera.obsNet) <-  ~X+Y+timedata
uppera.uk <- krige(formula=TEMP~MODIS.LST+dem,
uppera.obsNet,st.grid.s,model=vgm.st,debug.level=-1)
criterion <- mean(uppera.uk$var1.var,na.rm=TRUE)
p <- runif(1) # to allow accepting an inferior design
if (criterion <= oldcriterion) {
  oldpoints <- netPts; old.uppera.obsNet <- uppera.obsNet;
     oldDelPoints <- delPts
  old.uppera.obsDel <- uppera.obsDel
  oldcriterion <- criterion
  nr_designs <- nr_designs+1
  count <- 0
  cat("No improvement for",count,"iterations ")
  cat("[Will stop at",countMax,"iterations with no
       improvement]","\n")
  else if (criterion > oldcriterion & p <= (start_p*exp
   (-k/coolingFactor))) {
  oldpoints=netPts; old.uppera.obsNet=uppera.obsNet
  oldDelPoints=delPts
  old.uppera.obsDel=uppera.obsDel
  oldcriterion=criterion
  nr_designs=nr_designs+1
  count=count+1
  cat("No improvement for",count,"iterations  p=",p,"lim=",
    start_p*exp(-k/coolingFactor))
  cat("[Will stop at",countMax,"iterations with no
       improvement]","\n") }
else {
  criterion <- oldcriterion
  oldDelPt <- oldDelPoints
  old.uppera.obsDel <- old.uppera.obsDel
  netPts <- oldpoints
  uppera.obsNet <- old.uppera.obsNet
  nr_designs <- nr_designs
  count <- count + 1
  cat("No improvement for",count,"iterations ")
  cat("[Will stop at",countMax,"iterations with no
       improvement]","\n") }
criterionIterf[k] <- criterion
if (criterion < bestCriterion/1.0000001) {
  bestNetPts <- netPts
```

```
    best.uppera.obsNet <- uppera.obsNet
    bestCriterion <- criterion
    bestOldcriterion <- oldcriterion
    bestOldpoints <- oldpoints
    bestOld.uppera.obsNet <- old.uppera.obsNet
    bestOldDelPoints <- oldDelPoints
    bestOld.uppera.obsDel <- old.uppera.obsDel
    bestOldDelPt <- oldDelPt
    bestDelPts <- delPts
    bestUppera.obsDel <- uppera.obsDel }
  if (plotOptim) {
  plot(candidates)
    points(oldpoints,col=1,pch=19,cex=1.5)
    points(oldDelPoints,col=2,pch="X",cex=1.2)
    arrows(x0=120000, x1=120000, y0=260000, y1=275000,
        length = 0.15, angle = 30, code=2, lwd=2)
    text(124500, 263000, "N", cex=1.5)
    segments(x0=100000, x1=110000, y0=263000, lwd=5)
    text(105000, 260000, "10 km", cex=1)
    title('Spatial Simulated Annealing',
        xlab=(paste("Criterion=",
      signif(criterion,digits=5),"(Best Criterion=",
      signif(bestCriterion,digits=5),")")),
           ylab=(paste("Iterations=",k))) }
  if (count==countMax) {
  if (criterion > bestCriterion*1.000001) {
  oldpoints <- bestOldpoints
      old.uppera.obsNet <- bestOld.uppera.obsNet
      netPts <- bestNetPts
      uppera.obsNet <- best.uppera.obsNet
      oldcriterion <- bestOldcriterion
      criterion <- bestCriterion
      oldDelPoints <- bestOldDelPoints
      old.uppera.obsDel <- bestOld.uppera.obsDel
      oldDelPt <- bestOldDelPt
      delPts <- bestDelPts
      uppera.obsDel <- bestUppera.obsDel
      nr_designs <- nr_designs + 1
      count <- 0
      cat("Reached countMax with suboptimal design,
        restarting with previously best design \n")
      cat("No improvement for",count,"iterations ")
      cat("[Will stop at",countMax,"iterations with no
           improvement]","\n") }
    else
    break }}
```

Acknowledgment

We would like to acknowledge support provided by the Aquatic Research and Development Section of the Ontario Ministry of Natural Resources, Peterborough, Ontario, Canada.

References

Angelbratt J, Mellqvist J, Blumenstock T, Borsdorff T, Brohede S, Duchatelet P, Forster F, Hase F, Mahieu E, Murtagh D, Petersen K, Schneider M, Sussmann R and Urban J 2011 A new method to detect long term trends of methane (CH_4) and nitrous oxide (N_2O) total columns measured within the NDACC ground-based high resolution solar FTIR network. *Atmospheric Chemistry and Physics* **11**, 6167–6183.

Angulo JM, Ruiz-Medina MD, Alonso FJ and Bueso MC 2005 Generalized approaches to spatial sampling design. *Environmetrics* **16**, 523–534.

Baume OP, Gebhardt A, Gebhardt C, Heuvelink GBM and Pilz J 2011 Network optimization algorithms and scenarios in the context of automatic mapping. *Computers and Geosciences* **37**, 289–294.

Bivand RS, Pebesma EJ and Gómez-Rubio V 2008 *Applied Spatial Data Analysis with R*. Springer.

Brus DJ and de Gruijter JJ 2011 Design-based generalized least squares estimation of status and trend of soil properties from monitoring data. *Geoderma* **164**, 172–180.

Brus DJ and De Gruijter JJ 2012 A hybrid design-based and model-based sampling approach to estimate the temporal trend of spatial means. *Geoderma* **173**, 241–248.

Brus DJ and Heuvelink GBM 2007 Optimization of sample patterns for universal kriging of environmental variables. *Geoderma* **138**, 86–95.

Byrd RH, Lu P, Nocedal J and Zhu C 1995 A limited memory algorithm for bound constrained optimization. *SIAM Journal of Scientific Computing* **16**, 1190–1208.

Cameletti M 2012 *Spatio-temporal models in R*. Package vignettes. CRAN, University of Bergamo.

Cressie N and Wikle C 2011 *Statistics for Spatio-Temporal Data*. John Wiley & Sons, Ltd.

de Gruijter JJ, Brus DJ, Bierkens MFP and Knotters M 2006 *Sampling for Natural Resource Monitoring*. Springer.

Diggle P and Ribeiro PJ 2007 *Model-Based Geostatistics*. Springer.

Fonseca T and Steel M 2011 A general class of nonseparable space-time covariance models. *Environmetrics* **22**, 224–242.

Fuentes M, Chen L and Davis J 2008 A class of nonseparable and nonstationary spatial temporal covariance functions. *Environmetrics* **19**, 487–507.

Gething PW, Atkinson PM, Noor AM, Gikandi PW, Hay SI and Nixon MS 2007 A local space-time kriging approach applied to a national outpatient malaria data set. *Computers and Geosciences* **33**, 1337–1350.

Guedes L, Ribeiro P, Piedade S and Uribe-Opzao M 2011 Optimization of spatial sample configurations using hybrid genetic algorithm and simulated annealing. *Chilean Journal of Statistics* **2**, 39–50.

Heuvelink GBM and Griffith DA 2010 Space-time geostatistics for geography: a case study of radiation monitoring across parts of germany. *Geographical Analysis* **42**, 161–179.

Heuvelink GBM, Jiang Z, de Bruin S and Twenhöfel C 2010 Optimization of mobile radioactivity monitoring networks. *International Journal of GIS* **24**, 365–382.

Kirkpatrick S, Gelatt CD and Vecchi MP 1983 Optimization by simulated annealing. *Science* **220**, 671–680.

Kukkonena J, Pohjolaa M, Sokhib RS, Luhanab L, Kitwiroonb N, Fragkoub L, Rantamäkia M, Bergec E, Odegaardd V, Slordale L, Denby B and Finardif S 2005 Analysis and evaluation of selected local-scale PM10 air pollution episodes in four European cities: Helsinki, London, Milan and Oslo. *Atmospheric Environment* **39**, 2759–2773.

Ma C 2003 Families of spatio-temporal stationary covariance models. *Journal of Statistical Planning and Inference* **116**, 489–501.

Melles SJ, Heuvelink GBM, Twenhöfel CJW, Van Dijk A, Hiemstra PH, Baume O and Stöhlker U 2011 Optimizing the spatial pattern of networks for monitoring radioactive releases. *Computers and Geosciences* **37**, 280–288.

Müller W 2007 *Collecting Spatial Data* 3rd edn. Springer.

Müller W and Stehlik M 2010 Compound optimal spatial designs. *Environmetrics* **21**, 254–264.

Neteler M 2010 Estimating daily land surface temperatures in mountainous environments by reconstructed modis lst data. *Remote Sensing* **2**, 333–351.

Pebesma E 2012 *Spatio-temporal geostatistics using gstat*. Package vignettes. CRAN, University of Münster.

Pilz J and Spock G 2006 Spatial sampling design for prediction taking account of uncertain covariance structure. In *Proceedings of the 7th International Symposium on Spatial Accuracy Assessment in Natural Resources and Environmental Sciences* (eds Caetano M and Painho M). Portuguese Geographic Institute, Lisboan.

Rue H 2012 The r-inla project. Available at http://www.r inla.org/. Last accessed on 28 February 2012.

Snepvangers JJJC, Heuvelink GBM and Huisman JA 2003 Soil water content interpolation using spatio-temporal kriging with external drift. *Geoderma* **112**, 253–271.

Spadavecchia L 2008 *Estimation of Landscape Carbon Budgets: Combining Geostatistical and Data Assimilation Approaches*. PhD thesis, The University of Edinburgh, Edinburgh.

Braak CJ 2006 A Markov chain Monte Carlo version of the genetic algorithm differential evolution: easy Bayesian computing for real parameter spaces. *Statistics and Computing* **16**, 239–249.

van Groenigen JW, Siderius W and Stein A 1999 Constrained optimization of soil sampling for minimization of the kriging variance. *Geoderma* **87**, 239–259.

Vrugt JA, Robinson BA and Hyman JM 2008 Self-adaptive multimethod search for global optimization in real-parameter spaces. *IEEE Transactions on Evolutionary Computation* **13**, 243–259.

Wan Z, Zhang Y, Zhang Q and Li ZL 2004 Quality assessment and validation of the modis global land surface temperature. *International Journal of Remote Sensing* **25**, 261–274.

Xia G, Miranda M and Gelfand A 2006 Approximately optimal spatial design approaches for environmental health data. *Environmetrics* **17**, 363–385.

Yu HL, Kolovos A, Christakos G, Chen JC, Warmerdam S and Dev B 2007 Interactive spatiotemporal modelling of health systems: the Seks-Gui framework. *Stochastic Environmental Research and Risk Assessment* **21**, 555–572.

Zhu Z and Stein M 2006 Spatial sampling design for prediction with estimated parameters. *Journal of Agricultural, Biological and Environmental Statistics* **11**, 24–49.

10

Space-time adaptive sampling and data transformations

José M. Angulo[1], María C. Bueso[2] and Francisco J. Alonso[1]

[1]*Department of Statistics, University of Granada, Spain*

[2]*Department of Applied Mathematics and Statistics, Technical University of Cartagena, Murcia, Spain*

10.1 Introduction

In many applications, spatio-temporal sampling procedures allow for temporal adaptability, in the sense that the set of spatial sampling locations can be (totally or partially) modified between different observation times. Prior or inferential knowledge on structural properties, jointly with real-time sample-path observed evolution, provide the basic information sources for dynamic modeling and prediction in this context. In this chapter, some related approaches, mainly based on entropy criteria, are described and illustrated. The effect of variable transformations such as averages and/or maxima over given time periods or space regions, often considered in the formulation of critical indicators, is also evaluated in terms of entropy-based information measures.

Spatio-temporal design: Advances in efficient data acquisition, First Edition.
Edited by Jorge Mateu and Werner G. Müller.

We use Shannon's entropy, defined for a continuous random variable (or vector) X as

$$H(X) = E[-\log(f(X))] = \int_{\Omega} -\log(f(x))f(x)dx,$$

where f is the density function and Ω is the support of X. For a Gaussian random vector of dimension N, the entropy can be calculated as

$$H(X) = \frac{N}{2}(1 + \log(2\pi)) + \frac{1}{2}\log|C_X|,$$

where C_X is the covariance matrix of X. The entropy of X conditional on $Y = y_s$ is given by

$$\begin{aligned} H(X|Y = y_s) &= E_{X|Y=y_s}[-\log(f(X|Y = y_s))] \\ &= \int_{\Omega} -\log(f(x|Y = y_s))f(x|Y = y_s)dx. \end{aligned}$$

Thus, the mean conditional entropy of X given Y can be expressed as

$$H(X|Y) = E_Y[H(X|Y = y)] = E_{(X,Y)}[-\log(f(X|Y))].$$

The mutual information between X and Y is given by

$$\begin{aligned} I(X, Y) &= H(X) - H(X|Y) = H(Y) - H(Y|X) \\ &= H(X) + H(Y) - H(X, Y). \end{aligned}$$

If (X, Y) is jointly Gaussian, the conditional entropy can be expressed in terms of the determinants of covariance matrices as

$$H(X|Y) = \frac{N}{2}(1 + \log(2\pi)) + \frac{1}{2}\log\left[\frac{|C_{X,Y}|}{|C_Y|}\right],$$

and hence,

$$I(X, Y) = \frac{1}{2}\log|C_X| - \frac{1}{2}\log\left[\frac{|C_{X,Y}|}{|C_Y|}\right],$$

where N is the dimension of vector X, and C represents the covariance matrix of the subscript variable or vector. Note that the quotient of determinants is equal to $|C_{X|Y}|$, the determinant of the conditional covariance matrix.

Optimality criteria based on entropy and related quantities such as mutual information have been adopted for the design of spatial and spatio-temporal sampling strategies, particularly in the context of environmental applications. A comprehensive review can be found, for instance, in the books of Le and Zidek (2006), Müller (2007), and Gelfand *et al.* (2010). In particular, the pioneering work following this approach was introduced by Caselton and Zidek (1984). Bueso *et al.* (1998) formulate this approach in a state-space model framework, first in the spatial case, with later extensions to the spatial multivariate case (Angulo and Bueso 2001), to spatio-temporal configurations (Angulo *et al.* 2000), and to generalized processes defined in terms of functionals and indicator functions (Angulo *et al.* 2005). A dynamic spatio-temporal sampling network design approach is developed in Wikle and Royle (2005). Chang *et al.* (2007) study monitoring network design with the objective to measure extremes. A combined entropy-utility approach in a Bayesian framework is proposed by Fuentes *et al.* (2007).

In this chapter, we consider the sampling network real-time adaptation based on the observations, and the quantification of the information loss derived from transformation of the data. Specifically, in Section 10.2, the adaptive sampling design procedure is formulated, and illustrated with a simulated example. The effect of transformations on predictive information and, as a consequence, on the sampling designs is studied in Section 10.3. In Section 10.4, the temperature data given by 37 stations located in the Upper Austria region are analyzed in relation to the above mentioned aspects. A summary is presented in Section 10.5.

10.2 Adaptive sampling network design

Our objective is to predict a spatio-temporal process of interest, X, from the observation of a related spatio-temporal process, Y. The processes X and Y are assumed not to be mutually independent. For simplicity, we consider only one-step ahead forecasts: at each time t, predictions for X at time $t+1$ must be obtained on a set of sites, the 'region of interest', denoted Λ_{t+1}. Although, in theory, the region of interest could be continuous, the implementation of the optimization criterion requires finite discretization. Predictions are derived from the historical observations of process Y up to time t provided by the adaptive sampling network . The base criterion consists of maximizing the mutual information between the set of observation variables and the set of variables of interest, or, equivalently, minimizing the expected residual entropy conditional on those observations. In practice, changes in the region of interest can be motivated, for instance, by the consideration of the observed behavior of the process, information from covariates, or even factors external to the phenomenon but of interest to the investigator. In this section, real-time collected observations are used to redefine the region of interest, at each time, according to the actual sample-path evolution of the system. Assume, for instance, the case where interest is concentrated on areas with relatively higher values. Specifically, let Λ_1 be the region of interest for the unobservable process X at the initial time. Then, the set of locations to be observed at time 1, S_1, is selected by maximizing the mutual information between $\mathbf{X}_{\Lambda_1}$ (the random vector whose co-ordinates represent the spatio-temporal process of interest at Λ_1) and $Y_{(S,1)}$ (the observable process at set S, time 1):

$$S_1 = \arg \max_{S \subset \mathcal{S}_1} I(X_{\Lambda_1}, Y_{(S,1)}),$$

where $\mathcal{S}_1$ denotes the class of admissible sets considered from which S will be selected.

At each time $t \geq 1$, an adaptive design procedure may be performed according to the following steps:

1. Update the region of interest for the unobservable process X at time $t+1$, Λ_{t+1}, as follows:
   (a) Specify a distribution F to select the sites to be included in the set Λ_{t+1}. In particular, a Beta distribution will be considered here to generate points of interest in the neighborhood of critical values observed at time t.
   (b) By using the observations available at time t from $S_1, \ldots S_t$, obtain estimated values for the process of interest X at time t on a regular mesh. A function normalized to the interval [0, 1] is determined from these values to reflect the spatial distribution of the current risk.
   (c) Generate random values from the distribution F and assign each generated value to the closest value among the transformed values on the grid. The sites corresponding to these values determine the set Λ_{t+1}.
2. Select the set of locations to be observed at time $t+1$, S_{t+1}, by maximizing the mutual information between $\mathbf{X}_{\Lambda_{t+1}}$ (the random vector whose co-ordinates represent the spatio-temporal process of interest at Λ_{t+1}) and $Y_{(S_1,1),\ldots,(S_t,t),(S,t+1)}$ (the observable process at sets $S_1, \ldots, S_t, S$, times $1, \ldots, t, t+1$, respectively):

$$S_{t+1} = \arg \max_{S \subset \mathcal{S}_{t+1}} I(X_{\Lambda_{t+1}}, Y_{(S_1,1),\ldots,(S_t,t),(S,t+1)}),$$

   where $\mathcal{S}_{t+1}$ denotes the class of admissible sets considered from which S will be selected.

The function F specified in Step 1(a) must take into account both the characteristic to be detected and the structural properties of the process of interest. For instance, if we are interested in areas with very high values, a Beta distribution with a large value for the first parameter and a value lower than 1 for the second one would be an appropriate choice. Also, a high variability in the process of interest can lead to a large number of relative extremes, and hence the function F must be selected to make possible detection of such points eventually far from the absolute maximum which determines the normalization (in the case of the Beta distribution, the first parameter should be taken moderately large). On the other hand, in the case of lower variability, there will be a large number of higher values in the neighborhood of the global maximum (a larger value should be taken for the first parameter of the Beta distribution to get a higher density of points in that neighborhood in the region of interest).

Extension of the procedure to non integer or non equally spaced time intervals is straightforward.

10.2.1 A simulated illustration

In this section, the adaptive sampling criterion introduced is applied to network design in some cases where the underlying process has a spatio-temporal dependence structure depending on a parametric model. For illustration, here we only present the results obtained for two particular cases, where the effect of the structural characteristics on the network design is shown.

Let X be the spatio-temporal process of interest defined on the spatial domain $D \subset \mathbb{R}^2$ and modeled by convolution of the following spatio-temporal filter f with the innovation process ε, considered to be spatio-temporal Gaussian white noise with mean 0 and variance σ_ε^2:

$$f(s, t; \mathbf{x}, \mathbf{y}; \beta, \boldsymbol{\theta}) = \frac{\beta}{(1 + |t - s|^2 + \|\mathbf{x} - \mathbf{y}\|^2)^{h(\boldsymbol{\theta}, \mathbf{x})}}, \qquad \mathbf{x}, \mathbf{y} \in D, \tag{10.1}$$

with $\beta > 0$, $\boldsymbol{\theta} \in \boldsymbol{\Theta}$, and h defined as

$$h(\boldsymbol{\theta}, \mathbf{x}) = \frac{\theta_1}{\theta_2 + \|\mathbf{x}\|^{2/\theta_3}}, \tag{10.2}$$

where, for $\|\mathbf{x}\| > 0$, $\boldsymbol{\Theta}$ is defined with the restriction $h > 3/2$ (Ruiz-Medina and Angulo 2007). The covariance function of X is then defined by spatio-temporal self-convolution of f.

In all cases, we assume that the observation process Y consists of the process of interest affected by additive noise,

$$Y_t(\mathbf{s}) = X_t(\mathbf{s}) + \epsilon_t(\mathbf{s}), \quad \mathbf{s} \in D \subset \mathbb{R}^2, \ t \in \{1, \ldots, T\}, \tag{10.3}$$

where ϵ represents a spatio-temporal Gaussian white noise process, mutually uncorrelated with X, with mean 0 and variance σ_ϵ^2. In practice, at each time the process Y can be potentially observable on a set of spatial sites Π and our objective is to select the best subset $S \subset \Pi$ composed of a fixed number of sites or, alternatively, composed of the minimum number of sites required to achieve a target value of amount of information. Although the set Π could be a continuous set, as in the case of the region of interest, we define Π as a finite set of locations for computational considerations. For the case in which Π is a continuous set, Bueso *et al.* (2005) carried out an extensive empirical study showing the influence of the level of refinement of the discretization considered in the continuous region to be sampled on the final designs, as well as on the corresponding ratios of information.

In the design stage, all of the parameters involved are assumed to be known. For this model, we study two cases for different configurations of parameter values. In case A, we take $\beta = 25$, $\theta_1 = 100$, $\theta_2 = 0$, $\theta_3 = 3/2$, $\sigma_\varepsilon^2 = 3$, and $\sigma_\epsilon^2 = 0.001$. In case B, $\beta = 25$, $\theta_1 = 6$, $\theta_2 = 0$, $\theta_3 = 5$, $\sigma_\varepsilon^2 = 3$, and $\sigma_\epsilon^2 = 0.001$.

In both cases, we consider $D = [0, 1]^2 \subset \mathbb{R}^2$ as the common spatial domain and $T = 6$. In Figures 10.1 and 10.2 realizations generated from process X for

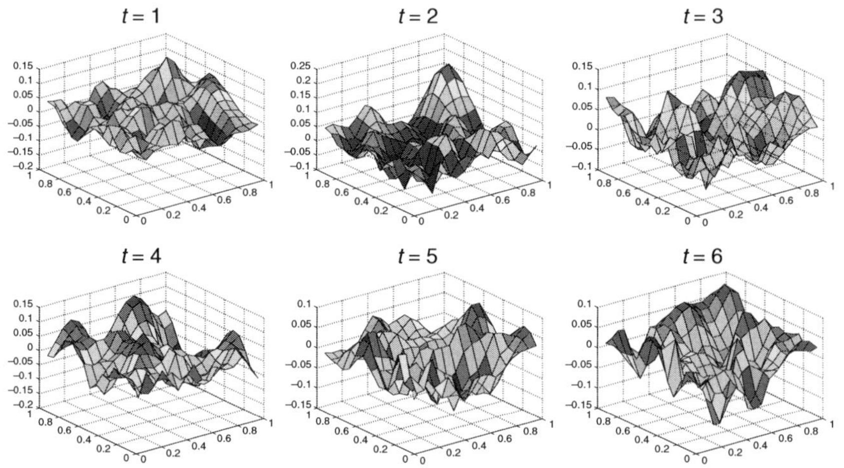

Figure 10.1 Case A. Simulated values of the spatio-temporal process of interest at times t $= 1, \ldots, 6$.

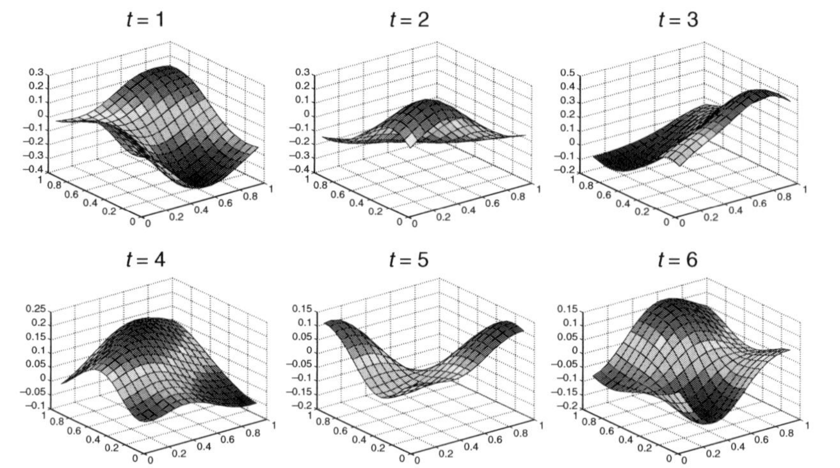

Figure 10.2 Case B. Simulated values of the spatio-temporal process of interest at times t $= 1, \ldots, 6$.

cases A and B on D and at times $t = 1, \ldots, 6$ are displayed, showing the different behavior in the evolution of the underlying process.

The set of potentially observable sites, Π, consists of 150 locations randomly generated from a regular 16×16 grid defined on D (Figure 10.3). The set of locations of interest is defined by 16 fixed points plus 26 additional points to be selected, at each time, from the 16×16 grid.

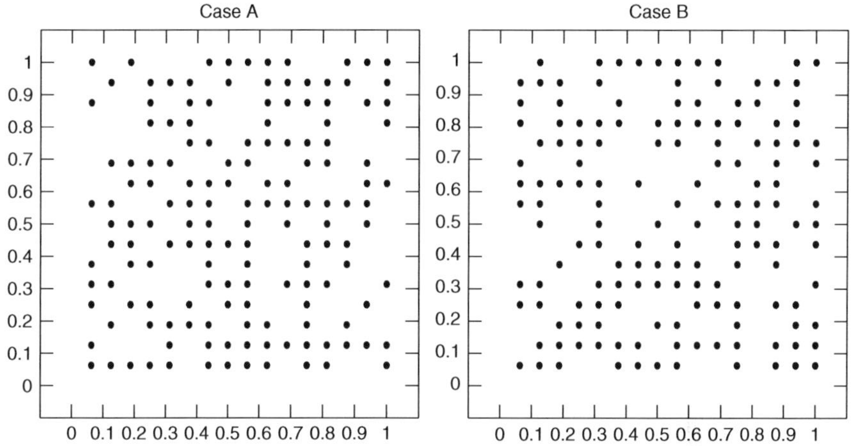

Figure 10.3 *Set of candidate sites for observation.*

For $t = 1, \ldots, 6$, the design criteria are applied to sequentially select, at each time t, 50 observation sites from the set of candidates. The starting set of locations of interest Λ_1, given by 42 fixed sites, and the locations included in the successive regions of interest obtained for $t = 2, \ldots, 6$ are shown in Figures 10.4 (case A) and 10.5 (case B), where we can observe relatively higher concentrations

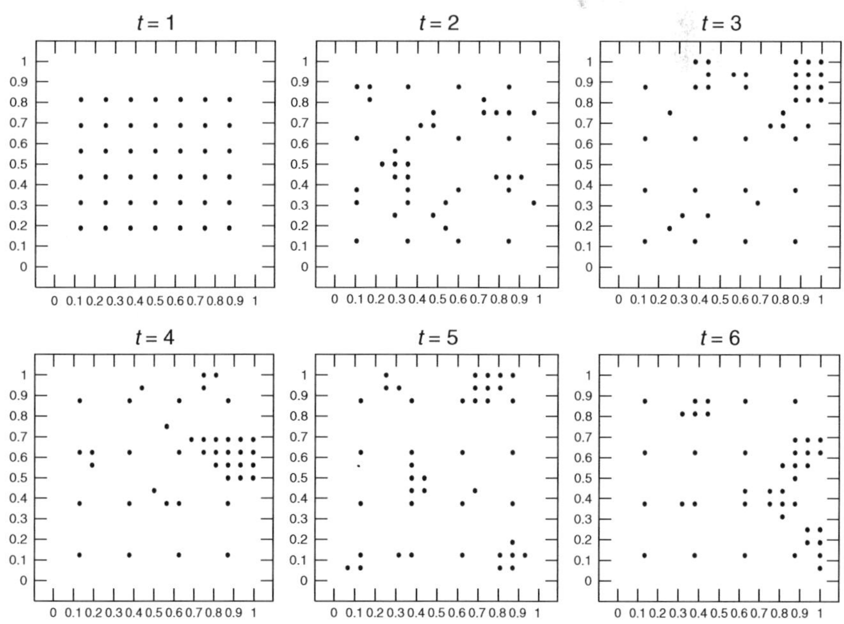

Figure 10.4 *Case A. Locations of interest at times* $t = 1, \ldots, 6$.

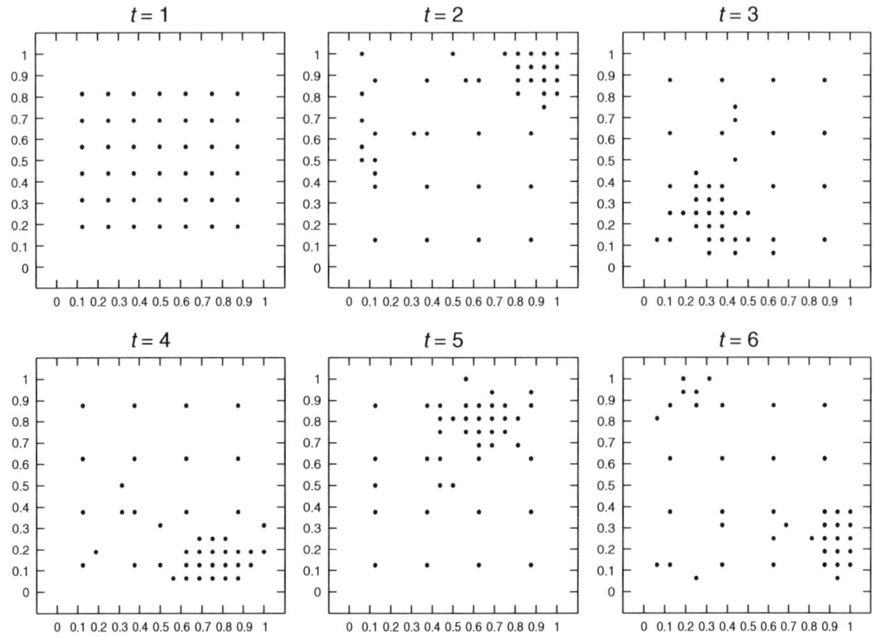

Figure 10.5 Case B. Locations of interest at times $t = 1, \ldots, 6$.

located in the areas with higher values for the simulated spatio-temporal process X. The resulting designs, displayed in Figures 10.6 and 10.7, show the dynamical adaptation of the network, clearly influenced by the model dependence structure assumed in each case. Thus, we can observe that the configuration of the region of interest is updated at each time with substantial changes for both cases, while the sampling network designs obtained in case A show changes in the configurations more evident than in case B.

Finally, Figures 10.8 and 10.9 show the results obtained when the first region of interest is maintained fixed for all subsequent times, that is, the network minor changes over time are due only to structural variations, again reflecting the dependence effect.

10.3 Predictive information based on data transformations

Another problem of interest we consider in this chapter consists of evaluating the effect of certain transformations of the variables used for predicting the configuration of the sampling network designs. Such transformations are usually related to the formulation of some indicators, and in most applications the same transformation considered to define the magnitude of interest (average or maximum

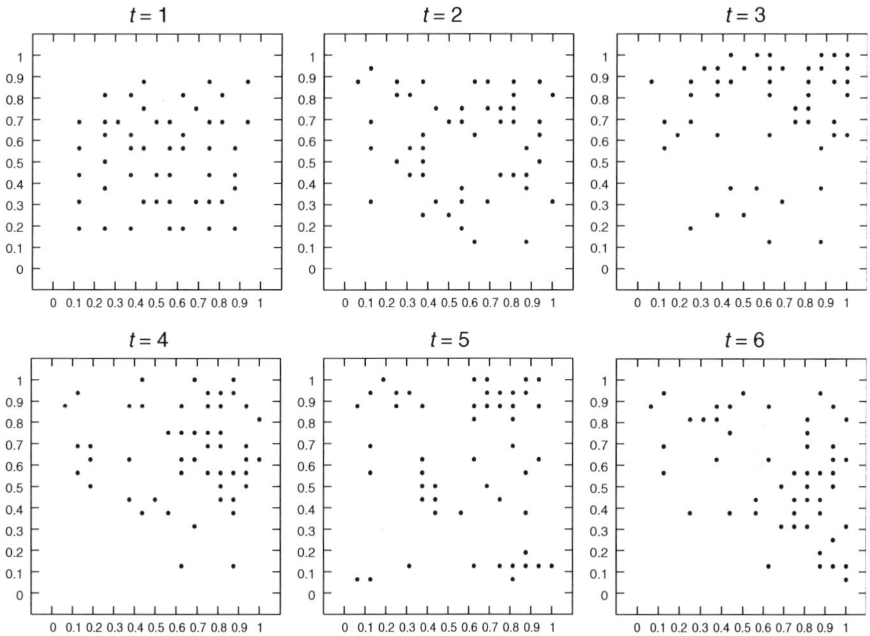

Figure 10.6 Case A. Sequentially selected network for times t = 1, ..., 6.

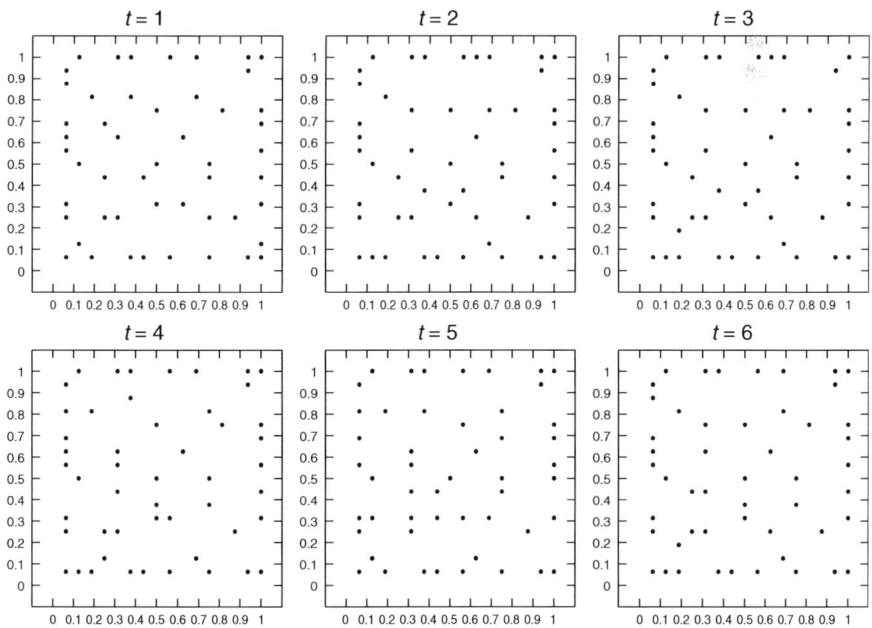

Figure 10.7 Case B. Sequentially selected network for times t = 1, ..., 6.

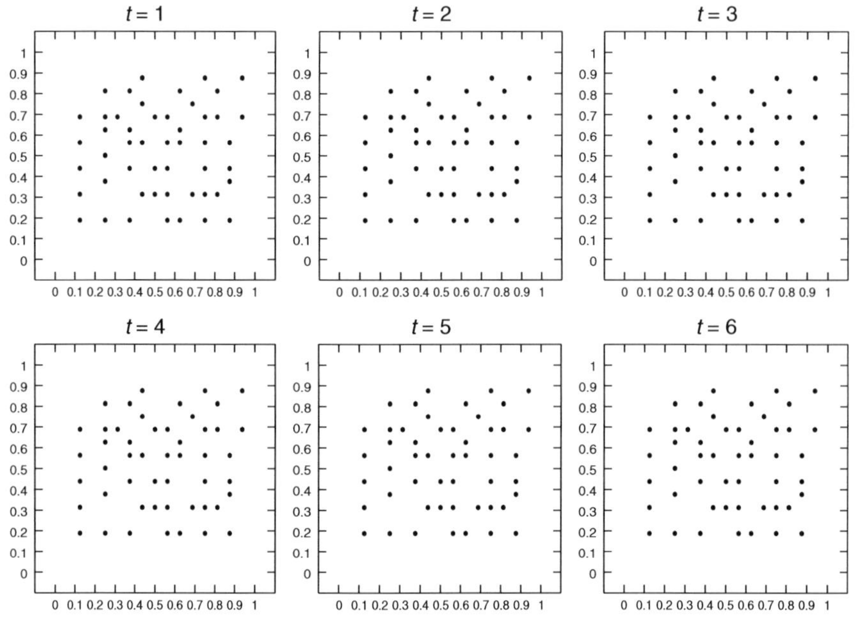

Figure 10.8 Case A. Sequentially selected network for times t = 1, . . . , 6*, for fixed region of interest.*

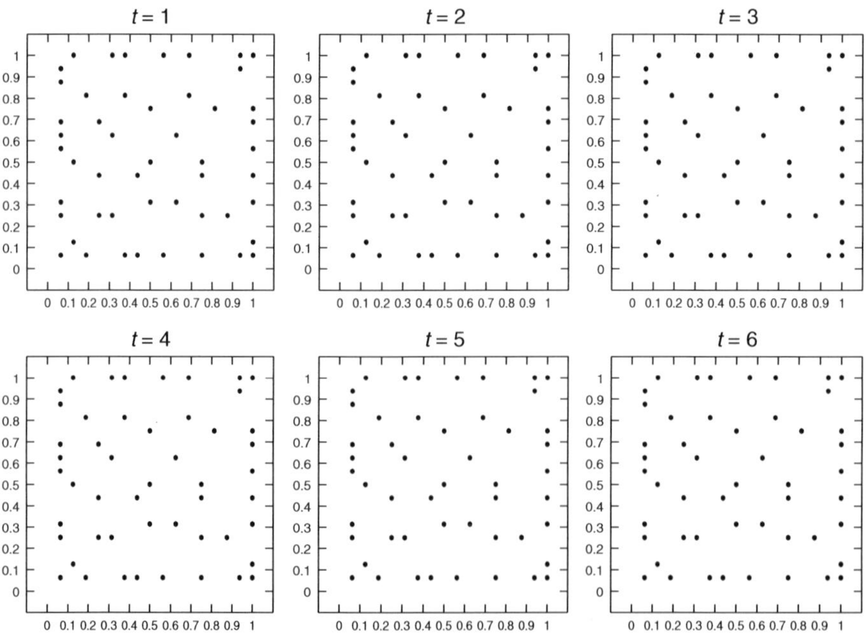

Figure 10.9 Case B. Sequentially selected network for times t = 1, . . . , 6*, for fixed region of interest.*

over a fixed period, exceedance over a given threshold, etc.) is used to redefine the sample information from the raw original data.

As a quantifier, to compare the relative information between different cases, we use again the mutual information $I(RI, SI)$, where RI represents the risk indicator of interest to be predicted, and SI is the random vector providing the set of sample information defined from the original observed variables Y, that is, the original observations or a transformation of them.

As can be expected, the information decreases or, at the most, remains unchanged when the data are transformed by any function g, $I(RI, Y) \geq I(RI, g(Y))$. However, the information loss depends on the particular scenario, in terms of the underlying model structure and the transformation considered (Alonso *et al.* 2012).

For illustration purposes, we consider the model defined by (10.1)–(10.3) under specifications for the parameter values given in case B. Now the objective is to predict the average of process X for times $t = 5, 6$ at the starting region of interest for X, Λ_1 (Figure 10.5). We study two cases. In the first case (case 1B), three alternative sets of sample information from Y related to observations at set Π, composed of 150 locations (Figure 10.3, case B), are analyzed: (a) the raw values of process Y for times $t = 1, \ldots, 4$ at all locations (SI_1); (b) the moving two-time averages, corresponding to times (1,2), (2,3) and (3,4) (SI_2); and (c) the disjoint two-time averages, corresponding to times (1,2) and (3,4) (SI_3). In the second case (case 2B), the following sets of sample information are considered: (a) the raw values of process Y for times $t = 1, \ldots, 6$, at all locations (SI_1); (b) the moving two-time averages, corresponding to times (1,2), (2,3), (3,4), (4,5) and (5,6) (SI_2); and (c) the disjoint two-time averages, corresponding to times (1,2), (3,4) and (5,6) (SI_3). Table 10.1 presents the absolute predictive information values and the corresponding ratios between the different sets of sample information considered. These quantities show the relative loss of information when the original data are transformed, reflecting the fact that SI_3 is obtained by transformation from SI_2, and SI_2 from SI_1. In particular, in the first case the information loss ranges from about 23 to 49%, while in the second case the maximum loss is only about 0.02%.

To study the influence of data transformations on the sampling configuration of the network designs, we have compared the resulting networks when 20 observation sites are sequentially selected from the set Π. For case 1B, Figure 10.10 displays the resulting networks for the sets of sample information considered; note

Table 10.1 Cases 1B and 2B. Amount of information values and the corresponding ratio of information values for the different sets of sample information.

	$I(RI, SI_1)$	$I(RI, SI_2)$	$I(RI, SI_3)$	$\frac{I(RI, SI_2)}{I(RI, SI_1)}$	$\frac{I(RI, SI_3)}{I(RI, SI_1)}$	$\frac{I(RI, SI_2)}{I(RI, SI_3)}$
Case 1B	0.024520	0.018755	0.012562	0.764886	0.512316	0.669795
Case 2B	46.040559	46.037334	46.031149	0.999929	0.999795	0.999865

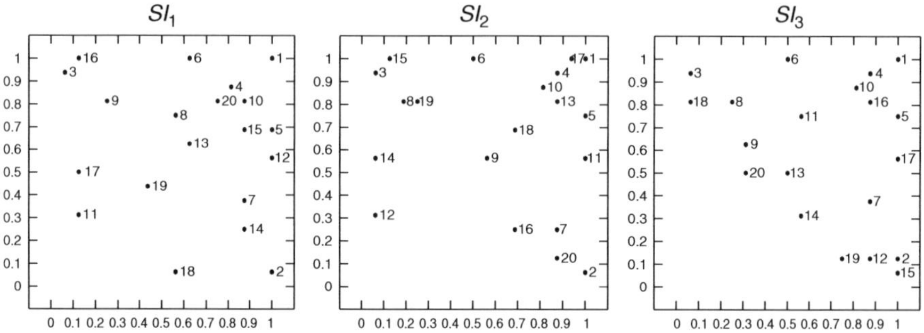

Figure 10.10 *Case 1B. Sequentially selected network for the different sets of sample information.*

the different configurations obtained. By contrast, for case 2B, Figure 10.11 shows that the resulting networks coincide (even in the selection ordering) for the three cases.

10.4 Application to Upper Austria temperature data

Temperature data are observed at 37 stations distributed in the Upper Austria state (Figure 10.12); in 35 of them the elevation ranges from 262 to 725 m, and for two specific stations the elevation values are 1618 and 2050 m. For each station, the temperature data are monthly recorded from January 1994 until December 2009, and in a large number of stations certain observations are missing (e.g., the time series at Weyregg station only has 44 observations available from the 192 possible ones). The number of stations without missing data is 18, including Feuerkogel station with an elevation of 1618 m.

Figure 10.13 displays, for illustration, the time evolution of the temperature data at 5 stations, showing a significant seasonal component. As expected, the time series corresponding to higher elevations present lower temperature values.

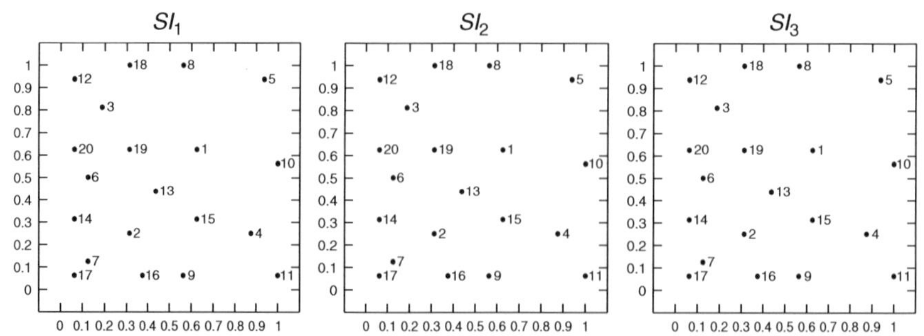

Figure 10.11 *Case 2B. Sequentially selected network for the different sets of sample information.*

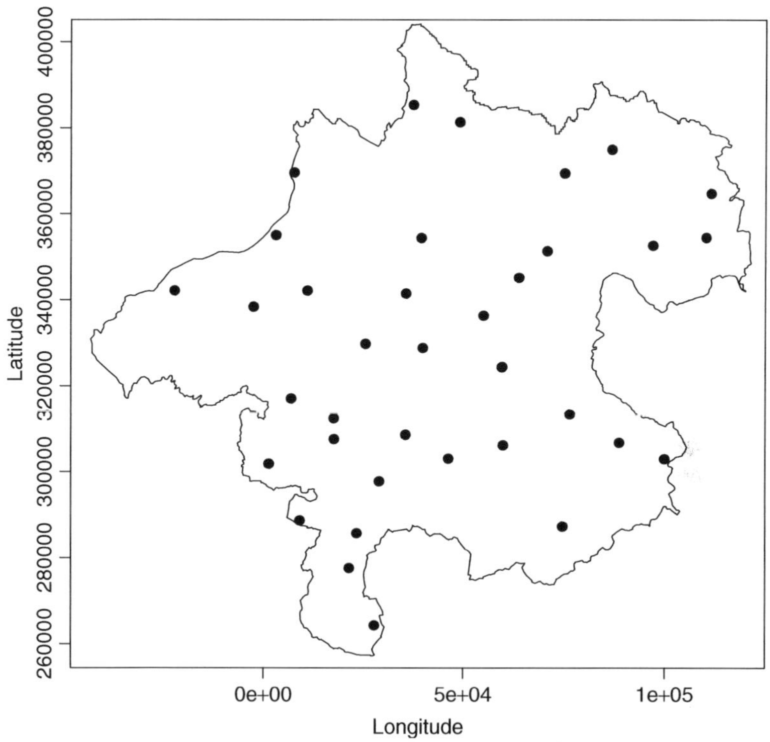

Figure 10.12 Spatial locations of Upper Austria stations.

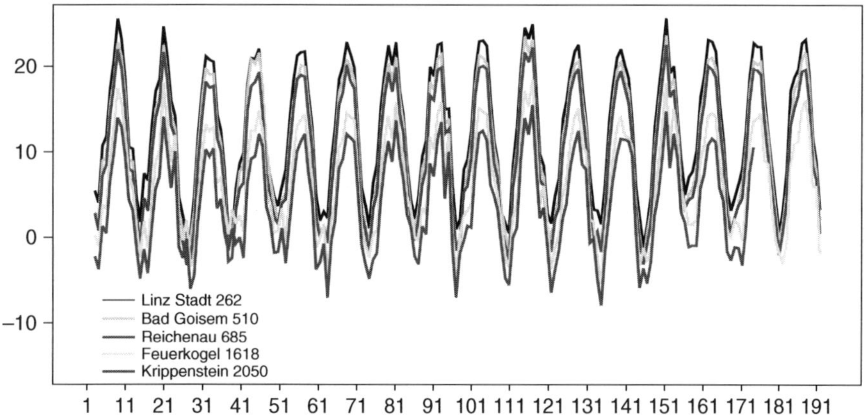

Figure 10.13 Time series observed at 5 stations.

Considering the seasonality of the data and the influence of the elevation together with latitude and longitude coordinates in temperature, in a first step, we have fitted the linear regression model

$$Temp(\mathbf{s},t) = \mu + \sum_{i=2}^{12} \alpha_i \delta_i(t) + \beta_1 Lat(\mathbf{s}) + \beta_2 Long(\mathbf{s}) + \beta_3 Elev(\mathbf{s}) + e(\mathbf{s},t),$$

where μ is a constant representing a global mean related to January; $\alpha_2, \ldots, \alpha_{12}$, are constants representing the deviations of the mean temperature in the ith month with respect to μ; $\beta_1, \beta_2, \beta_3$ are parameters that account for the effect of the station position. The functions $\delta_i(t)$ are defined as

$$\delta_i(t) = \begin{cases} 1 & \text{if } t \text{ corresponds to the } i\text{th month} \\ 0 & \text{otherwise} \end{cases}.$$

The term $e(\mathbf{s},t)$ is the random error component. The parameter estimates are presented in Table 10.2. It can be observed that all the coefficients are statistically significant.

From the residuals obtained after fitting the regression model, in a second step, we model the stochastic spatio-temporal dependence structure. First, we analyze the $e(\mathbf{s},t)$ residual for each fixed $\mathbf{s}$, as individual time series, using a SARIMA model. We observe that, except for the two highest elevation stations, the seasonal effect has been removed, and only two autoregressive (AR) terms, at maximum, are needed to explain the ordinary temporal dependence (in some cases, the second-order coefficient, and even the first-order one, are not significant).

Table 10.2 Parameter estimates for the regression model fitted to temperature data.

Parameter	Estimate	t-value	p-value
μ	2.61549072	8.653	0.000
α_2	1.92159533	16.199	0.000
α_3	5.16892397	43.594	0.000
α_4	9.98552975	84.094	0.000
α_5	15.34853809	129.259	0.000
α_6	18.31967230	153.977	0.000
α_7	19.89720773	167.483	0.000
α_8	19.27921681	162.361	0.000
α_9	14.66750203	123.523	0.000
α_{10}	10.33730804	86.885	0.000
α_{11}	5.05760370	42.530	0.000
α_{12}	0.96187586	8.085	0.000
β_1	−1.9533593E-06	−2.633	0.008
β_2	−5.1375160E-06	−6.095	0.000
β_3	−0.00467499	−62.457	0.000

We assume that the residuals can be modeled according to

$$e(\mathbf{s}, t) = X(\mathbf{s}, t) + \varepsilon(\mathbf{s}, t),$$

where $\varepsilon(\mathbf{s}, t)$ represents observation white noise, assumed to be Gaussian, with mean 0 and variance σ_ε^2, uncorrelated with X.

From the residuals, we try to estimate the spatio-temporal covariance of process $e(\mathbf{s}, t)$. Unfortunately, in this case, the scale of distances between stations is significantly larger than the scale at which supposedly spatial dependence occurs. Once the seasonal and positional effects have been removed, the residuals then do not allow us to estimate further spatial dependence.

In the case where such a spatio-temporal dependence structure would exist at the spatial scale of interest, a dynamic network design could be performed according to the following procedure. Our aim is to select at each time, the stations providing optimal predictive information for a region of interest which changes over time. Let us suppose that we are concerned about the occurrence of high residuals. The region of interest for time $t + 1$ is defined from the kriging surface based on the residuals at time t and their estimated covariance, identifying those areas where extremal episodes are detected.

At time $t = 1$, the region of interest consists of a certain regular mesh plus the locations of the available stations. The amount of information provided by X on the residual $e(\mathbf{s}, t)$ at these locations is then maximized, giving the observation set S_1:

$$I(e(\Lambda_1, 1), X(S_1, 1)).$$

Based on this observation set, a new region of interest is specified for $t = 2$, Λ_2. A new observation set S_2 is now defined, such that the information provided jointly with S_1 is maximized:

$$I(e(\Lambda_2, 2), (X(S_2, 2), X(S_1, 1))).$$

The procedure is iterated for all subsequent times.

Concerning the problem of the effect of data transformations on predictive information, we consider here as objective to predict the annual average temperature from the monthly temperature data at Linz/Stadt station. We define the following sample information sets: (a) the monthly temperature data at the station (SI_1); (b) the moving 12-month average temperature data (SI_2); and (c) the annual average temperature data (SI_3). At the first stage, we obtain the parameter estimates for the AR(2) model fitted to the residuals $e(\mathbf{s}, t)$ for this station, shown in Table 10.3. This model is suggested by the corresponding sample autocorrelation and sample partial autocorrelation functions (ACF and PACF), as can be observed in Figure 10.14. All the coefficients are significant at the level 0.05. The AR(2) model residuals have white noise structure, with estimated variance 2.798.

Using this model, we compute the covariances involved in the three scenarios considered, from which we obtain the corresponding values of predictive information, displayed in Table 10.4. The ratios with respect to the maximum

Table 10.3 Parameter estimates for the AR(2) model fitted to the residuals obtained after removing the trend from the original time series at Linz/Stadt station.

Parameter	Estimate	t-value	p-value
AR1	0.156	2.172	0.031
AR2	0.145	2.015	0.045
Const.	0.624	3.622	0.000

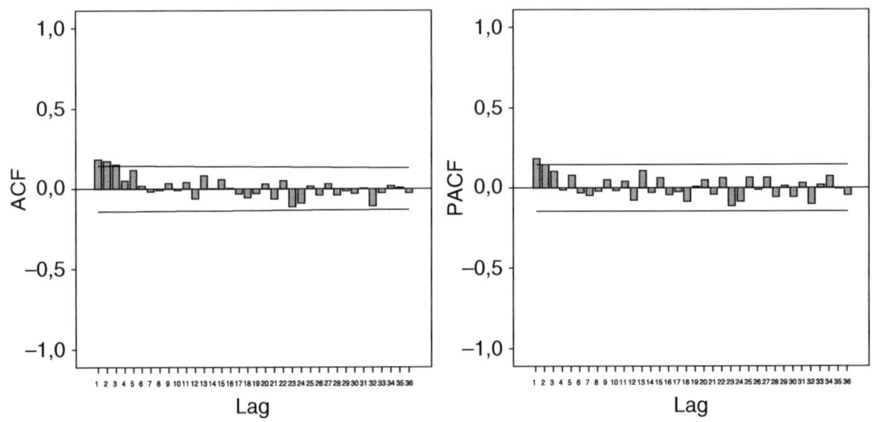

Figure 10.14 ACF and PACF for the Linz/Stadt station.

information, given by the monthly data residuals, are 0.9528 and 0.1714 for cases (b) and (c), respectively. Hence, the loss of information when using the moving 12-month average data is less than 5%, while the loss using the annual averages is higher than 82%.

10.5 Summary

In this chapter, we study and illustrate two aspects of interest in the context of spatio-temporal sampling design. First, we propose a methodological strategy for dynamic network sampling design based on the knowledge of the spatio-temporal

Table 10.4 Amount of information values and the corresponding ratio of information values for the different sets of sample information.

$I(RI, SI_1)$	$I(RI, SI_2)$	$I(RI, SI_3)$	$\frac{I(RI, SI_2)}{I(RI, SI_1)}$	$\frac{I(RI, SI_3)}{I(RI, SI_1)}$	$\frac{I(RI, SI_2)}{I(RI, SI_3)}$
0.0062	0.0059	0.0011	0.9528	0.1714	0.1864

dependence structure of the underlying model jointly with real-time observations. In this context, changes in the region of interest are allowed dynamically in time. These changes can be motivated, for instance, by the realization of the process, changes in covariates affecting the process, or even a deterministic *ad hoc* dynamical planning. These modifications in the region of interest jointly with the dependence structure of the process lead to the selection at each time of the sampling region.

The methodology introduced is illustrated with a simulation study in which the underlying process is represented by a parametric model defining the spatio-temporal dependence structure. The proposed procedure is applied to different scenarios and the results show, in consecutive times, the adaptation of the spatial configuration of the region of interest as well as of the sampling points. For computational implementation of the optimal solution searching methods, we have considered that the region of interest as well as the region to be sampled are given by finite sets. Although in general these may refer to continuous regions, some form of discretization will be used for practical purposes, with the level of refinement having a certain effect on the final network design configurations.

Secondly, we address the problem of the effect of data transformations on predictive information and, as a consequence, on the sampling designs. Assuming, as before, the underlying model structure is known, we show for different scenarios how such data transformations affect the final network designs, as well as the substantial information loss in some cases, an important aspect to take into consideration when the available sample information is given in terms of certain risk indicators. In some applications, the values of the transformed variables are the only information that the researcher can obtain even when the values of the variables themselves are directly observed; for example, if only some levels or functions of the data are the object of interest for legal regulations.

Acknowledgments

This work has been supported in part by project P08-FQM-3834 of the Andalusian CICE, and project MTM2009-13250 of the SGPI, MICINN, Spain.

References

Alonso FJ, Bueso MC and Angulo JM 2012 Effect of data transformations on predictive risk indicators. *Methodology and Computing in Applied Probability* **14**, 705–716.

Angulo JM and Bueso MC 2001 Random perturbation methods applied to multivariate spatial sampling design. *Environmetrics* **12**, 631–646.

Angulo JM, Bueso MC and Alonso FJ 2000 A study on sampling design for optimal prediction of space-time stochastic processes. *Stochastic Environmental Research and Risk Assessment* **14**, 412–427.

Angulo JM, Ruiz-Medina MD, Alonso FJ and Bueso MC 2005 Generalized approaches to spatial sampling design. *Environmetrics* **16**, 523–534.

Bueso MC, Angulo JM and Alonso FJ 1998 A state-space model approach to optimum spatial sampling design based on entropy. *Environmental and Ecological Statistics* **5**, 29–44.

Bueso MC, Angulo JM, Alonso FJ and Ruiz-Medina MD 2005 A study on sensitivity of spatial sampling designs to a priori discretization schemes. *Environmental Modelling and Software* **20**, 891–902.

Caselton WF and Zidek JV 1984 Optimal monitoring network designs. *Statistics and Probability Letters* **2**, 281–298.

Chang H, Fu AQ, Le ND and Zidek JV 2007 Designing environmental monitoring networks to measure extremes. *Environmental and Ecological Statistics* **14**, 301–321.

Fuentes M, Chaudhuri A and Holland DM 2007 Bayesian entropy for spatial sampling design of environmental data. *Environmental and Ecological Statistics* **14**, 323–340.

Gelfand AE, Diggle PJ, Fuentes M and Guttorp P 2010 *Handbook of Spatial Statistics*. Chapman & Hall/CRC.

Le ND and Zidek JV 2006 *Statistical Analysis of Environmental Space-Time Processes*. Springer.

Müller WG 2007 *Collecting Spatial Data: Optimum Design of Experiments for Random Fields*. Springer.

Ruiz-Medina MD and Angulo JM 2007 Functional estimation of spatio-temporal heterogeneities. *Environmetrics* **18**, 775–792.

Wikle CK and Royle JA 2005 Dynamic design of ecological monitoring networks for non-Gaussian spatio-temporal data. *Environmetrics* **16**, 507–522.

11

Adaptive sampling design for spatio-temporal prediction

Thomas R. Fanshawe[1] **and Peter J. Diggle**[1,2]

[1]*Lancaster Medical School, Lancaster University, UK*

[2]*Institute of Infection and Global Health, University of Liverpool, UK*

11.1 Introduction

Environmental monitoring provides a typical setting that gives rise to spatio-temporal design problems, and motivates the work in this chapter. Suppose public health authorities wish to monitor levels of particulate air pollution in a large city, with a view to constructing real-time, city-wide maps of particulate concentrations. Where should the devices be sited?

Fixed configurations of monitoring devices have been the design of choice for most epidemiological studies of the effects of air pollution on human health, because exposure modelling typically relies on data collected from permanent air monitoring networks (Cox 2000). Choosing the monitoring locations in this setting is the *fixed* sampling design problem, which is reviewed in detail by Zidek and Zimmerman (2010).

As exposure assessment technology improves, it is likely that mobile monitoring networks will be used instead of, or in addition to, permanent networks, possibly in conjunction with alternative technologies such as leaf sampling

Spatio-temporal design: Advances in efficient data acquisition, First Edition.
Edited by Jorge Mateu and Werner G. Müller.

(Maher *et al.* 2008). This gives rise to the *adaptive* or *dynamic* sampling design problem.

In this chapter, we consider model-based design, in which the optimal design problem requires two key features to be specified:

(i) a statistical or mathematical model for the process under consideration;

(ii) a criterion with respect to which the design is required to be optimized.

In practice, a good choice of model may not be known until data collection has begun. This highlights another benefit of adaptive designs: model fitting can take place as data become available, and future design locations chosen accordingly.

We demonstrate how both the choice of model and design criterion may affect the adaptive choice of a fixed number of sampling locations over a sequence of sampling times. A possible extension, which we do not consider, would be to allow the number of sampling locations to vary over the data collection period and to treat this as an additional component of the design problem, as is commonplace in analogous adaptive designs used in pharmacological studies (Schäfer *et al.* 2006).

We shall assume that the phenomenon under consideration can be modelled as a realisation of a stochastic process $S(x, t)$, and that data y_{it} are generated at a set of time-dependent locations $x_{it} : i = 1, ..., n$. We write $\mathcal{H}_t$ for the data up to time t, hence $\mathcal{H}_t = \{y_{is} : i = 1, ..., n; s = 1, ..., t\}$, write $S_t = \{S(x, t) : x \in A\}$, where A denotes the spatial region of interest, and define the predictive distribution of S_t to be its conditional distribution given $\mathcal{H}_t$, which we write as $[S_t|\mathcal{H}_t; \theta]$ where θ denotes the set of model parameters. We use this notation acknowledging that this distribution is defined only on a finite-dimensional set of locations, which in practice is used to approximate the region A.

The *plug-in* predictive distribution of S_t is then $[S_t|\mathcal{H}_t; \hat{\theta}]$ where $\hat{\theta}$ is any point estimate of θ, whilst the *Bayesian* predictive distribution is

$$[S_t|\mathcal{H}_t] = \int [S_t|\mathcal{H}_t; \theta] p(\theta|\mathcal{H}_t) d\theta$$

where $p(\cdot)$ is the Bayesian posterior for θ.

Any sample drawn from a predictive distribution for S_t automatically produces a sample from any target functional $T(S_t)$ of interest by direct calculation, i.e. if S_t^* is a draw from $[S_t|\mathcal{H}_t; \theta]$ then $T^* = T(S_t^*)$ is a draw from $[T(S_t)|\mathcal{H}_t; \theta]$, and likewise for plug-in or Bayesian predictive distributions. In practice, all but the simplest of spatio-temporal prediction problems are solved by Monte Carlo sampling from the required predictive distribution. For this reason, the few theoretical results that are available concerning optimal spatio-temporal design are confined to simple linear targets such as $T(S_t) = \int_A S(x, t) dx$ and pragmatic design criteria such as integrated mean square error,

$$IMSE_t = \int_A \mathrm{E}[\{\hat{S}(x, t) - S(x, t)\}^2] dx.$$

For other targets and/or design criteria, we will suggest algorithms for finding reasonable, but not necessarily optimal, adaptive designs. Two design criteria that we will consider in detail are to minimize, at each time t:

(i) $\max_{x \in A}[\text{Var}\{S(x,t)\}]$

(ii) $-\int_A \{\text{P}(S(x,t) > c) - p_0\}^2 dx$, where c and p_0 are fixed values, to be specified in advance.

The first of these is a minimax criterion that controls the maximum prediction variance of S_t across the whole of A, and is a common choice; see, for example, Wikle and Royle (1999) and Zimmerman (2006). The second is motivated by problems in which we seek to delineate regions that lie above or below a policy-relevant threshold. For example, if c is the maximum acceptable concentration of a pollutant, the aim is to discriminate between regions that are or are not in compliance. By choosing $p_0 = 0.5$, the criterion becomes a measure of how well the design discriminates, the ideal being that all predictive probabilities are 0 or 1.

After reviewing spatial and spatio-temporal adaptive designs in Section 11.2, we will consider the performance of adaptive design-finding algorithms with respect to these criteria for two different models for S: the stationary Gaussian model in Section 11.3; and a dynamic process convolution model in Section 11.4. In Section 11.5 we use the second of these models to consider adaptive designs for the Upper Austria rainfall data. In Section 11.6 we give a concluding discussion.

Before proceeding further, we define more formally what we mean by an adaptive design. We use *adaptive* to mean any design in which the design points x_{is} and x_{it} are allowed to differ for distinct times s and t. We use k*–adaptive* to mean any design in which at most k of the design points x_{is} and x_{it} are allowed to differ for consecutive time points s and $t = s + 1$. In most of the work below we consider *1–adaptive* designs.

11.2 Review of spatial and spatio-temporal adaptive designs

Early work in spatial design is summarized by Royle and Nychka (1998), who consider intuitively appealing designs based on space-filling, thereby ensuring even coverage of the sampling region. A greedy 'point-swapping' algorithm is available as the `cover.design` function in the `fields` library (Furrer *et al.* 2010) of R (R Development Core Team 2008) for selecting a near-optimal subset of sampling locations from a discrete set of candidate locations in an irregularly shaped region.

This criterion uses purely geometric considerations unrelated to modelling assumptions, and typically leads to regular lattices of sampling locations that cover the design region. Johnson *et al.* (1990) show that, in a regularly shaped region A such as a square, and ignoring edge effects, the optimal set of design

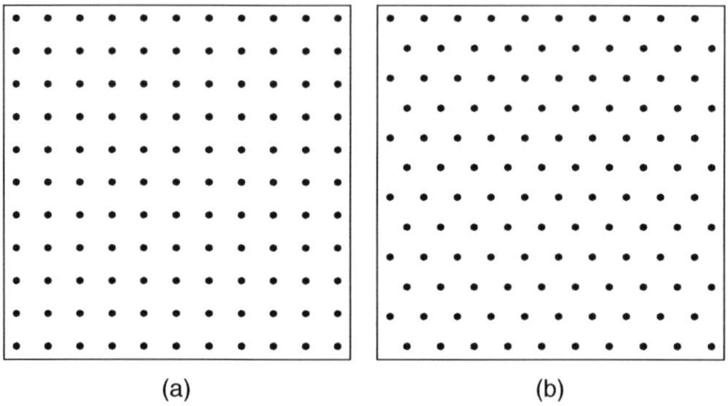

Figure 11.1 Regular (a) rectangular and (b) hexagonal lattices.

points $x_1, ..., x_n$ to minimize $\max_i[\max_{x\in A}\|x - x_i\|]$ forms a hexagonal lattice (Figure 11.1). We will return to this result in Section 11.3.

van Groenigen *et al.* (2000) consider another distance-based optimality criterion, which they term a 'weighted means of shortest distances'. Under this criterion, for given design points $x_1, ..., x_n$, define $d(x) = \min_i \|x - x_i\|$, the distance between a point $x \in A$ and its nearest design point, and let $w(x)$ be a weight function. The optimal design is the one that minimizes

$$\int_A d(x)w(x)dx.$$

This approach generalizes the simple space-filling criterion, for which $w(x)$ is constant. In these space-filling criteria, sampling design considerations rely not on the observed data Y, but only on the locations where sampling takes place.

Most recent work has focused on optimising a design criterion under a set of modelling assumptions relating to the phenomenon to be modelled. In the purely spatial setting the stationary Gaussian model is a popular choice. Zhu and Stein (2005, 2006) consider this model in redesigning an air monitoring network, noting the computational complexity of evaluating the predictive distribution at a large number of potential design points in order to compare different designs. Jardim and Ribeiro Jr (2007) use a stationary, isotropic log-Gaussian model to compare twelve proposed sampling designs for fish abundance surveys.

The design criteria we consider here are based on prediction of S. Other proposed criteria use the precision of parameter estimates, or combined summary measures of precision and predictive performance. Lark (2002) and Zimmerman (2006) consider optimal designs based on spatial prediction whilst allowing for uncertainty in parameter estimates. A related concept is the *entropy* of the predictive distribution, i.e. for a target T with predictive density $f(t)$, the quantity $\int -f(t)\log(f(t))dt$; see, for example, Guttorp *et al.* (1993), Zidek *et al.* (2000), Fuentes *et al.* (2007) and Ruiz-Cárdenas *et al.* (2010).

The designs that emerge vary widely according to the context, assumptions and criteria under consideration. The conclusions of Zimmerman (2006) are more general than most. He shows that prediction with known parameters results in near-optimal designs that consist of regular lattices or clumps of design points close to the boundary of A, whereas criteria in which parameters are estimated from the data tend to produce regularly spaced clusters of design points. These conclusions broadly agree with those of Diggle and Lophaven (2006), who describe an example of the latter design as 'lattice plus close pairs', in which a regular lattice is supplemented by pairs of points with small spatial separation. Further details of much of the work outlined above are provided by other chapters in this volume.

There is relatively little work on adaptive designs in the spatio-temporal setting, although recent years have seen the development of rich classes of models for analysing spatio-temporal data. Gneiting and Guttorp (2010) provide a summary and Finkelstadt *et al.* (2006) a book-length treatment. Wikle and Royle (1999, 2005) consider the problem of determining adaptive designs for spatio-temporal processes, and approach the problem using both Gaussian and non-Gaussian state-space models. They use $\max_{x \in A}[\text{Var}\{S(x,t)\}]$ as a design criterion, and conclude that for Gaussian models with a separable space-time structure, lattice designs are good choices.

To ease computation, in design studies of this kind a finite set of potential design points is usually pre-specified. This reflects practical limitations in environmental studies for which only a relatively small number of locations are suitable for siting monitoring equipment; see, for example, Fanshawe *et al.* (2008). Consequently, all but the simplest spatial design problems require the continuous region A to be approximated by a discrete grid of possible locations.

11.3 The stationary Gaussian model

11.3.1 Model specification

In this section, we assume that the phenomenon under investigation can be modelled as a realisation of a stationary spatio-temporal Gaussian process $S(x, t)$ with mean zero and variance σ^2, and that at each of m times $t = 1, ..., m$, data $Y_{it} : i = 1, ..., n$ are generated as

$$Y_{it} = d(x_{it})'\beta + S(x_{it}, t) + Z_{it} \tag{11.1}$$

where $d(x_{it})$ is a matrix of spatially and temporally varying covariates and the $Z_{it} \sim \text{N}(0, \tau^2)$ are mutually independent. We assume also that $S(x, t)$ has a *separable* correlation structure, so the correlation between $S(x, t)$ and $S(x - u, t - v)$ is of the form $\rho_x(||u||)\rho_t(|v|)$, where $\rho_x(\cdot)$ and $\rho_t(\cdot)$ are valid monotone, nonincreasing spatial and temporal correlation functions, respectively.

For model (11.1), let

$$Y = (Y_{11}, ..., Y_{n1}, Y_{12}, ..., Y_{n2}, ..., Y_{1m}, ..., Y_{nm})'$$

be the stacked vector of all observations. This has a multivariate Gaussian distribution with mean $\mathbf{d}'\beta$ and variance matrix $\sigma^2 V$, say. For any fixed (x, t), let r_{ij} denote the correlation between Y_{ij} and $S(x, t)$, and write $\mathbf{r}$ for the stacked vector of all of the r_{ij}. It is straightforward to show (Diggle and Ribeiro Jr 2007) that the minimum mean square error predictor for $T = S(x, t)$ has a Gaussian distribution with mean

$$\mathrm{E}[T|Y] = \mathbf{d}'\beta + \mathbf{r}'V^{-1}(Y - \mathbf{d}'\beta) \tag{11.2}$$

and variance

$$\mathrm{Var}[T|Y] = \sigma^2(1 - \mathbf{r}'V^{-1}\mathbf{r}). \tag{11.3}$$

11.3.2 Theoretically optimal designs

The prediction variance (11.3) does not depend on the data values Y, and the spatio-temporal pattern of prediction variances is therefore determined solely by the locations and times at which measurements are taken. If we assume that parameter values are known, the design that minimizes $\max_{x \in A}[\mathrm{Var}\{S(x, t)\}]$ will therefore be the one that minimizes the maximum distance between points in A and the data locations. The resulting design is a hexagonal lattice across the spatio-temporal design space (Johnson *et al.* 1990); rectangular lattices are also commonly used. For a purely spatial design, these are shown in Figure 11.1. For 1–adaptive designs, for each time point t the optimum location x_t of the new design point can be found as the one that maximizes $\min_i(||x_t - x_i||)$, where the minimum is taken with respect to all existing design points x_i.

For our second design criterion, $-\int_A \{\mathrm{P}(S(x, t) > c) - p_0\}^2 dx$, the predictive distribution of $\mathrm{P}(S(x, t) > c)$ is intractable, and it is therefore not possible to provide theoretical results regarding optimal designs. We proceed instead by evaluating the predictive distribution using Monte Carlo simulation.

11.3.3 A comparison of design strategies

In an initial empirical investigation of different design criteria, we let $c = 1$, $\sigma^2 = 1$, $\mathbf{d}'\beta = 0$, and suppose that $\rho_x(\cdot)$ is the exponential correlation function with range parameter $\phi = 0.2$, and that the process is temporally constant. For each of 1000 simulations, we generated a realisation of this process on the unit square and compared the performance of four competing design strategies. Although we do not permit S to change over time, we do consider 1–adaptive strategies: data points are sampled sequentially, and predictions are repeatedly updated in the light of the new observations to guide decisions about where to sample S subsequently.

11.3.3.1 Design criteria

Criterion 1 Minimize $\max_{x \in A}[\mathrm{Var}\{S(x)\}]$.
Criterion 2 Minimize $-\int_A \{(\mathrm{P}(S(x) > 1) - 0.5)^2\} dx$.

11.3.3.2 Design strategies

(a) Forty-nine design points were located on the rectangular grid $(i/14, j/14)$, where $i, j = 1, 3, 5, .., 13$.

(b) Nine design points were located on the rectangular grid $(i/6, j/6)$, where $i, j = 1, 3, 5$, and a further 40 further design points were added sequentially (in a 1−adaptive design) at the point x where the estimated value of $\mathrm{P}\{S(x) > 1\}$ was closest to $p_0 = 0.5$.

(c) As (b), with $p_0 = 1 - \Phi(1) = 0.159$, the marginal probability of exceeding the threshold in the absence of any data.

(d) As (b), but the final 40 points were added sequentially as points generated uniformly on the region where the estimated value of $\mathrm{P}\{S(x) > 1\}$ fell in the interval $[0.1, 0.9]$.

11.3.3.3 Summary statistics

We present the results of the simulations using summary statistics. For criterion 1, we used the numerical value of the criterion. For criterion 2, we used the following statistics:

$D_1 = -\int_A \{(\mathrm{P}(S(x) > 1) - 0.5)^2\} dx.$
$D_2 = -\int_A \{(\mathrm{P}(S(x) > 1) - 0.159)^2\} dx.$
$D_3 = \int_A \{(\mathrm{P}(S(x) > 1) - \mathrm{I}(S(x) > 1))^2\} dx$, where $\mathrm{I}(S(x) \geq 1)$ is an indicator function of whether the true value of $S(x)$ exceeds the threshold.

D_1 and D_2 are summaries of the performance relative to the design criterion, and D_3 a summary of the predictive performance relative to the true value of S. All of these criteria are based on data available at the design locations, which we have omitted from the definitions for clarity of notation.

Results are shown in Table 11.1 and Figures 11.2 and 11.3. Under criterion 2, strategies (b) and (d) perform similarly, and are preferred to strategies (a) and (c).

Table 11.1 Table of results of simulation for the design strategies (a)–(d) and criteria 1 and 2 described in the text. Figures shown are mean (standard deviation). The summary statistics D_1, D_2 and D_3 are defined in the text. Lower values of all statistics indicate improved performance.

Strategy	Criterion 1	Criterion 2		
		D_1	D_2	D_3
(a)	0.40 (0)	−0.188 (0.032)	−0.779 (0.150)	0.062 (0.033)
(b)	0.85 (0.04)	−0.195 (0.037)	−0.784 (0.158)	0.054 (0.039)
(c)	0.83 (0.05)	−0.184 (0.036)	−0.786 (0.166)	0.066 (0.046)
(d)	0.80 (0.08)	−0.195 (0.044)	−0.772 (0.152)	0.055 (0.037)

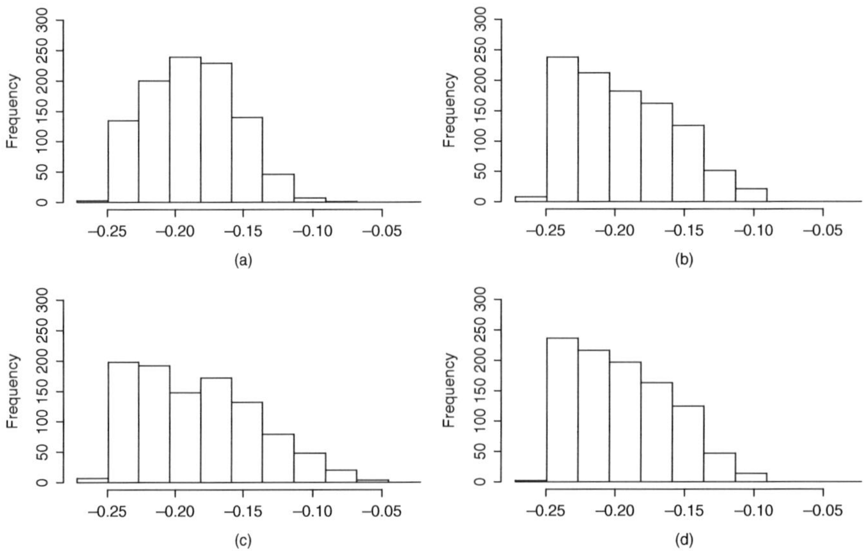

Figure 11.2 Results of simulation, criterion 2, summary statistic D_1, *for design strategies (a)–(d).*

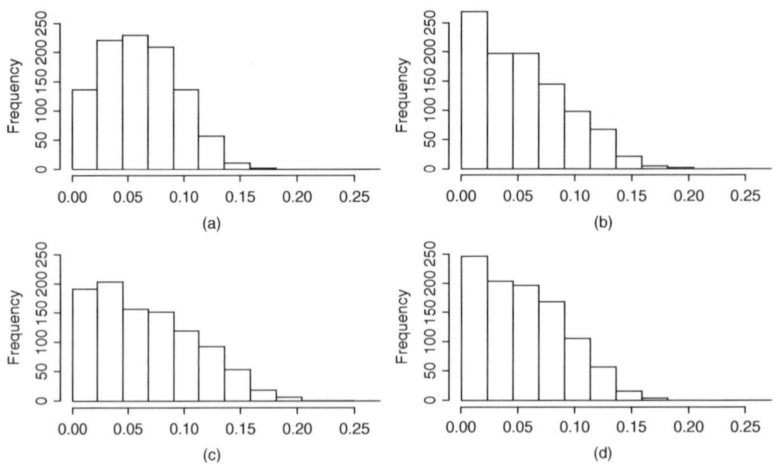

Figure 11.3 Results of simulation, criterion 2, summary statistic D_3, *for design strategies (a)–(d).*

As expected from theoretical results, the regular lattice strategy (a) is by far the best for criterion 1.

Further comparisons between the design strategies for criterion 2 can be made by examining the sets of points generated by the different criteria for a given realisation of S. Figure 11.4 provides a typical example. As the figure illustrates, the 1–adaptive strategy (b) tends to give new design points at close spatial

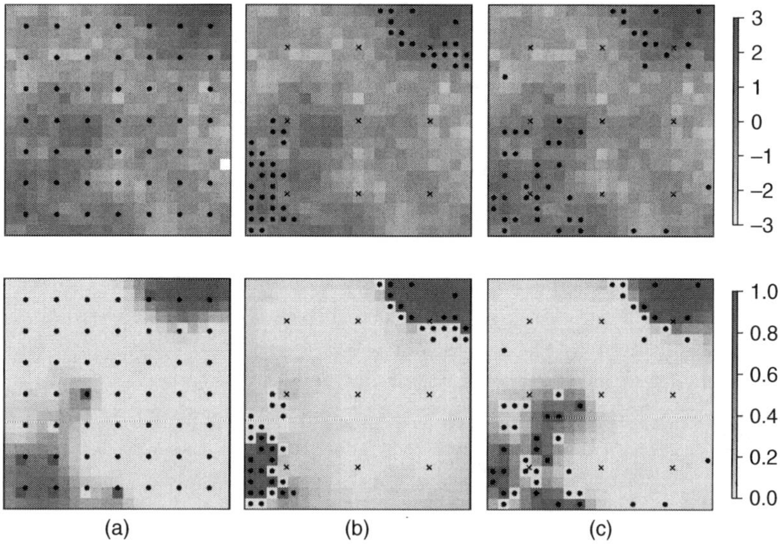

Figure 11.4 Design points selected by three strategies for a single set of simulated data. Top row: simulated data; bottom row: estimate of P(S(x) > 1). *Columns refer to the design strategies indicated, as described in the text.*

separation in order to delineate the threshold between regions in space for which $S(x) > z$ and those for which $S(x) < z$. As a consequence, such thresholds are more finely resolved by an adaptive strategy than by the regular lattice design (a). The cost is that regions elsewhere in the design space may remain unexplored, so some local maxima may be missed. Strategy (d) provides a compromise between strategies (a) and (b).

11.4 The dynamic process convolution model

In order to investigate promising strategies in the spatio-temporal setting, it is instructive to widen the modelling framework beyond the simple Gaussian model described above. A broad class of models for spatio-temporal processes provided by Calder (2007), and based on earlier work by Higdon (2002) and Xu *et al.* (2005), may be described as *dynamic process convolution models*. In this section we introduce this model class and use it to examine how the design strategies introduced in Section 11.3 perform in the spatio-temporal setting.

11.4.1 Model specification

The model is specified in two parts. First, an unobserved *knot process* $\mathbf{z}_t$ is defined on a fixed set of M locations or *knots*, $w_1, ..., w_M$, within the spatial region of interest. This process $\mathbf{z}_t$ [with components $\mathbf{z}_t(w_j)$, $j = 1, ..., M$] is typically correlated over time. For example, in the spatially independent first-order random walk we have

$$\mathbf{z}_t = \mathbf{z}_{t-1} + \nu_t, \tag{11.4}$$

where ν_t is a multivariate zero-mean Gaussian random variable with variance matrix $\lambda_\nu I_M$.

In Section 11.4.2 we use instead

$$\mathbf{z}_t = \alpha \mathbf{z}_{t-1} + \nu_t, \tag{11.5}$$

where we assume that the variance matrix of $\mathbf{z}_0$ is $\sigma^2/(1-\alpha^2)$ and set $\lambda_\nu = \sigma^2$, so that, importantly for our purposes, the variance of each component of $\text{Var}(\mathbf{z}_t)$ assumes a constant value $\sigma^2/(1-\alpha^2)$ for all $t \geq 1$. This corresponds to a stationary first-order autoregressive model for the knot process.

Secondly, the model specification is completed with an *observation process* defined by

$$\mathbf{y}_t = \mu_t + K\mathbf{z}_t + \epsilon, \tag{11.6}$$

where μ is a mean term, K is an $N \times M$ matrix whose entries $K_{ij} = k(\|x_i - w_j\|)$ are defined by a *kernel function* k, and ϵ is a multivariate zero-mean Gaussian random variable with variance matrix $\lambda_\epsilon I_N$. Spatially and/or temporally varying covariates may also be added to the right-hand side of (11.6).

Likelihood-based inference for this model is computationally challenging. However, after specifying priors for λ_μ, λ_ϵ and $\mathbf{x}_0$, it is possible to use Gibbs sampling, based on the full conditional distributions of the model parameters, to carry out Bayesian inference.

The likelihood can be factorised as

$$[\mathbf{y}_1, \ldots, \mathbf{y}_t | \mathbf{z}_0, \ldots, \mathbf{z}_t, \lambda_\epsilon, \lambda_\nu] = \prod_{i=1}^{t} [\mathbf{y}_i | \mathbf{z}_i, \lambda_\epsilon]$$

and the posterior distribution is given by

$$[\mathbf{z}_0, \ldots, \mathbf{z}_t, \lambda_\epsilon, \lambda_\nu | \mathbf{y}_1, \ldots, \mathbf{y}_t] = \prod_{i=1}^{t} [\mathbf{y}_i | \mathbf{z}_i, \lambda_\epsilon][\mathbf{z}_i | \mathbf{z}_{i-1}, \lambda_\nu][\mathbf{z}_0][\lambda_\epsilon][\lambda_\nu].$$

Conjugate priors are $\mathbf{z}_0 \sim \text{MVN}(0, v_0 I_M)$, $\lambda_\epsilon \sim \text{IG}(\alpha_\epsilon, \beta_\epsilon)$ and $\lambda_\nu \sim \text{IG}(\alpha_\nu, \beta_\nu)$, where $\text{IG}(\cdot, \cdot)$ denotes the inverse Gamma distribution and all hyper-parameters are specified numerically. Details of all relevant conditional distributions are given in Appendix 11.1. Calder (2007) provides further information about the properties of the model.

11.4.2 A comparison of design strategies

To evaluate properties of design strategies similar to those in Section 11.3, we conducted another simulation experiment with parameter values assumed known, consistent with our focus on prediction of (functionals of) $S(x, t)$.

Specifically, we set $\mu_t = 0$ for all t, $\upsilon_0 = 1$, $\alpha = 0.95$ and $\lambda_\upsilon = \sigma^2 = 1 - \alpha^2$ [so $\text{Var}(\mathbf{z}_t) = I_M$ for each t], and $\lambda_\epsilon = 0.2^2$. We used the Gaussian kernel, $k(s) = 2\sigma_k^2/(\phi_k\sqrt{\pi})\exp(-0.5(s/\phi_k)^2)$, with $\sigma_k^2 = 0.3$ and $\phi_k = 0.3$. We generated data on a fine grid covering the unit square by simulation from the model using a static 5×5 rectangular grid of knot locations, for $t = 1, \ldots, 10$.

In each simulation, we took as the observed data the simulated values of $\mathbf{y}_t$ at nine observation locations at each time point t. Observation locations were allowed to change over time for adaptive design strategies, as described below. We re-fit the model to this observed data based on $\mathcal{H}_t$ for each of $t = 3, \ldots, 10$, To ensure stability of the estimates of $\mathbf{z}_t$, we did not fit the model at times 1 and 2, although data at these time points did contribute to $\mathcal{H}_t$. A static 3×3 rectangular grid of knots was sufficient to estimate values of $\mathbf{z}_t$ using Monte Carlo simulation. Draws from the distributions of each $\mathbf{z}_t$ then allowed functions of the predictive distribution to be evaluated.

11.4.2.1 Design criteria

For comparison with the results of Section 11.3.3, at each time t we used $-\int_A\{\text{P}(S(x,t) > 1) - 0.5\}^2 dx$ as the quantity to be minimized. Because of the relatively poor performance of design strategies based on $p_0 = 1 - \Phi(1)$ in Section 11.3, we exclude these from consideration here.

11.4.2.2 Design strategies

We compared the following strategies:

(a) Nine design points were located on a static rectangular grid on the unit square and observations taken at this fixed set of locations at each time t;

(b) For $t = 1, \ldots, 3$, observations were taken as in (a), after which a 1–adaptive design was used in which at each time t, the current design point x with the estimated value of $\text{P}(S(x,t) > 1)$ furthest from 0.5 was replaced by the point x' on the unit square with the estimated value of $\text{P}(S(x',t) > 1)$ closest to 0.5.

(c) As (b), but at each time $t \geq 4$ one of the current design points x with estimated value of $\text{P}(S(x,t) > 1) \notin [0.1, 0.9]$, was replaced by a point x' on the unit square with the estimated value of $\text{P}(S(x',t) > 1) \in [0.1, 0.9]$. Candidate points for removal and addition were selected at random (with equal probability) if more than one satisfied the criterion.

11.4.2.3 Summary statistics

At each time t, we calculated D_1 and D_3 in the same way as in Section 11.3, based on all data recorded at or before time t. We also considered the integrated mean square error

$$D_4 = IMSE_t = \int_A \mathrm{E}[\{\hat{S}(x,t) - S(x,t)\}^2]dx.$$

As the total number of observations contributing to $\mathcal{H}_t$ increases linearly with t, it is computationally demanding to fit this model for a large number of simulations. We therefore considered 500 randomly generated datasets, and evaluated the three design strategies for each.

11.4.2.4 Results

Figure 11.5 shows how the average estimated summary statistics vary with t for each design strategy, using the 500 simulated spatio-temporal processes. The two adaptive strategies offer improved performance over the fixed design in estimating exceedance probabilities across A. Under these strategies, exceedance probabilities are on average further from 0.5 (as measured by D_1) and closer in magnitude to the binary indicator of exceedance, $I(S(x,t) > 1)$ (as measured by D_3). Decreases in D_3 over time indicate, as would be expected, that each strategy becomes better at identifying regions above the threshold as time passes.

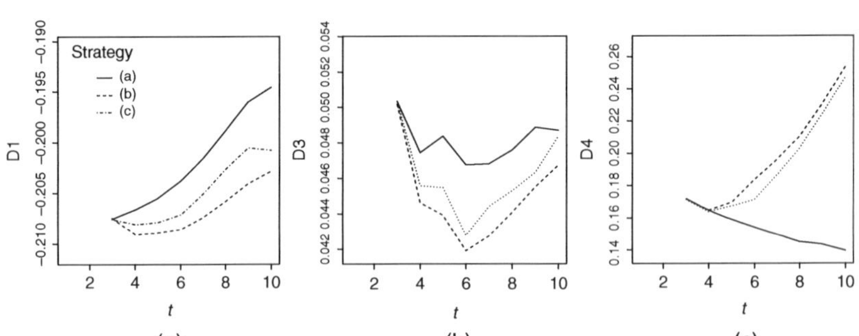

Figure 11.5 Comparison of 1–adaptive design strategies (a), (b) and (c) using three summary statistics D_1 *(left),* D_3 *(middle) and* D_4 *(right). See text for details.*

In contrast, the integrated mean square error is substantially worse for the adaptive strategies than for the static lattice design. This is not unexpected: the adaptive strategies are intended to concentrate observation locations in regions of the design space close to the exceedance threshold. Parts of A for which S is far from the threshold are sampled infrequently, and so predictions in these regions are poor.

We repeated the simulation to consider k–adaptive strategies. Here, for strategies (b) and (c), all nine points were permitted to change location between consecutive sampling times. The results are shown in Figure 11.6, and the effect is similar to that seen in the 1–adaptive case, albeit more pronounced. These 9–adaptive strategies lead to points congregating in certain parts of A much more rapidly than the 1–adaptive strategies.

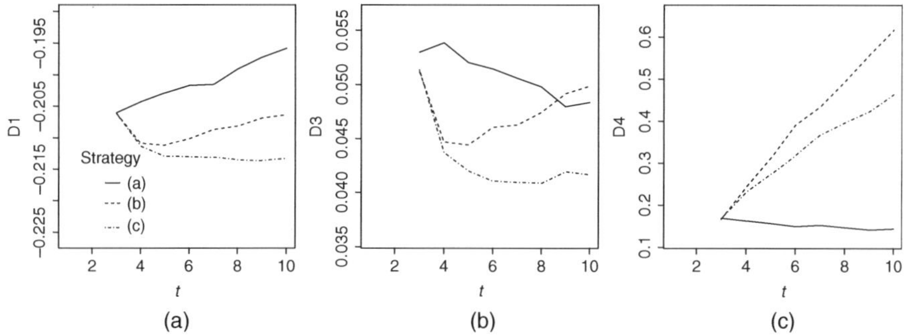

Figure 11.6 Comparison of 9–adaptive design strategies (a), (b) and (c) using three summary statistics D_1 *(left),* D_3 *(middle) and* D_4 *(right). See text for details.*

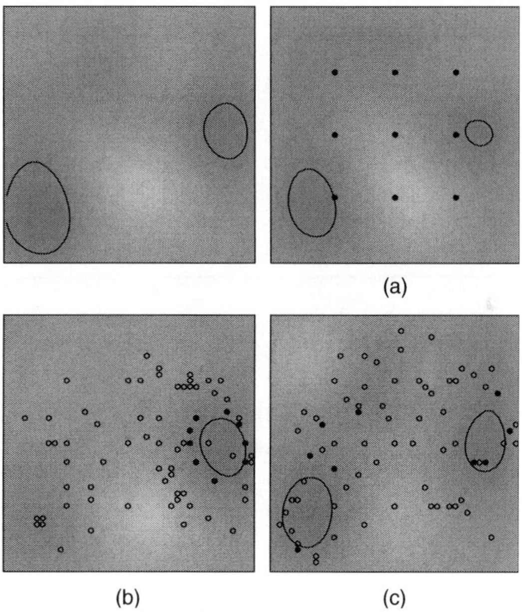

Figure 11.7 Illustration of design points used by strategies (a), (b) and (c) for a 9–adaptive design, 10 time points. Upper left: simulated realisation of S*; upper right: regular lattice design strategy (a); lower left: strategy (b); lower right: strategy (c). Predicted values of* S *are shown in each case. Solid circles indicate design points at the current time* ($t = 10$)*, and open time points design points for previous time points. Contour lines indicate regions where* $\hat{S} > 1$. *(Please see plate section for color version of the figure.)*

The types of design resulting from these design strategies are illustrated in Figure 11.7. An animation of a similar design problem can be viewed at www.lancs.ac.uk/staff/fanshawe/adaptive. It shows a typical example of how the set of design points changes over time for each strategy. Under the adaptive strategies, design points tend to cluster in certain parts of A. In the example, strategy (b) oversamples to the right of the unit square and, by not sampling towards the bottom left at later times, fails to identify that S exceeds the threshold in this region. The more flexible strategy (c), in contrast, performs better than the lattice design as it allows occasional visits to be made to locations more widely spread across the square. Both adaptive strategies perform worse than the lattice design in predicting S in regions where S is small.

11.5 Upper Austria rainfall data example

Here we illustrate our design strategies using the Upper Austria rainfall data. A common objective using monitoring data such as this is to estimate the probability of exceedance of a threshold, possibly after accounting for seasonal and/or long-term trends.

Let Y_{it} be the log-transformed value of the observed rainfall at site i ($i = 1, ..., 37$) in month t, and $d(x_i)$ be the altitude of site i. First, using all available data from 1994–2009, we fit a linear model of the form

$$Y_{it} = \alpha + \beta_1 d(x_i) + \beta_2 \cos(2\pi i/12) + \beta_3 \sin(2\pi i/12) + Z_{it}, \tag{11.7}$$

where $Z_{it} \sim \mathrm{N}(0, \tau^2)$ are mutually independent. This model recognizes both the strong seasonal component to rainfall levels and their relationship with altitude. We then 'de-seasonalize' the original data by subtracting the fitted values $\hat{Y}_{it}$ from (11.7) from the original data Y_{it}, giving residuals R_{it}, and suppose that interest lies in determining whether R_{it} exceeds a threshold of $c = 0.5$, akin to determining months and locations in which there is unseasonably high rainfall.

As an expository example, we present results on adaptive design strategies using only the values of R_{it} for 1995 and supposing that only 12 of the 37 monitoring stations are available for use during any one month, with the design configuration to be determined. First, we obtained the set of 12 maximally spatially regular sites using the `cover.design` function of R. This configuration is shown in Figure 11.8. Figure 11.9 shows the average, maximum and minimum of the R_{it} across all sites for each month of 1995.

Next, we fit the dynamic process convolution model to data from these 12 sites for January–May 1995, a period of 'run-in'. For the remaining months of this year, we compared the following three 1–adaptive strategies, chosen to mimic the strategies used in Section 11.4.2:

(a) The static configuration of 12 maximally regular sites was used for June–December.

(b) After the run-in period, a 1–adaptive design was used in which at each subsequent time t, the current site x with the estimated value of $\mathrm{P}(S(x, t) > 0.5)$ furthest from 0.5 was replaced by the site x', from

Figure 11.8 Upper Austria sampling locations. The twelve filled circles indicate the maximally spatially regular design configuration.

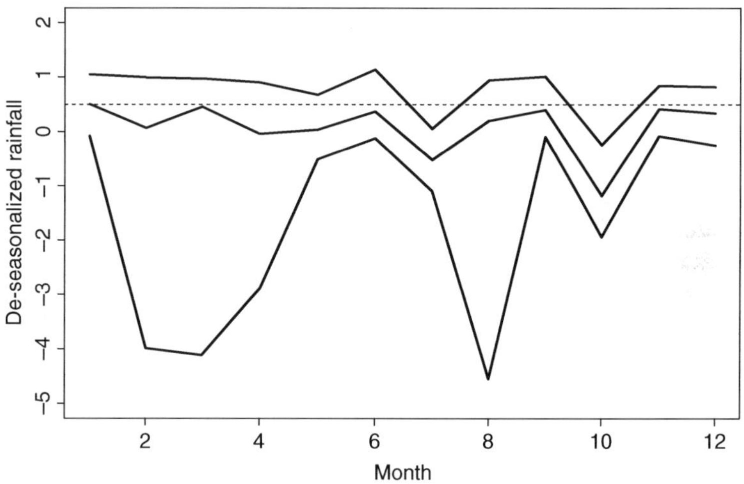

Figure 11.9 Maximum, average and minimum monthly de-seasonalized log-rainfall data for 1995. The dotted line represents the threshold value of 0.5.

the set of available monitor locations, with the estimated value of $P(S(x', t) > 0.5)$ closest to 0.5.

(c) As (b), but after the run-in period, at each subsequent time t a site x with estimated value of $P(S(x, t) > 0.5) \notin [0.1, 0.9]$, was replaced by a site x', from the set of available monitor locations, with the estimated value of $P(S(x', t) > 0.5) \in [0.1, 0.9]$. The design was left unchanged if either of the sets was empty, and candidate points for removal and addition were selected at random (with equal probability) if more than one satisfied the criterion.

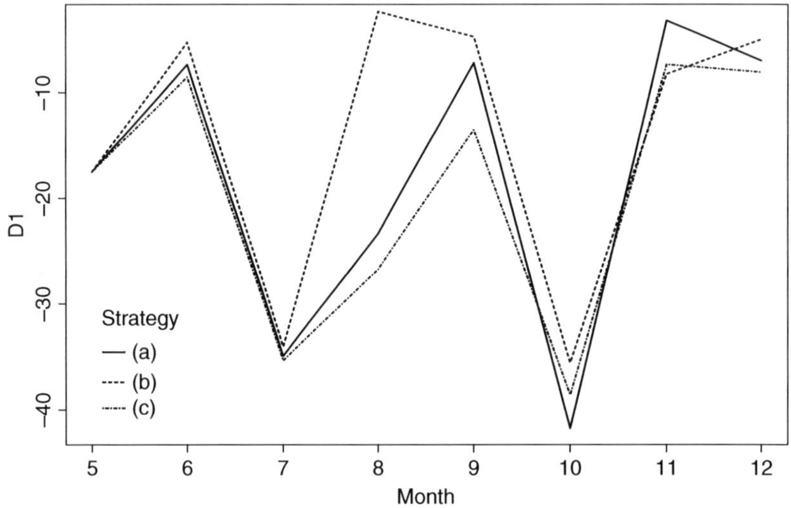

Figure 11.10 Design criterion evaluated for each month in 1995 under three design strategies.

Figure 11.10 shows the estimated value of $-\int_A \{P(S(x,t) > 0.5) - 0.5\}^2 dx$ at each time t between May and December 1995 for each design strategy. None of the three strategies performs consistently better than the other two, although strategies (a) and (c) appear to perform the best on average. The overall performance of the three strategies is similar, with the exception of strategy (c) for August, when the design configuration chosen left a large part of the centre of the study region unsampled. Nevertheless, picking sites in regions of residual rainfall close to the threshold value leads strategies (b) and (c) to outperform strategy (a) in particular months.

All three strategies give similarly low predicted probabilities in months of comparatively low rainfall, such as October. It should be noted that the performance of strategies (b) and (c) in this example is hampered by the relatively low temporal correlation in the data: sites with high rainfall in one month do not have a high probability of high rainfall in the next. There might be more reason to consider adaptive designs with data with a stronger temporal correlation, as might be expected from rainfall data at weekly, rather than monthly, resolution at this temporal resolution.

11.6 Discussion

In this chapter we have compared several strategies for geostatistical design in spatio-temporal problems. Our results show that there is value in considering alternative designs to the regular lattice if one is able to sample adaptively, changing one or more of the design locations over time. At least in the model-based design framework, a suitable choice of design will depend both on the assumed model for the data under consideration and the objectives of the study, which we have expressed in this chapter as a design criterion.

We chose to focus on a design criterion based on the probability of the modelled surface $S(x, t)$ exceeding a fixed threshold c, although the methods we have used are easily adapted to other criteria. For example, the amount by which the threshold is exceeded might also be considered. Threshold exceedance is often important in addressing public health concerns, for example in identifying regions of unusually high disease prevalence (Diggle *et al.* 2007) or areas in which air pollution levels are untypically high, and thereby liable to have an adverse impact on human health (Axelrod *et al.* 2001). In this latter context, extreme pollution exposures may have a disproportionate impact on respiratory conditions, and statistically neutral criteria such as integrated mean square error may lead to poor design choices.

When integrated mean square error *is* an appropriate design criterion, the adaptive design strategies considered in this chapter should not be considered. However, adaptive designs are better suited than lattice designs to delineating regions where $S(x, t)$ takes values within specified ranges, and may therefore play an increasingly important role in monitoring network design as the quality of mobile monitoring devices improves.

The geostatistical design literature has tended to make a distinction between design criteria based on prediction of $S(x, t)$ (Zhu and Stein 2005) and those based on estimation of the model parameters (Zhu and Stein 2006). We have concentrated on the former, assuming known parameters. It would also be useful to investigate the performance of the design strategies used in this chapter after incorporating uncertainty in parameter estimates, but this is computationally a more challenging problem.

We make no claim that the design strategies we have considered are theoretically optimal. The complexity of the design problem for correlated spatio-temporal data suggests that it is unlikely that analytic solutions to the adaptive design problem can be found in realistic settings. Also, the gains in practical terms made by improving a near-optimal design to one that is optimal may be so small as to make the endeavour of questionable value, especially if it is not possible to record data at the precise locations required for theoretical optimality (Fanshawe and Diggle 2011).

In many contexts there is only a relatively small set of possible sampling locations, in which case systematic evaluation of each possible design configuration may be possible. When the number of possible design locations is larger, it may be implausible to find the optimal configuration, but well-developed methods such as simulated annealing may allow a solution very close to the optimum to be found (Fuentes *et al.* 2007).

For these reasons, we have suggested intuitively appealing adaptive design strategies and have demonstrated their potential for greater predictive accuracy than would be attained by the default option of a spatially regular static design. Our specific suggestions surely leave room for further improvement. For example, they may lead to an over-concentration of points in a particular region of the design space, to the extent that the threshold may be exceeded in some other parts of the design space without being detected, especially if the temporal correlation of $S(x,t)$ is weak. This phenomenon was apparent in our analysis of the Upper Austria rainfall data.

In the 9–adaptive case considered in Section 11.4.2, a strategy that allowed new design points to be distributed according to a range of probability values $[0.1, 0.9]$ gave more promising results than the greedy strategy that expressly targeted areas with exceedance probability close to 0.5. A hybrid design of a fixed lattice supplemented by additional moving design points might achieve a similar effect.

Our most important single conclusion is that adaptive designs should be constructed by a criterion that directly measures the extent to which the primary scientific goal of the analysis is being met, and should therefore be strongly context-dependent.

Appendix 11.1

Under the specification of the dynamic process convolution model and the conjugate priors given in Section 11.4, it can be shown that, given data $\mathbf{y}_1, \ldots, \mathbf{y}_t$, the full conditional distributions are given by:

$$\mathbf{x}_i|\cdot \sim \text{MVN}(M_i, W_i) \qquad (i = 0, \ldots, t)$$

$$\lambda_\epsilon|\cdot \sim \text{IG}(\alpha_\epsilon + Nt/2 + 1, \frac{1}{2}\sum_{i=1}^{n}(\mathbf{y}_i - K\mathbf{z}_i)^T(\mathbf{y}_i - K\mathbf{z}_i) + \beta_\epsilon)$$

$$\lambda_\nu|\cdot \sim \text{IG}(\alpha_\nu + Mt/2 + 1, \frac{1}{2}\sum_{i=1}^{n}(\mathbf{z}_i - \mathbf{z}_{i-1})^T(\mathbf{z}_i - \mathbf{z}_{i-1}) + \beta_\nu)$$

where

$$M_0 = \frac{W_0^T \mathbf{z}_1}{\lambda_\nu}$$

$$W_0^{-1} = \frac{(v_0 + \lambda_\nu)I_M}{\lambda_\nu v_0}$$

$$M_i = W_i^T\left(\frac{\mathbf{z}_{i-1} + \mathbf{z}_{i+1}}{\lambda_\nu} + \frac{K^T y_i}{\lambda_\epsilon}\right) \qquad (i = 1, \ldots, t-1)$$

$$W_i^{-1} = \frac{\lambda_\nu K^T K + 2\lambda_\epsilon I_M}{\lambda_\epsilon \lambda_\nu} \qquad (i = 1, \ldots, t-1)$$

$$M_t = W_t^T\left(\frac{x_{t-1}}{\lambda_\nu} + \frac{K^T y_t}{\lambda_\epsilon}\right)$$

$$W_t^{-1} = \frac{\lambda_\nu K^T K + \lambda_\epsilon I_M}{\lambda_\epsilon \lambda_\nu}$$

References

Axelrod D, Lee Davis D, Hajek R and Jones L 2001 It's time to rethink dose: the case for combining cancer and birth and developmental defects. *Environmental Health Perspectives* **109**, A246–A249.

Calder C 2007 Dynamic factor process convolution models for multivariate space-time data with application to air quality assessment. *Environmental and Ecological Statistics* **14**, 229–247.

Cox L 2000 Statistical issues in the study of air pollution involving airborne particulate matter. *Environmetrics* **11**, 611–626.

Diggle P and Lophaven S 2006 Bayesian geostatistical design. *Scandinavian Journal of Statistics* **33**, 53–64.

Diggle P and Ribeiro Jr P 2007 *Model-based Geostatistics*. New York: Springer.

Diggle P, Thomson M, Christensen O, Rowlingson B, Obsomer V, Gardon J, Wanji S, Takougang I, Enyong P, Kamgno J, Remme J, Boussinesq M and D.H. M 2007 Spatial modelling and the prediction of loa loa risk: decision making under uncertainty. *Annals of Tropical Medicine and Parasitology* **101**, 499–509.

Fanshawe T and Diggle P 2011 Spatial prediction in the presence of prediction error (with discussion). *Environmetrics* **22**, 109–131.

Fanshawe T, Diggle P, Rushton S, Sanderson R, Lurz P, Glinianaia S, Pearce M, Parker L, Charlton M and Pless-Mulloli T 2008 Modelling spatio-temporal variation in exposure to particulate matter: a two-stage approach. *Environmetrics* **19**, 549–566.

Finkelstadt B, Held L and Isham V (eds) 2006 *Statistical Methods for Spatio-temporal Systems*. Boca Raton, FL: Chapman and Hall/CRC.

Fuentes M, Chaudhuri A and Holland D 2007 Bayesian entropy for spatial sampling design of environmental data. *Environmental and Ecological Statistics* **14**, 323–340.

Furrer R, Nychka D and Sain S 2010 *Fields: Tools for spatial data*. R package version 6.3.

Gneiting T and Guttorp P 2010 Continuous parameter spatio-temporal processes. In: *Handbook of Spatial Statistics*. Boca Raton, FL: CRC Press, pp. 427–436.

Guttorp P, Le N, Sampson P and Zidek J 1993 Using entropy in the redesign of an environmental monitoring network. In: *Multivariate Environmental Statistics*. New York: North-Holland, pp. 175–202.

Higdon D 2002 Space and space-time modeling using process convolutions. In: *Quantitative Methods for Current Environmental Issues*. New York: Springer-Verlag, pp. 37–56.

Jardim E and Ribeiro Jr P 2007 Geostatistical assessment of sampling designs for Portuguese bottom trawl surveys. *Fisheries Research* **85**, 239–247.

Johnson M, Moore L and Ylvisaker D 1990 Minimax and maximin distance designs. *Journal of Statistical Planning and Inference* **26**, 131–148.

Lark R 2002 Optimized spatial sampling of soil for estimation of the variogram by maximum likelihood. *Geoderma* **105**, 49–80.

Maher B, Moore C and Matzka J 2008 Spatial variation in vehicle-derived metal pollution identified by magnetic and elemental analysis of roadside tree leaves. *Atmospheric Environment* **42**, 364–373.

R Development Core Team 2008 *R: A Language and Environment for Statistical Computing*. Vienna: R Foundation for Statistical Computing.

Royle J and Nychka D 1998 An algorithm for the construction of spatial coverage designs with implementation in SPLUS. *Computers & Geosciences* **24**, 479–488.

Ruiz-Cárdenas R, Ferreira M and Schmidt A 2010 Stochastic search algorithms for optimal design of monitoring networks. *Environmetrics* **21**, 102–112.

Schäfer H, Timmesfeld N and Müller HH 2006 An overview of statistical approaches for adaptive designs and design modifications. *Biometrical Journal* **48**, 507–520.

van Groenigen J, Pieters G and Stein A 2000 Optimizing spatial sampling for multivariate contamination in urban areas. *Environmetrics* **11**, 227–244.

Wikle C and Royle J 1999 Space-time dynamic design of environmental monitoring networks. *Journal of Agricultural, Biological, and Environmental Statistics* **4**, 489–507.

Wikle C and Royle J 2005 Dynamic design of ecological monitoring networks for non-gaussian spatio-temporal data. *Environmetrics* **16**, 507–522.

Xu K, Wikle C and Fox N 2005 A kernel-based spatio-temporal dynamical model for nowcasting radar precipitation. *Journal of the American Statistical Association* **100**, 1133–1144.

Zhu Z and Stein M 2005 Spatial sampling design for parameter estimation of the covariance function. *Journal of Statistical Planning and Inference* **134**, 583–603.

Zhu Z and Stein M 2006 Spatial sampling design for prediction with estimated parameters. *Journal of Agricultural, Biological, and Environmental Statistics* **11**, 24–44.

Zidek J and Zimmerman D 2010 Monitoring network design In: *Handbook of Spatial Statistics*. Boca Raton, FL: CRC Press, pp. 131–148.

Zidek J, Sun W and Le N 2000 Designing and integrating composite networks for monitoring multivariate Gaussian pollution fields. *Applied Statistics* **49**, 63–79.

Zimmerman D 2006 Optimal network design for spatial prediction, covariance parameter estimation, and empirical prediction. *Environmetrics* **17**, 635–652.

12

Semiparametric dynamic design of monitoring networks for non-Gaussian spatio-temporal data

Scott H. Holan and Christopher K. Wikle
Department of Statistics, University of Missouri, Columbia, USA

12.1 Introduction

Environmental processes and ecosystem phenomena have received considerable attention over the past several decades as a result of potential climate change, as well as issues surrounding endangered species, sustainable agriculture and renewable energy. As a result, both scientists and those involved in making public-policy decisions have realized the importance of data collected from environmental monitoring networks. In particular, one may have a need to add more locations to a monitoring network, remove stations due to cost, or move stations through time. Choosing sampling locations in the future, based on information observed in the past, is known as the 'design' problem, and model-based design in this dynamic spatio-temporal context frequently requires a non-Gaussian spatio-temporal dynamic model.

Spatio-temporal design: Advances in efficient data acquisition, First Edition.
Edited by Jorge Mateu and Werner G. Müller.

The problem of spatial design has been the subject of extensive research. In particular, the issue of obtaining spatial designs for environmental processes has culminated into a large body of substantive work in both the statistics and nonstatistics literature (Fedorov and Müller 1989; Haas 1992; Guttorp *et al.* 1993; Oehlert 1996; Bueso *et al.* 1998; Nychka and Saltzman 1998; Sansó and Müller 1998; Müller and Pázman 2003; Wikle and Royle 2005; Zidek and Zimmerman 2010). For a comprehensive discussion regarding optimal spatial design see Müller (2007) and the references therein. It is important to note, typically, the same optimal design strategies used in the pure spatial setting can be adapted to the more general setting of spatio-temporal correlation, if the design remains static over time (Le and Zidek 1994). However, in dynamic settings, where the underlying process evolves in space and time, static designs based on spatial correlation alone are typically less efficient than designs that consider the joint spatio-temporal dependence structure. In fact, Wikle and Royle (1999, 2005) and Hooten *et al.* (2009) demonstrated that allowing the design to change with time can lead to significant efficiency gains, in terms of average prediction variance, in the spatio-temporal context. This is typically referred to as *adaptive* (or *dynamic*) design, which is not the same as adaptive sampling (Thompson and Seber 1996), although the notions are similar. The primary distinction is that the former is model-based, whereas the latter arises in the context of traditional sample surveys and modifies sample unit selection as observations are made.

There are several alternative approaches to the design problem, including design criteria based on different characteristics of the semivariogram that are related to Kriging (Müller and Zimmerman 1999). Alternatively, Nychka and Saltzman (1998) use a space-filling design, whereas Müller (1999) proposes a simulation-based strategy for obtaining optimal designs based on maximizing a utility function. Other authors have considered design criteria based on entropy-based methods. For example, Zidek *et al.* (2000) proposed a maximum entropy approach which uses multivariate time series techniques to model observations collected at various monitoring sites in order to maximize the entropy (e.g., expected information) of potential locations not being monitored. Additionally, while accounting for environmental cost and statistical information, (Fuentes *et al.* 2007) developed a framework that ranks various subnetworks. Their approach obtains an optimal subnetwork using a Bayesian algorithm in an entropy setting. In contrast, to choose optimal designs for a monitoring network, Ruiz-Cárdenas *et al.* (2010) propose a new hybrid genetic algorithm to maximize entropy.

Motivated by the problem of selecting a subset of sampling locations in which to measure precipitation from monitoring locations in Upper Austria, we consider a dynamic model-based spatio-temporal design. Specifically, extending the approach of Wikle and Royle (2005), we introduce a low-rank semiparametric spatio-temporal dynamic Gamma model that allows important environmental factors to be related to the process mean (or location parameter) nonlinearly. Due to the excessive computation involved in estimating the expected utility function associated with various designs from this model we utilize the extended Kalman filter approximation to aid in implementation. The goal of our design is to select a small number of sampling locations (e.g., 20) at time $T + 1$, from the 37 possible locations, using an approach that yields maximal information about the spatial distribution of total precipitation.

The remainder of this chapter proceeds as follows. Section 12.2 provides details surrounding our low-rank semiparametric spatio-temporal dynamic Gamma model-based design approach. Application of our approach to the motivating problem of choosing sampling locations in Upper Austria for measuring precipitation is presented in Section 12.3. Finally, Section 12.4 provides concluding discussion.

12.2 Semiparametric non-Gaussian space-time dynamic design

12.2.1 Semiparametric spatio-temporal dynamic Gamma model

Following Wikle and Royle (2005), let $\mathbf{y}_t \equiv (y_t(r_1^t), \ldots, y_t(r_{m_t}^t))'$ denote spatio-temporal data for spatial locations $r_i^t \in \{s_1, \ldots, s_n\}$ in $\mathbb{R}^2$ and times $t = 1, \ldots, T$. In this context, the subscript notation on the spatial locations r_i is needed because not all possible sampling locations will necessarily include data for a given time t. Further, let $\boldsymbol{\gamma}_t \equiv (\gamma_t(s_1), \ldots, \gamma_t(s_n))'$ denote a spatio-temporal process at spatial locations $\{s_1, \ldots, s_n\}$ and for each time t define a $m_t \times p$ matrix of covariates $\mathbf{X}_t$ and an associated matrix of random effects $\mathbf{Z}_t^{(1)}$ (e.g., matrix containing low-rank basis expansion; Wikle 2010) of dimension $m_t \times q$. In our case the dimension q will directly depend on the number of covariates being modeled nonlinearly. Finally, let $\mathbf{Y}_1^t \equiv \{\mathbf{y}_t, \mathbf{y}_{t-1}, \ldots, \mathbf{y}_1\}$ and $\boldsymbol{\gamma}_1^t \equiv \{\boldsymbol{\gamma}_t, \boldsymbol{\gamma}_{t-1}, \ldots, \boldsymbol{\gamma}_1\}$. Importantly, in the models considered here, the spatio-temporal process, $\boldsymbol{\gamma}_t$, can also be represented using a low-rank basis function expansion.

In the total precipitation design problem of interest here, we have monthly precipitation data at times $t = 1, \ldots, T$ from January 1994 through December 2010. The total precipitation data are not assumed to be available for all locations s_i, $i = 1, \ldots, n$, at all times $t = 1, \ldots, T$. Additionally, the potential covariates in the example presented here are elevation, latitude and longitude. As previously alluded to, the design problem of interest is to select a subset of the spatial monitoring locations $s_1, \ldots, s_n$ at which to take samples at time $T + 1$. In other words, we wish to find a subset of monitoring locations at time $T + 1$, $\{r_i^{T+1} \in \{s_1, \ldots, s_n\}\}$.

Assuming conditional independence for the measurement process, we have

$$p(\mathbf{y}_t | \boldsymbol{\gamma}_1^t, \mathbf{Y}_1^{t-1}, \boldsymbol{\beta}, \mathbf{u}, \phi) = p(\mathbf{y}_t | \boldsymbol{\gamma}_t, \boldsymbol{\beta}, \mathbf{u}, \phi), \quad \text{for } t = 1, 2, \ldots,$$

where $\boldsymbol{\beta}$ and $\boldsymbol{u}$ denote vectors of parameters associated with $\mathbf{X}_t$ and $\mathbf{Z}_t^{(1)}$, respectively, contained in the model for the location parameter, and ϕ denotes a scale parameter. In this setting $\mathbf{Z}_t^{(2)}$ denotes an $m_t \times n$ incidence matrix. Further, assume that $p(\mathbf{y}_t|\cdot)$ is from the m_t-dimensional Gamma distribution; i.e., $p(\mathbf{y}_t|\boldsymbol{\gamma}_t, \boldsymbol{\beta}, \mathbf{u}, \phi) \sim Gam(\mathbf{c}_t, \phi)$, where $\log(\mathbf{c}_t) = \mathbf{X}_t\boldsymbol{\beta} + \mathbf{Z}_t^{(1)}\mathbf{u} + \mathbf{Z}_t^{(2)}\boldsymbol{\gamma}_t$, which implies that $\boldsymbol{\mu}_t = \phi\mathbf{c}_t$. The general case of the natural exponential family is considered in Wikle and Royle (2005) and illustrated using a Poisson data model.

Rather than modeling $\mathbf{c}_t$ or the mean directly, we model the mean by specifying a link function $g(\cdot)$ such that

$$g(\boldsymbol{\mu}_t) = \log(\boldsymbol{\mu}_t) = \log(\phi) + \mathbf{X}_t\boldsymbol{\beta} + \mathbf{Z}_t^{(1)}\mathbf{u} + \mathbf{Z}_t^{(2)}\boldsymbol{\gamma}_t. \tag{12.1}$$

This implies that $h(\cdot) = g^{-1}(\cdot)$ is given by

$$\begin{aligned} h\{\log(\phi) + \mathbf{X}_t\boldsymbol{\beta} + \mathbf{Z}_t^{(1)}\mathbf{u} + \mathbf{Z}_t^{(2)}\boldsymbol{\gamma}_t\} &= \exp\{\log(\phi) + \mathbf{X}_t\boldsymbol{\beta} + \mathbf{Z}_t^{(1)}\mathbf{u} + \mathbf{Z}_t^{(2)}\boldsymbol{\gamma}_t\} \\ &= \phi\exp(\mathbf{X}_t\boldsymbol{\beta} + \mathbf{Z}_t^{(1)}\mathbf{u} + \mathbf{Z}_t^{(2)}\boldsymbol{\gamma}_t) = \boldsymbol{\mu}_t. \end{aligned}$$

Thus, the dispersion parameter ϕ acts as an offset when modeling the log mean function.

In principal, $\mathbf{Z}_t^{(1)}$ can be specified using any complete (orthonormal) basis expansion. However, our exposition and analysis in Section 12.3 uses a low-rank penalized spline basis function; e.g., see Ruppert *et al.* (2009) and Wikle (2010) and the references therein. For ease of exposition, and effective illustration, we temporarily consider a special case of (12.1) having two nonlinear functions (one univariate and one bivariate). In particular, let

$$g(\boldsymbol{\mu}_t) = \beta_0 + \beta_1\boldsymbol{x}_{1,t} + f_2(\boldsymbol{x}_{2,t}) + f(\mathbf{x}_t) + \mathbf{Z}_t^{(2)}\boldsymbol{\gamma}_t, \tag{12.2}$$

where, for $\ell = 1, 2$ $\boldsymbol{x}_{\ell,t}$ are vectors of covariates at time t with elements $\boldsymbol{x}_{\ell,tj}$ and, in this context, $\mathbf{x}_t$ is an $m_t \times 2$ vector with elements, $\mathbf{x}_{tj}$, equal to *latitude* and *longitude* (or positions in x-direction and y-direction on a Cartesian coordinate system).

The model given in (12.2) can be equivalently expressed using matrix notation as

$$g(\boldsymbol{\mu}_t) = \mathbf{X}_t\boldsymbol{\beta} + \mathbf{Z}_{t,K_1}^{(1)}\mathbf{b}_1 + \mathbf{Z}_{t,K_2}^{(1)}\mathbf{b}_2 + \mathbf{Z}_t^{(2)}\boldsymbol{\gamma}_t, \tag{12.3}$$

where $\boldsymbol{\beta} = (\beta_0, \beta_1, \beta_2)'$, $\mathbf{b}_1 = (b_{11}, \ldots, b_{1M_1})'$, $\mathbf{b}_2 = (b_{21}, \ldots, b_{2M_2})'$ and

$$\mathbf{X}_t = \begin{bmatrix} 1 & x_{1tj} & x_{2tj} & \mathbf{x}'_{tj} \end{bmatrix}_{1\le j\le m_t}; \quad \mathbf{Z}_{tK_1}^{(1)} = [|x_{2tj} - \kappa_k|^3]_{\substack{1\le j\le m_t;\\ 1\le k\le M_1}}$$
$$\mathbf{Z}_{tK_2}^{(1)} = [W(||\mathbf{x}_{tj} - \boldsymbol{\kappa}_k||)]_{\substack{1\le j\le m_t,\\ 1\le k\le M_2}}$$

with

$$W(\mathbf{r}) = ||\mathbf{r}||^2 \log ||\mathbf{r}||, \tag{12.4}$$

$||\mathbf{r}|| = \sqrt{\mathbf{r}'\mathbf{r}}$ and κ_k, $\boldsymbol{\kappa}_k$ denoting knot points in $\mathbb{R}^1$ and $\mathbb{R}^2$, respectively. Here, K_1 and K_2 refer to the untransformed Z matrices associated with the matrix formulation of (12.2) defined above; M_1 and M_2 denote the number of knot points associated with the low-rank representations expressed using the Z matrices. Additionally, note that $W(\mathbf{r})$ can also be chosen as radial basis functions corresponding to a proper covariance structure (Ruppert *et al.* 2003).

Further, let

$$\boldsymbol{\Omega}_{K_1} = [|\kappa_k - \kappa_{k'}|^3]_{1\le k,k'\le M_1}, \quad \boldsymbol{\Omega}_{K_2} = [||\boldsymbol{\kappa}_k - \boldsymbol{\kappa}'_k||^2 \log ||\boldsymbol{\kappa}_k - \boldsymbol{\kappa}'_k||]_{1\le k,k'\le M_2},$$

in (12.3), the elements of $\boldsymbol{\beta}$ are interpreted as 'fixed' effects and $\mathbf{b}_\ell$ ($\ell = 1, 2$) are interpreted as 'random' effects, with $\mathrm{E}(\mathbf{b}_\ell) = \mathbf{0}$, $\mathrm{Cov}(\mathbf{b}_\ell) = \sigma^2_{u\ell}\boldsymbol{\Omega}^{-1}_{K_\ell}$; see Holan *et al.* (2008) and the references therein for a comprehensive discussion. Following Crainiceanu *et al.* (2005), define $\mathbf{Z}^{(1)}_{t,\ell} = \mathbf{Z}^{(1)}_{t,K_\ell}\boldsymbol{\Omega}^{-1/2}_{K_\ell}$ and $\mathbf{b}_\ell = \boldsymbol{\Omega}^{-1/2}_{K_\ell}\mathbf{u}_\ell$ (for $\ell = 1, 2$). Then, (12.3) can equivalently be rewritten as

$$g(\boldsymbol{\mu}_t) = \mathbf{X}_t\boldsymbol{\beta} + \mathbf{Z}^{(1)}_{t,1}\mathbf{u}_1 + \mathbf{Z}^{(1)}_{t,2}\mathbf{u}_2 + \mathbf{Z}^{(2)}_t\boldsymbol{\gamma}_t, \tag{12.5}$$

where $\mathrm{E}(\mathbf{u}_\ell) = \mathbf{0}$ and $\mathrm{Cov}(\mathbf{u}_\ell) = \sigma^2_{u\ell}\mathbf{I}_{K_\ell}$. Note that, for further economy of notation, (12.5) can be expressed as

$$g(\boldsymbol{\mu}_t) = \mathbf{X}_t\boldsymbol{\beta} + \mathbf{Z}^{(1)}_t\mathbf{u} + \mathbf{Z}^{(2)}_t\boldsymbol{\gamma}_t,$$

where $\mathbf{u} = (\mathbf{u}'_1, \mathbf{u}'_2)'$ and $\mathbf{Z}^{(1)}_t = [\mathbf{Z}^{(1)}_{t,1}\ \mathbf{Z}^{(1)}_{t,2}]$. Although notationally cumbersome, it is relatively straightforward to include any number of additional linear and/or nonlinear covariates to (12.5). Nevertheless, for simplicity of exposition, we defer migration to the general case here.

Importantly, a model for the latent process $\boldsymbol{\gamma}_t$ must still be specified. In environmental processes, like the one considered in Section 12.3, the latent spatio-temporal process, $\boldsymbol{\gamma}_t$, tends to be dynamic, frequently motivated by an underlying partial differential equation that describes a real world physical process (Wikle 2003; Wikle and Hooten 2010; Cressie and Wikle 2011). We assume that the spatio-temporal process $\boldsymbol{\gamma}_t$ can be written using the low-rank representation $\boldsymbol{\Phi}\boldsymbol{\alpha}_t$ (Wikle and Cressie 1999; Cressie and Wikle 2011), where $\boldsymbol{\Phi}$ ($n \times p_\alpha$) is a set of spatial basis function (e.g., Empirical Orthogonal Functions, EOFs) and $\boldsymbol{\alpha}_{t,}$ ($p_\alpha \times 1$) has a first-order Markov structure [i.e., a vector autoregressive model of order one – VAR(1)]

$$p(\boldsymbol{\alpha}^t_1, \boldsymbol{\alpha}_0, \boldsymbol{\nu}) = p(\boldsymbol{\alpha}_0)\prod_{t=1}^{T} p(\boldsymbol{\alpha}_t|\boldsymbol{\alpha}_{t-1}, \boldsymbol{\nu}),$$

where $\boldsymbol{\nu}$ is a vector of parameters describing the dynamical evolution. More specifically, we assume that

$$\boldsymbol{\alpha}_t|\boldsymbol{\alpha}_{t-1}, F(\boldsymbol{\nu}), Q(\boldsymbol{\nu}) \sim N_n(F(\boldsymbol{\nu})\boldsymbol{\alpha}_{t-1}, Q(\boldsymbol{\nu})), \tag{12.6}$$

where the transition matrix $\mathbf{F}$ and noise covariance matrix $\mathbf{Q}$ depend on $\boldsymbol{\nu}$. Typically, $\boldsymbol{\nu}$ is given a prior distribution $p(\boldsymbol{\nu})$ and estimated as part of the model.

Additionally, although we assume a distribution for the initial condition $\boldsymbol{\alpha}_0$, $p(\boldsymbol{\alpha}_0)$, this parameter is also frequently assumed to be known. It is important to note that, while $\boldsymbol{\alpha}_t$ is not spatio-temporal, the low-rank representation $\boldsymbol{\Phi}\boldsymbol{\alpha}_t$ is spatio-temporal.

12.2.2 Simulation-based dynamic design

Now, given $\mathbf{Y}_1^T$, the design problem of interest is to find the design $\mathbf{d}_{T+1}$ that optimizes some aspect of the distribution

$$p(\boldsymbol{\alpha}_{T+1}, \boldsymbol{\beta}, \mathbf{u}, \mathbf{y}_{T+1}|\mathbf{Y}_1^T, \mathbf{d}_{T+1}).$$

In this context, 'design' is meant to denote a subset of the n possible spatial locations $\{r_1^{T+1}, \ldots, r_{m_{t+1}}^{T+1}\}$ at which data are collected at time $T+1$. To simplify notation, let $\boldsymbol{\theta}_t = (\boldsymbol{\beta}', \mathbf{u}', \boldsymbol{\alpha}_t')'$ and define $u(\boldsymbol{\theta}_{T+1}, \mathbf{y}_{T+1}, \mathbf{d}_{T+1})$ to be some utility function of interest. Then, given the data $\mathbf{Y}_1^T$, the goal is to select the design, $\mathbf{d}_{T+1}$ that maximizes the expected utility with respect to $\mathbf{y}_{T+1}$ and $\boldsymbol{\theta}_{T+1}$. In other words, we seek the design $\mathbf{d}_{T+1}$ from the set of possible designs D that maximizes

$$U(\mathbf{d}_{T+1}|\mathbf{Y}_1^T) = \int u(\boldsymbol{\theta}_{T+1}, \mathbf{y}_{T+1}, \mathbf{d}_{T+1}) \times p_d(\mathbf{y}_{T+1}|\boldsymbol{\theta}_{T+1}) p(\boldsymbol{\theta}_{T+1}|\mathbf{Y}_1^T) d\boldsymbol{\theta}_{T+1} d\mathbf{y}_{T+1}, \quad (12.7)$$

where $p(\boldsymbol{\theta}_{T+1}|\mathbf{Y}_1^T)$ is the posterior predictive distribution of $\boldsymbol{\theta}_{T+1}$ and $p_d(\mathbf{y}_{T+1}|\boldsymbol{\theta}_{T+1})$ is the data model given the design $\mathbf{d}_{T+1}$ [e.g., $p(\mathbf{y}_t|\boldsymbol{\alpha}_t, \boldsymbol{\beta}, \mathbf{u}, \phi)$]. For convenience of notation, we temporarily suppress the possible dependence on $\boldsymbol{v}$ and the subscript d on $p_d(\cdot)$ is used to denote the dependence on $\mathbf{d}_{T+1}$.

Typically, (12.7) is intractable; therefore, in practice, this integral will need to be approximated. In this case we can obtain samples from the distribution $\{\mathbf{y}_{d,T+1}^j, \boldsymbol{\theta}_{T+1}^j\}_{j=1}^N$ by sampling from the posterior predictive distribution of $\boldsymbol{\theta}_{T+1}$; then, assuming design $\mathbf{d}_{T+1}$, using this sample in the data model. In other words, given the design $\mathbf{d}_{T+1}$ and the data $\mathbf{Y}_1^T$, $\{\mathbf{y}_{d,T+1}^j, \boldsymbol{\theta}_{T+1}^j\}$ denotes the jth sample from the joint distribution at time $T+1$.

In theory, a Monte Carlo (MC) estimate of the expected utility can be obtained using

$$\widehat{U}(\mathbf{d}_{T+1}|\mathbf{Y}_1^T) = \frac{1}{N}\sum_{j=1}^{N} u(\boldsymbol{\theta}_{T+1}^j, \mathbf{y}_{d_{T+1}}^j); \quad (12.8)$$

however, evaluating (12.8) will often prove challenging in practice. For example, in the precipitation example in Section 12.3 the objective is to estimate the total precipitation produced over the monitored region. Therefore, if $y_t(s)|c_t(s), \phi \sim Gam(c_t(s), \phi)$ then the total observed precipitation at time $T+1$, $\sum_s y_{T+1}(s)$, is distributed as $Gam(c_t^*(s), \phi)$, where $c_t^*(s) = \sum_s c_t(s)$. Therefore, it seems sensible to find a design that minimizes $\text{Var}(\sum_s \phi c_t(s)|\mathbf{Y}_1^T, \mathbf{d}_{T+1})$. Specifically, we are interested in minimizing the variance of the sum, over space, of the predicted mean total precipitation for $T+1$ (e.g., the mean of y at time $T+1$).

Define the mean total precipitation at location s_i and time t as $\mu_t(s_i) = \phi c_t(s_i)$. Then, following Wikle and Royle (2005), finding the design that minimizes $\text{Var}(\sum_s \mu_{T+1}(s)|\mathbf{Y}_1^T, \mathbf{d}_{T+1})$ is equivalent to maximizing the following expected utility

$$
\begin{aligned}
&U(\mathbf{d}_{T+1}|\mathbf{Y}_1^T)\\
&= -\text{E}\left(\left(\sum_{i=1}^{n} \mu_{T+1}(s_i) - \text{E}\left(\sum_{i=1}^{n} \mu_{T+1}(s_i)|\mathbf{y}_{d,T+1}, \mathbf{Y}_1^T\right)\right)^2 |\mathbf{y}_{d,T+1}, \mathbf{Y}_1^T\right)\\
&= -\text{E}\left(\left(\sum_{i=1}^{n} h(\theta_{T+1}(s_i)) - \text{E}\left(\sum_{i=1}^{n} h(\theta_{T+1}(s_i))|\mathbf{y}_{d,T+1}, \mathbf{Y}_1^T\right)\right)^2 |\mathbf{y}_{d,T+1}, \mathbf{Y}_1^T\right),
\end{aligned}
$$

which, in theory, can be estimated using MC, given design dependent samples $\{\theta_{T+1}^j, \mathbf{y}_{d,T+1}^j\}$, $j = 1, \ldots, N$,

$$
\begin{aligned}
&\widehat{U}(\mathbf{d}_{T+1}|\mathbf{Y}_1^T)\\
&= -\frac{1}{N}\sum_{j=1}^{N}\left(\sum_{i=1}^{n} h^j(\theta_{T+1}(s_i)) - \text{E}\left(\sum_{i=1}^{n} h(\theta_{T+1}(s_i))|\mathbf{y}_{d,T+1}, \mathbf{Y}_1^T\right)\right)^2. \qquad (12.9)
\end{aligned}
$$

The problem with Equation (12.9) is that calculating the second expectation using an MC approach is computationally prohibitive [cf. Equation (10) in Wikle and Royle (2005)]. Nevertheless, an approximate estimate of the expected value $\text{E}(\sum_{i=1}^n h(\theta_{T+1}(s_i))|\mathbf{y}_{d,T+1}^j, \mathbf{Y}_1^T)$ can be obtained through the dynamical properties of the spatio-temporal process by using an extended Kalman filter (EKF) (Grewal and Andrews 1993; Wikle and Royle 2005; Cressie and Wikle 2011).

12.2.3 Extended Kalman filter for dynamic Gamma models

We arrive at an approximation estimate for the expected value $\text{E}(\sum_{i=1}^n h(\theta_{T+1}(s_i))|\mathbf{y}_{d,T+1}, \mathbf{Y}_1^T)$, by implementing an EKF for dynamic Gamma models. In particular, Fahrmeir and Kaufmann (1991) and Fahrmeir (1992) demonstrated that an EKF can be used for multivariate exponential families and dynamic processes. However, their exposition is provided in terms of the natural exponential family. Although the Gamma distribution can be written in natural exponential form, this formulation is less advantageous for the dynamic generalized linear model proposed here. Therefore, the EKF they present cannot be used directly; instead, a straightforward modification is needed to accommodate the extra parameter.

Recall, $\boldsymbol{\theta}_t = (\boldsymbol{\beta}', \mathbf{u}', \boldsymbol{\alpha}_t')'$; therefore, the dynamic model (12.6) can be expressed as

$$\boldsymbol{\theta}_t = \widetilde{F}\boldsymbol{\theta}_{t-1} + \widetilde{\boldsymbol{\eta}}_t,$$

where

$$\widetilde{\mathbf{F}} = \begin{bmatrix} \mathbf{I} & \mathbf{0} \\ \mathbf{0} & \mathbf{F} \end{bmatrix}, \quad \widetilde{\eta}_t = \begin{bmatrix} \mathbf{0} \\ \eta_t \end{bmatrix} \text{ where } \widetilde{\mathbf{Q}} = \begin{bmatrix} \mathbf{0} & \mathbf{0} \\ \mathbf{0} & \mathbf{Q} \end{bmatrix}.$$

Now, let $\boldsymbol{\mu}_t = h(\widetilde{\mathbf{X}}_t \boldsymbol{\theta}_t)$ and $\widetilde{\mathbf{X}}_t = [\mathbf{X}_t \; \mathbf{Z}_t] = [\mathbf{X}_t \; \mathbf{Z}_t^{(1)} \; \mathbf{Z}_t^{(2)}]$, where, again, $\mathbf{Z}_t^{(2)}$ is a $m_t \times n$ model matrix of random effects associated with $\boldsymbol{\alpha}_t$. Next, define the notation $\boldsymbol{\theta}_{t|t} \equiv \mathrm{E}(\boldsymbol{\theta}_t|\mathbf{Y}_1^t)$, $\boldsymbol{\Sigma}_{t|t} \equiv \mathrm{var}(\boldsymbol{\theta}_t|\mathbf{Y}_1^t)$, $\boldsymbol{\theta}_{t|t-1} \equiv \mathrm{E}(\boldsymbol{\theta}_t|\mathbf{Y}_1^{t-1})$ and $\boldsymbol{\Sigma}_{t|t-1} \equiv \mathrm{var}(\boldsymbol{\theta}_t|\mathbf{Y}_1^{t-1})$. Then, analogous to Wikle and Royle (2005) we have the following iterative EKF algorithm:

1. *Initialization step.*

$$\widehat{\boldsymbol{\theta}}_{0|0} = \boldsymbol{\theta} \qquad \widehat{\boldsymbol{\Sigma}}_{0|0} = \widetilde{\mathbf{Q}},$$

   where $\boldsymbol{\theta}_0$ and $\widetilde{\mathbf{Q}}$ are specified.

2. *Prediction step.*

$$\begin{aligned} \boldsymbol{\theta}_{t|t-1} &= \widetilde{\mathbf{F}}\widehat{\boldsymbol{\theta}}_{t-1|t-1} \\ \widehat{\boldsymbol{\Sigma}}_{t|t-1} &= \widetilde{\mathbf{F}}\widehat{\boldsymbol{\Sigma}}_{t-1|t-1}\widetilde{\mathbf{F}}' + \widetilde{\mathbf{Q}}, \end{aligned}$$

   where $\widetilde{\mathbf{F}}$ is specified.

3. *Correction step.*

$$\begin{aligned} \widehat{\boldsymbol{\theta}}_{t|t} &= \widehat{\boldsymbol{\theta}}_{t|t-1} + \mathbf{G}_t(\mathbf{y}_t - \boldsymbol{\mu}_t) \\ \widehat{\boldsymbol{\Sigma}}_{t|t} &= (\mathbf{I} - \mathbf{G}_t\widehat{\mathbf{H}}_t')\widehat{\boldsymbol{\Sigma}}_{t|t-1}, \end{aligned} \tag{12.10}$$

   where

$$\mathbf{G}_t = \widehat{\boldsymbol{\Sigma}}_{t|t-1}\widehat{\mathbf{H}}_t[\widehat{\mathbf{H}}_t'\widehat{\boldsymbol{\Sigma}}_{t|t-1}\widehat{\mathbf{H}}_t + \widehat{\boldsymbol{\Sigma}}_t]^{-1}$$

   is the Kalman gain, $\widehat{\boldsymbol{\Sigma}}_t = \boldsymbol{\Sigma}_t(\widehat{\boldsymbol{\theta}}_{t|t-1})$ is the conditional covariance matrix associated with our model under the Gamma distribution evaluated at $\widehat{\boldsymbol{\theta}}_{t|t-1}$ and $\widehat{\boldsymbol{\mu}}_t$ is equal to $\boldsymbol{\mu}_t$ evaluated at $\widehat{\boldsymbol{\theta}}_{t|t-1}$; i.e., the conditional mean vector associated with our model under the Gamma distribution evaluated at $\widehat{\boldsymbol{\theta}}_{t|t-1}$. Finally, the matrix $\widehat{\mathbf{H}}_t$ is given by

$$\widehat{\mathbf{H}}_t = \left. \frac{\partial h(\widetilde{\mathbf{X}}_t \boldsymbol{\theta}_t)}{\partial \boldsymbol{\theta}_t} \right|_{\boldsymbol{\theta}_t = \widehat{\boldsymbol{\theta}}_{t|t-1}}.$$

   In the Gamma data model case we have

$$\begin{aligned} \boldsymbol{\mu}_t &= h(\widetilde{\mathbf{X}}_t \boldsymbol{\theta}_t), \\ \boldsymbol{\Sigma}_t &= \mathrm{diag}\left[\phi^2 \exp\left\{\mathbf{X}_t\boldsymbol{\beta} + \mathbf{Z}_t^{(1)}\mathbf{u} + \mathbf{Z}_t^{(2)}\boldsymbol{\alpha}_t\right\}\right] \\ &= \mathrm{diag}\left[\exp\left\{2\log(\phi) + \mathbf{X}_t\boldsymbol{\beta} + \mathbf{Z}_t^{(1)}\mathbf{u} + \mathbf{Z}_t^{(2)}\boldsymbol{\alpha}_t\right\}\right], \end{aligned}$$

and $\mathbf{H}_t = \widetilde{\mathbf{X}}_t \text{diag}(\boldsymbol{\mu}_t)$. The EKF is an approximation based on local linearization and it is assumed that the matrices $\widetilde{\mathbf{F}}$ and $\widetilde{\mathbf{Q}}$ and the starting value $\boldsymbol{\theta}_0$ are known. In practice, these quantities need to be estimated and we will do so using a Bayesian approach.

12.2.4 Extended Kalman filter design algorithm

Frequently, the design problem suffers from the presence of a large design space, the so-called *curse of dimensionality*. Therefore, in this context, computationally efficient approaches to optimization are essential. For example, based on an augmented model that contains the design as a random variable, Müller (1999) considers a simulation-based design strategy to explore the design space. This approach is shown to be equivalent to the design problem considered in Wikle and Royle (2005) and hence the one considered above. As is well known, when using any practical optimization algorithm, the solution is not guaranteed to be the 'optimal' design; nevertheless, one can be assured that such an algorithm will explore 'good' designs. An additional drawback to this type of simulation-based approach is the need to find the mode of the design distribution. To circumvent these difficulties, we adopt the strategy of Wikle and Royle (2005) and approximate the expectation in (12.9) through an EKF.

The approach we follow combines MC and EKF methodology. Due to the high dimensionality frequently encountered in spatio-temporal statistics and for reasons associated with uncertainty quantification, it is advantageous to consider the models in our analysis from a Bayesian hierarchical perspective (e.g., Cressie and Wikle 2011). As a result, we use Markov chain Monte Carlo (MCMC) methods to obtain the posterior predictive distribution $p(\boldsymbol{\theta}_{T+1}|\mathbf{Y}_1^T, \boldsymbol{\nu})$. Then, for a given design $\mathbf{d}_{T+1}$, samples of 'data' can be generated from the posterior predictive data distribution and, subsequently, MCMC can be used to obtain estimates of $\mathbf{F}$, $\mathbf{Q}$ and $\boldsymbol{\theta}_0$. These estimates are then plugged into the EKF to arrive at an estimate of the needed expectation from (12.9). More specifically, using MCMC we obtain an estimate of $\mathrm{E}(\boldsymbol{\theta}_{T+1}|\mathbf{Y}_1^T, \mathbf{F}, \mathbf{Q})$, make the necessary transformation and arrive at an estimate of the expectation of interest, $\mathrm{E}(\sum_{i=1}^n h(\theta_{T+1}(s_i))|\mathbf{y}_{d,T+1}^j, \mathbf{Y}_1^T)$. This estimate is then plugged in to a MC approximation to obtain the expected utility. Following Wikle and Royle (2005), the MC–EKF dynamic design algorithm can be summarized as follows:

Step 1. Use MCMC to obtain samples $\boldsymbol{\theta}_{T+1}^j$ from the posterior predictive distribution $p(\boldsymbol{\theta}_{T+1}|\mathbf{Y}_1^T, \boldsymbol{\nu})$.

Step 2. Select a design $\mathbf{d}_{T+1}$ from the set of candidate designs D.

Step 3. Given the design $\mathbf{d}_{T+1}$ generate 'data' $\mathbf{y}_{d,T+1}^j$ from the posterior predictive data distribution $p_d(\mathbf{y}_{T+1}|\boldsymbol{\theta}_{T+1}^j)$.

Step 4. Use the posterior mean estimates of $\mathbf{F}$, $\mathbf{Q}$, $\boldsymbol{\theta}_0$, obtained from the MCMC, and samples of the 'data' generated in Step 3 in the EKF algorithm to obtain estimates of $\mathrm{E}(\sum_{i=1}^{n} h(\theta_{T+1}(s_i))|\mathbf{y}_{d,T+1}^{j}, \mathbf{Y}_1^T)$ for each $j = 1, \ldots, N$.

Step 5. Use samples obtained in Steps 1 and 3, along with the corresponding estimated means from Step 4, to calculate the expected utility, $U(\mathbf{d}_{T+1}|\mathbf{Y}_1^T)$, given in (12.9).

Step 6. Repeat Steps 2–5 until the 'optimal' design is selected (i.e., the design having maximum expected utility).

The matter of how to select the design, $\mathbf{d}_{T+1}$, in Step 2 has not yet been described. To facilitate selection, we use a basic *exchange* algorithm. These types of algorithms have been used extensively in practice and many alternatives have been proposed (Cook and Nachtsheim 1980; Atkinson and Fedorov 1988; Nychka and Saltzman 1998). Such algorithms are known to be 'greedy' and often converge to local optima. Nevertheless, for moderately sized problems, like the one considered in Section 12.3, these algorithms typically find 'good' designs that either achieve the true optimum or a value close to it.

12.3 Application: Upper Austria precipitation

In this section we examine the problem of choosing sampling locations from a monitoring network for efficiently measuring total precipitation (in mm). The monitoring network consists of $n = 37$ monitoring sites in Upper Austria and data are collected monthly from January 1994 through December 2009 (i.e., $T = 192$); see Figure 12.1. Note, not all monitoring sites contain data at each time (i.e., some of the data are missing). In addition to precipitation, information on elevation and monitoring location (latitude and longitude) is available.

The Gamma distribution is well-suited for modeling total precipitation and has been effectively used in previous analyses; e.g., see Husak *et al.* (2007) and Zhai and Feng (2009) among others. Therefore, based on this precedent and exploratory analysis we implement a low-rank semiparametric Gamma EKF-assisted design. The advantage of the low-rank semiparametric representation will be demonstrated through the posterior distribution of the elevation predictor.

We assume a dynamic Gamma model for total precipitation of the form

$$\mathbf{y}_t|\boldsymbol{\beta}, \boldsymbol{u}, \phi, \boldsymbol{\alpha}_t \sim Gam(\exp\{\mathbf{X}_t\boldsymbol{\beta} + \mathbf{Z}_t^{(1)}\boldsymbol{u} + \mathbf{Z}_t^{(2)}\boldsymbol{\Phi\alpha}_t\}, \phi) \tag{12.11}$$

where $\mathbf{y}_t$ is an m_t-dimensional vector of precipitation observations (measured on 0.1 mm increments) and $\mathbf{X}_t$ is an $m_t \times 2$ matrix containing a column of ones and an indicator for high elevation stations. Specifically, the stations located at Krippenstein and Feuerkogel are considered 'high' elevation stations whereas the remainder are considered 'low' elevation. Additionally, $\boldsymbol{\beta}$ is 2×1 with $\beta_1 = \log(\phi)$ assumed known. In particular, we assume that ϕ is known and fix it, based on a pilot analysis, to be $\phi = \exp\{\beta(1)\} = \exp(1.7) = 5.474$. The matrix $\mathbf{Z}_t^{(1)}$ is $m_t \times 4$ with 4 knot points for the 35 lower elevation sites (where elevation

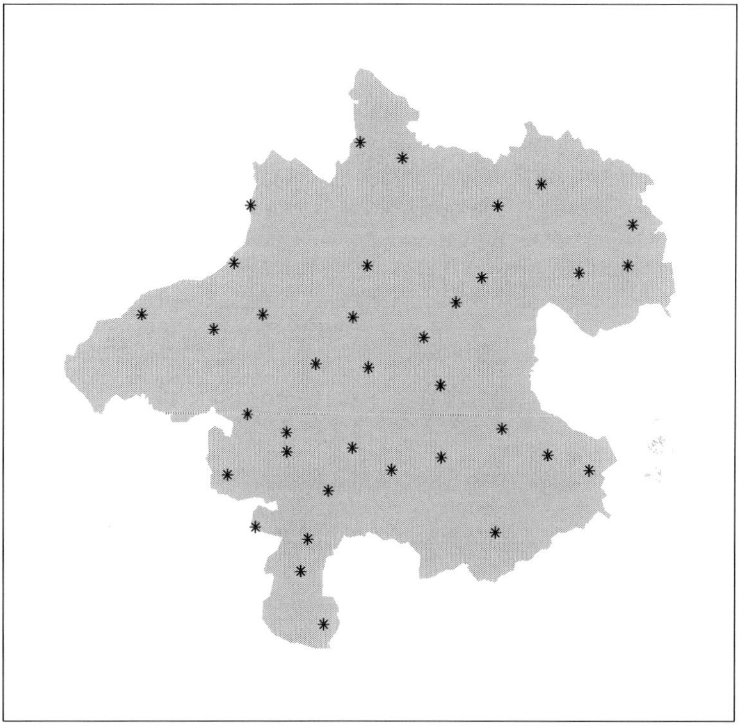

Figure 12.1 Upper Austria monitoring network: the Upper Austria monitoring network consists of n = 37 *locations marked by an asterisk.*

is standardized and the one-dimensional knot points are equally spaced in the quantile domain of the 35 lower elevation sites). Further, $\mathbf{Z}_t^{(2)}$ is an $m_t \times n$ incidence matrix, and $\mathbf{\Phi}$ is an $n \times 6$ matrix of basis functions [EOFs obtained from a pilot run and determined from the differences: $\widehat{\mathbf{k}}_t = \log(\mathbf{y}_t/\widehat{\phi}) - (\mathbf{X}_t\widehat{\boldsymbol{\beta}} + \mathbf{Z}_t^{(1)}\widehat{\boldsymbol{u}})$]. Finally, $\boldsymbol{\alpha}_t$ is a 6×1 vector of random coefficients, assumed to follow a vector AR(1) dynamic process. It is important to note that there exists significant seasonality in the precipitation response and this seasonality is accounted for through the $\boldsymbol{\alpha}_t$ process dynamics.

The low-rank dynamic process is given by

$$\boldsymbol{\alpha}_t = \boldsymbol{F}\boldsymbol{\alpha}_{t-1} + \boldsymbol{\eta}_t, \quad \boldsymbol{\eta}_t \sim Gau(\mathbf{0}, \mathbf{Q}),$$

with $vec(\boldsymbol{F}) \sim N(\boldsymbol{\mu}_f, \boldsymbol{\Sigma}_f)$, $\mathbf{Q}^{-1} \sim W((\nu_Q\mathbf{C}_Q)^{-1}, \nu_Q)$ and $\boldsymbol{\alpha}_0 \sim Gau(\boldsymbol{\alpha}_0, \boldsymbol{\Sigma}_0)$. Additionally, for $\boldsymbol{\beta}$ and $\mathbf{u}$ from (12.11), $\boldsymbol{\beta} \sim N(\boldsymbol{\beta}_0, \boldsymbol{\Sigma}_\beta)$ and $\boldsymbol{u} \sim N(\boldsymbol{u}_0, \sigma_u^2\mathbf{I})$, with $\sigma_u^2 \sim IG(q_u, r_u)$. Our hyperparamters reflect relatively vague priors and are specified as follows: $\boldsymbol{\beta}_0 = \mathbf{0}$, $\boldsymbol{\Sigma}_\beta = 100\,\mathbf{I}$, $\boldsymbol{\alpha}_0 = \mathbf{0}$, $\boldsymbol{\Sigma}_0 = 100\,\mathbf{I}$, $\boldsymbol{u}_0 = \mathbf{0}$, $\boldsymbol{\mu}_f = 0.9\,\mathbf{1}$, $\boldsymbol{\Sigma}_f = 10\,\mathbf{I}$, $\nu_Q = 6$, $\mathbf{C}_Q = \mathbf{I}$. Lastly, $q_u = 2.01$ and $r_u = 0.99$; in our parameterization this corresponds to an Inverse Gamma prior for σ_u^2 with mean 1 and variance 100.

The MCMC was comprised of a relatively straightforward Metropolis within Gibbs sampling algorithm. The MCMC was run for 110 000 iterations, the first 10 000 of which were discarded for burn-in. Convergence of the MCMC algorithm was evaluated through visual inspection of trace plots of the sample chains and showed no evidence of nonconvergence. For purposes of the MC estimation of the expected utility 1000 samples were used, based on every 100th sample of the 100 000 post burn-in samples. The initial design in the exchange algorithm was selected randomly from the 37 possible sampling locations and the exchange algorithm considered the 15 closest sampling locations for exchange.

In our analysis, we consider choosing various different numbers of monitoring stations to sample among the 37 possible locations at time $T+1$ (i.e., January 2010). Specifically, we consider designs having 20, 15, 10 and 5 sampling locations (Figure 12.2). For comparison purposes, we also generated 20000 random designs having 20 locations.

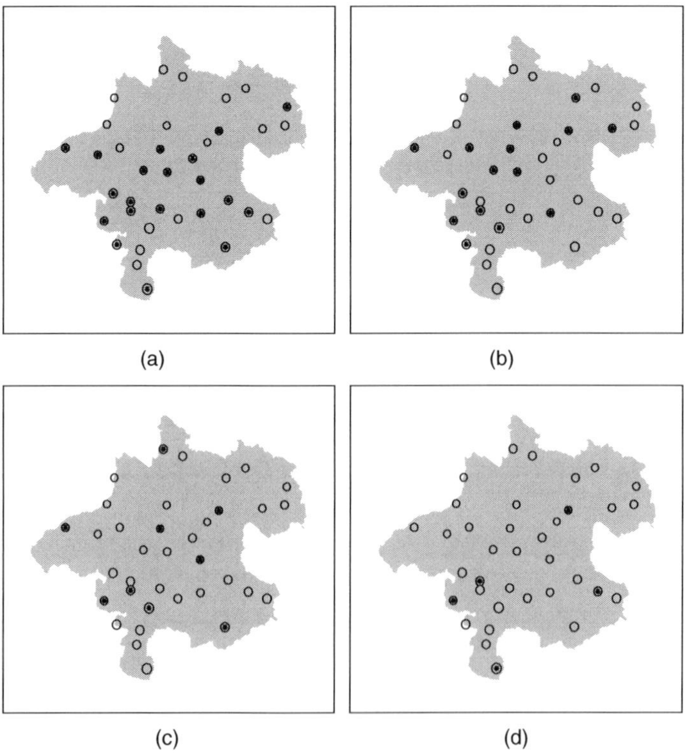

Figure 12.2 'Optimal' T + 1 *(January 2010) design (asterisks) out of* n = 37 *possible sampling stations (circles): (a) 20 sampling stations; (b) 15 sampling stations; (c) 10 sampling stations; and (d) 5 sampling stations. Note that the circles are proportional to the posterior predicted precipitation and that, although (c) appears to have only 9 stations selected, the true number selected is 10 (this occurs because two stations – Kremsmuenster1 and Kremsmuenster2 – are located very close to one another).*

Out of the 20000 random designs having 20 locations, the best design criterion was 3.85×10^5. In contrast, the 'optimal' designs for the various numbers of stations yielded the following design criterion: 20 sampling locations $= 3.15 \times 10^5$; 15 sampling locations $= 3.34 \times 10^5$; 10 sampling locations $= 3.45 \times 10^5$; and 5 sampling locations $= 3.86 \times 10^5$. A couple of things are important to note. First, with 10, 15, and 20 sampling locations we did better (in terms of our design criterion) choosing a design based on our approach rather than randomly choosing 20 locations. Secondly, even with only 5 locations the performance is on par with the best randomly chosen 20 location design out of 20000. Although our approach may not actually choose the 'best' design, certainly as evidenced by the performance of our method under different sampling scenarios, our approach provides a method of effectively choosing 'good' designs.

Another noteworthy component to our approach is the low-rank semiparametric model placed on the covariates. More specifically, using only 4 knot points we were able to capture the nonlinear structure displayed by the lower 35 elevation

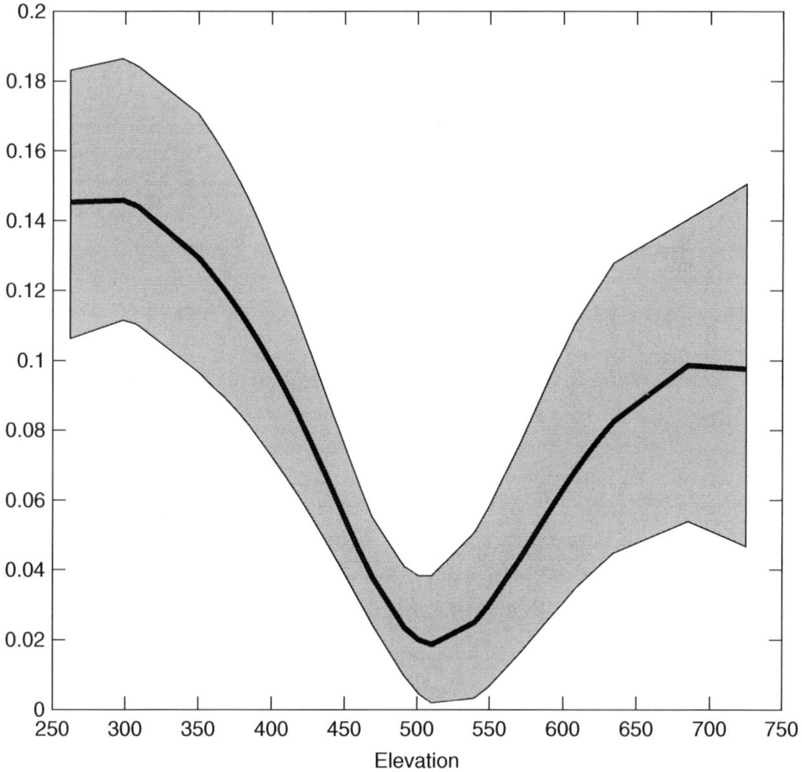

Figure 12.3 Low-rank nonparametric spline representation for elevation *(applied to the lower 35 stations). The black line represents the posterior mean curve whereas the shaded region reflects the pointwise 95% posterior credible interval.*

locations (i.e., excluding Krippenstein and Feuerkogel) when modeling the log mean precipitation; see Figure 12.3. This type of approach has the potential to improve estimated designs by reducing the possibility of model misspecification. Lastly, as demonstrated in Figure 12.3, elevation has a more pronounced effect on the log mean precipitation for both lower and higher elevations.

12.4 Discussion

The dynamic design problem for spatial sampling has grown tremendously over the past decade, with many different methods now available. The approach taken here is to maximize an expected utility function related to some important characteristic of the data. In our specific case we have chosen to minimize the variance of the estimated total precipitation. In practice, simulation-based approaches can be computationally intensive and to reduce the computational burden we have implemented a hybrid MC–EKF simulation-based design approach.

Precipitation has long been modeled using non-Gaussian data distributions and for this purpose we have chosen to use a dynamic Gamma regression model. One advantage of the model we propose is that it contains a low-rank semi-parametric component that can effectively capture structure in the covariates. Additionally, this representation illustrates that lower and higher elevations have a larger effect on the log mean precipitation.

More importantly, our approach effectively chooses designs with fewer sampling locations. Specifically, in terms of our design criterion, we are able to choose designs with 5 locations that perform on par with randomly choosing 20 locations. Additionally for designs of 10, 15, and 20 locations our approach typically chooses better designs (in terms of our design criterion) than choosing 20 sampling locations at random.

Although we have chosen a design criterion based on minimizing the total variation in the estimated precipitation, our framework has the flexibility to choose designs based on other aspects of interest for a particular analysis. For example, if the goal was to minimize cost related to sampling one could taylor the criterion for this purpose (Hooten *et al.* 2012). Alternatively, if one were interested in prediction, the utility function could be adapted to minimize forecast error. Depending on the goals of the monitoring program our approach can provide a flexible computationally efficient means of obtaining 'good' designs in nonlinear and non-Gaussian settings.

Acknowledgments

S.H.H.'s research was partially supported by a University of Missouri Research Board grant. C.K.W.'s research was partially supported by National Science Foundation grant DMS-1049093 Finally, both authors were partially supported by the NSF through the NSF-Census Research Network (NCRN) under grant SES-1132031.

References

Atkinson AC and Fedorov VV 1988 Optimum design of experiments. In *Encyclopedia of Statistics* (Supplemental volume) (eds Kotz S and Johnson N). John Wiley & Sons, Ltd, pp. 107–114.

Bueso JM, Angulo JM and Alonso FJ 1998 A state-space model approach to optimum spatial sampling design based on entropy. *Environmental and Ecological Statistics* **5**(1), 29–44.

Cook RD and Nachtsheim CJ 1980 A comparison of algorithms for constructing exact D-optimal designs. *Technometrics* **22**(3), 315–324.

Crainiceanu CM, Ruppert D and Wand MP 2005 Bayesian analysis for penalized spline regression using WinBUGS. *Journal of Statistical Software* **14**(14), 1–24.

Cressie N and Wikle CK 2011 *Statistics for Spatio-Temporal Data*. John Wiley & Sons, Ltd.

Fahrmeir L 1992 Posterior mode estimation by extended Kalman filtering for multivariate dynamic generalized linear models. *Journal of the American Statistical Association* **87**(418), 501–509.

Fahrmeir L and Kaufmann H 1991 On Kalman filtering, posterior mode estimation and Fisher scoring in dynamic exponential family regression. *Metrika* **38**(1), 37–60.

Fedorov VV and Müller W 1989 Comparison of two approaches in the optimal design of an observation network. *Statistics* **20**(3), 339–351.

Fuentes M, Chaudhuri A and Holland D 2007 Bayesian entropy for spatial sampling design of environmental data. *Environmental and Ecological Statistics* **14**(3), 323–340.

Grewal M and Andrews A 1993 *Kalman Filtering, Theory and Practice*. Prentice Hall.

Guttorp P, Le N, Sampson P and Zidek J 1993 Using entropy in the redesign of an environmental monitoring network. In *Multivariate Environmental Statistics* (eds Patil GP and Rao CR). North Holland, pp. 175–202.

Haas T 1992 Redesigning continental-scale monitoring networks. *Atmospheric Environment. Part A. General Topics* **26**(18), 3323–3333.

Holan SH, Wang S, Arab A, Sadler EJ and Stone K 2008 Semiparametric geographically weighted response curves with application to site-specific agriculture. *Journal of Agricultural, Biological, and Environmental Statistics* **13**(4), 424–439.

Hooten MB, Ross BE and Wikle CK 2012 Optimal spatio-temporal monitoring designs for characterizing population trends. In *Design and Analysis of Long-Term Ecological Monitoring Studies* (eds Gitzen RA, Millspaugh JJ, Cooper AB and Licht DS). Cambridge University Press, pp. 443–459

Hooten MB, Wikle CK, Sheriff S and Rushin J 2009 Optimal spatio-temporal hybrid sampling designs for ecological monitoring. *Journal of Vegetation Science* **20**(4), 639–649.

Husak G, Michaelsen J and Funk C 2007 Use of the gamma distribution to represent monthly rainfall in africa for drought monitoring applications. *International Journal of Climatology* **27**(7), 935–944.

Le N and Zidek J 1994 Network designs for monitoring multivariate random spatial fields. In *Recent Advances in Statistics and Probability: Proceedings of the 4th International Meeting of Statistics in the Basque Country, San Sebastián, Spain, 4–7 August, 1992*. VSP Intl Science, pp. 191–206.

Müller P 1999 Simulation-based optimal design. In *Bayesian Statistics 6: Proceedings of the Sixth Valencia International Meeting*. Oxford University Press, vol. 6, pp. 459–474.

Müller W 2007 *Collecting Spatial Data: Optimum Design of Experiments for Random Fields*. Springer Verlag.

Müller W and Pázman A 2003 Measures for designs in experiments with correlated errors. *Biometrika* **90**(2), 423–434.

Müller W and Zimmerman D 1999 Optimal designs for variogram estimation. *Environmetrics* **10**(1), 23–37.

Nychka D and Saltzman N 1998 Design of air-quality monitoring networks. In *Case Studies in Environmental Statistics* (eds Nychka D, Piegorsch W and Cox L). Springer, pp. 51–76.

Oehlert G 1996 Shrinking a wet deposition network. *Atmospheric Environment* **30**(8), 1347–1357.

Ruiz-Cárdenas R, Ferreira MAR and Schmidt AM 2010 Stochastic search algorithms for optimal design of monitoring networks. *Environmetrics* **21**(1), 102–112.

Ruppert D, Wand MP and Carroll RJ 2003 *Semiparametric Regression*. Cambridge University Press.

Ruppert D, Wand MP and Carroll RJ 2009 Semiparametric regression during 2003-2007. *Electronic Journal of Statistics* **3**, 1193.

Sansó B and Müller P 1998 Redesigning a network of rainfall stations. In *Case Studies in Bayesian Statistics, vol. IV*. Springer Verlag, pp. 383–394.

Thompson S and Seber A 1996 *Adaptive Sampling*. John Wiley & Sons, Ltd.

Wikle CK 2003 Hierarchical models in environmental science. *International Statistical Review* **71**(2), 181–199.

Wikle CK 2010 Low-rank representations for spatial processes. In *Handbook of Spatial Statistics* (eds Gelfand A, Diggle P, Fuentes M and Guttorp P). Chapman & Hall/CRC, pp. 107–118.

Wikle CK and Cressie N 1999 A dimension-reduced approach to space-time Kalman filtering. *Biometrika* **86**(4), 815–829.

Wikle CK and Hooten MB 2010 A general science-based framework for nonlinear spatio-temporal dynamical models. *Test* **19**, 417–451.

Wikle CK and Royle JA 1999 Space-time models and dynamic design of environmental monitoring networks. *Journal of Agricultural, Biological, and Environmental Statistics* **4**, 489–507.

Wikle CK and Royle JA 2005 Dynamic design of ecological monitoring networks for non-Gaussian spatio-temporal data. *Environmetrics* **16**(5), 507–522.

Zhai L and Feng Q 2009 Spatial and temporal pattern of precipitation and drought in Gansu Province, Northwest China. *Natural Hazards* **49**(1), 1–24.

Zidek J and Zimmerman D 2010 Monitoring network design. In *Handbook of Spatial Statistics* (eds Gelfand A, Diggle P, Fuentes M and Guttorp P). Chapman & Hall/CRC, pp. 131–148.

Zidek J, Sun W and Le N 2000 Designing and integrating composite networks for monitoring multivariate Gaussian pollution fields. *Journal of the Royal Statistical Society: Series C (Applied Statistics)* **49**(1), 63–79.

13

Active learning for monitoring network optimization

Devis Tuia[1,2], Alexei Pozdnoukhov[3], Loris Foresti[4] and Mikhail Kanevski[4]

[1]*Image Processing Laboratory, University of Valencia, Spain*
[2]*Laboratory of Geographic Information Systems, Lausanne Institute of Technology EPFL, Switzerland*
[3]*National Centre for Geocomputation, National University of Ireland, Maynooth, Ireland*
[4]*Institute of Geomatics and Analysis of Risk (IGAR), University of Lausanne, Switzerland*

13.1 Introduction

Machine learning (Vapnik 1998; Shawe-Taylor and Cristianini 2004; Bishop 2006; Cherkassky and Mulier 2007; Hastie *et al.* 2009; Haykin 2009) is a branch of artificial intelligence concerned with the design and development of algorithms that allow computers to reproduce dependencies observed in empirical data. A major focus of machine learning research is to design methods to automatically recognize complex patterns and make intelligent decisions based on data. Machine learning adopts many methods from nonparametric and semiparametric

Spatio-temporal design: Advances in efficient data acquisition, First Edition.
Edited by Jorge Mateu and Werner G. Müller.

statistics, but keeps the focus on the development of algorithms solving tasks where the data distribution is unknown and where no assumptions about it can be made. Moreover, machine learning algorithms are flexible since they can adapt to many nonlinear mappings, handle increased data dimensionality and noise types efficiently and thus model functions of great complexity. Machine learning is a very powerful framework for data analysis, pattern recognition and modeling, and is rapidly developing in different fields of application such as image processing (Camastra and Vinciarelli 2007), remote sensing (Camps-Valls and Bruzzone 2009) or spatially distributed environmental modeling (Dubois 2005; Kanevski *et al.* 2009). In these fields, the major problems considered by machine learning algorithms include temporal, spatial and spatio-temporal predictions (classification, regression, probability density estimation).

Success of machine learning algorithms strongly depends on the quality of the monitoring network used for building the prediction model. In this optic, monitoring network optimization, including design and redesign, is a complex fundamental problem. The configuration of an optimal network depends on the quality and quantity of available data, objectives of the study and complexity of the phenomena to be detected, modeled and predicted (Lovejoy *et al.* 1986; Tuia and Kanevski 2008).

In spatial statistics there is a long and successful history of approaches for monitoring networks optimization using kriging models, geostatistical conditional simulations and annealing algorithms (see for example Fedorov 1972; Fedorov and Hackl 1997; de Gruijter *et al.* 2006; Le and Zidek 2006, Müller 2007, Pronzato and Müller 2012). These approaches can be split into two main families: optimal experimental design and model-based design. Machine learning has adapted and developed ideas from both methodologies. Among the different applications proposed, great effort has been paid to model-based sequential re-design, referred to as *active learning* in the community. Contrarily to kriging-based monitoring network optimization, which optimizes the input space by 'filling' the geographical space, active learning depends strongly on the specific function that is being modeled: it uses uncertainty of the predictions to define input configurations that the current model cannot handle properly. The locations where these conditions occur are those that are the most beneficial to sample. In this sense, active learning is closer to an optimization done using sequential Gaussian simulations. In contrast to space filling criteria (that optimize exploration), active learning adds measurements where they are supposed to bring more information to the specific model (optimizing exploitation).

The chapter is organized as follows. Section 13.2 presents the general ideas of statistical learning from data. Then, Section 13.3 derives the basics of kernel-based support vector algorithms. The active learning framework is presented in Section 13.4, while machine learning extensions for active learning are described in Section 13.5. Kernel-based active learning strategies are tested on real case studies in Section 13.6. Section 13.7 concludes the chapter.

13.2 Statistical learning from data

Given a set of N input data samples, $\{\mathrm{x}_i\}_{i=1}^N \in \mathcal{X}$ and a set of N output data samples $\{y_i\}_{i=1}^N \in \mathcal{Y}$, in machine learning one aims to learn the mapping from the input $\mathcal{X}$ to the output space $\mathcal{Y}$. Typically, one considers input samples $\mathbf{x}$ in a d-dimensional feature space $\mathcal{X} \subseteq \mathbb{R}^d$ and outputs $\mathbf{y} \in \mathcal{Y}$ whose form depends on the type of problem (usually, $y_i \in \mathbb{R}^1$ in univariate regression and $y_i \in \{-1, +1\}$ in binary classification). Multioutput problems, where $\mathbf{y}_i \in \mathbb{R}^m$, also exist, even if not considered in this chapter.[1] Empirical examples used for this task are assumed to be drawn from a fixed but unknown distribution $P(\mathbf{x}, y) = P(y|\mathbf{x})P(\mathbf{x})$. In this sense, machine learning is closely related to nonparametric statistical modeling, and extends it by including ideas on model complexity to construct the mapping and provide distribution-free bounds to guarantee its consistency (Vapnik 1998; Cherkassky and Mulier 2007).

In the general case, the number of available input/output couples is limited, i.e., N is small. Typically one has unlimited access to $P(\mathbf{x})$ and can sample M additional *unlabeled* data $\{\mathbf{x}_i\}_{i=N+1}^{N+M}$ i.e., those data samples which consist of the input values only, while the desired output value is unknown. After sampling the unlabeled data the number of available samples is $Q = N + M$. This situation is common for most real-life learning problems and usually $M \gg N$. A simplified taxonomy of machine learning algorithms (MLAs) can then be devised in accordance with different learning tasks:

- Unsupervised learning: the only information available is the input space (main tasks: clustering, dimensionality reduction, modeling of the distribution of samples in the input space, visualization of high dimensional data): $N = 0,\ M > 0,\ Q = M$.

- Supervised learning: both input and output information is available, $N > 0$, $M = 0,\ Q = N$. Supervised learning is mainly considered in the present chapter.

- Semi-supervised learning: input data are available and for some input measurements also output information is available. $N > 0,\ M > 0,\ M \gg N$, $Q = N + M$. Semi-supervised learning combines both labeled and unlabeled data.

- Transduction: a particular semi-supervised setting where the emphasis is given to learning from particular to particular, trying to avoid induction and deduction phases of building a generalized mapping $\mathcal{X} \rightarrow \mathcal{Y}$. In this case $Q > N$, but the model is only derived to predict a particular set of samples in the input space, typically M.

[1] It is however not unusual to consider nonvectorial spaces $\mathcal{X}$ and $\mathcal{Y}$ in modern machine learning applications in computer science and bioinformatics: these complex outputs are referred to as structured outputs (Bakir *et al.* 2007), and are not covered by this chapter.

13.2.1 Algorithmic approaches to learning

The most general description of the data is based on a joint distribution function $P(\mathbf{x}, y)$. The knowledge of $P(\mathbf{x}, y)$ provides a full description of the process, as well as marginal and conditional distributions $P(\mathbf{x})$, $P(y|\mathbf{x})$. The problem of estimating $P(\mathbf{x}, y)$ is very challenging, but in many real life applications it is not even necessary as the task is to learn the dependencies between inputs and outputs and not $P(\mathbf{x}, y)$.

The process of learning the dependencies is set as the task to choose the 'best' available model from a given set $\mathbf{F} = \{f(\mathbf{x}, \alpha)\}$ defined by hyper-parameters α from some admissible set $\alpha \in \Lambda$. One measures the *loss* or discrepancy $L(\mathbf{y}, f(\mathbf{x}, \alpha))$ between the response $\mathbf{y}$ to a given input $\mathbf{x}$ and the response $f(\mathbf{x}, \alpha)$ provided by the model. Therefore, the problem of learning can be considered as a risk minimization problem with respect to α, with the risk defined as the expected value of the loss:

$$R[f] = \int L(\mathbf{y}, f) dP(\mathbf{x}, y). \tag{13.1}$$

The goal of learning is to find the function $f(\mathbf{x}, \alpha_0)$ which minimizes the risk in the situation where the joint probability distribution function is unknown and the only available information is contained in the training set (available measurements). For the three main learning tasks of classification, regression and density modeling, loss functions correspond to the classification error, mean square error, and Kullback–Leibler distance between the desired and modeled functions (Hastie *et al.* 2009). Typical assumptions are that the unknown dependency is 'smooth' in $\mathcal{X}$ and thus the family of functions $\mathbf{F}$ is chosen in accordance with this belief. Statistical learning theory (Vapnik 1998) aims at avoiding assumptions on particular data models and develops distribution-free bounds describing when the chosen algorithmic approach to derive f is consistent.

13.2.2 Over-fitting and model selection

Recall a univariate regression problem, with measured outputs $y_i = f(\mathbf{x}_i, \alpha) + \varepsilon_i$, where $f(\mathbf{x}_i, \alpha)$ is a structured part of data and ε_i is a normally distributed noise with $\mathrm{E}(\varepsilon) = 0$, $\mathrm{Var}(\varepsilon) = \sigma_\varepsilon^2$. The following expression can be derived for the expected prediction error of a regression at an input point $\mathbf{x}_0$ using squared-error loss:

$$\begin{aligned} \mathrm{Err}(\mathbf{x}) &= \mathrm{E}\left[(y - f(\mathbf{x}, \alpha))^2\right] \\ &= \sigma_\varepsilon^2 + \mathrm{Bias}^2[f(\mathbf{x}, \alpha)] + \mathrm{Var}[f(\mathbf{x}, \alpha)]. \end{aligned}$$

The first term is the variance of the target around its true mean, and cannot be avoided no matter how well we estimate the true function generating the data, unless $\sigma_\varepsilon^2 = 0$. The second term is the squared bias, the amount by which the average of our estimate differs from the true mean. The last term is the variance, the expected squared deviation of $f(\mathbf{x}, \alpha)$ around its mean.

As the family of functions **F** is usually very rich, a serious problem that arises in machine learning is the overfitting of the provided training examples. This means that the learned function fits too closely the training data (low bias) but, because of the high variance, it can not generalize well when used to predict new unseen data (high generalization error). Overfitting occurs when the function fits too closely the unique and noisy realization of the data, and is thus unable to generalize to new observations. One of the possible solutions is to balance the statistical bias and statistical variance when learning in order to achieve the smallest expected prediction error. In practice, this problem is usually solved by splitting data into training (used to train a model), validation (used to find optimal hyper-parameters) and testing (used to estimate generalization error) subsets. If the residual variance (noise) in data can be estimated independently from the selected model, the complexity can be controlled by extracting only structured information and neglecting residual variance. This approach, which in geostatistics corresponds to estimation of the nugget, was recently proposed and tested in machine learning context (gamma-test and others; Liitiainen *et al.* 2009) to account for higher dimensional spaces. It should be mentioned that the gamma-test was found to be efficient in feature selection when estimating directional noise along the input dimensions (Tsui *et al.* 2002).

There are also other regularization techniques to control the complexity of the model and to improve predictability (Hastie *et al.* 2009; Haykin 2009). The problem of bias-variance dilemma arises from the empirical risk minimization principle, or minimization of the error based on training data. A more general principle used in statistical learning theory is the structural risk minimization (SRM), where structural risk is composed of the empirical risk and a confidence term depending on the complexity of the model (Vapnik 1998). Given some regularity conditions (Vapnik 1998; Jiang *et al.* 2008), this principle enables finding optimal models with a finite number of observations. Many nonlinear learning procedures can be interpreted in terms of structural risk minimization, which means minimizing the empirical error, while keeping model complexity low. The most successful methodology based on SRM is the support vector machine (SVM), which has been developed to solve classification tasks efficiently. In the next section we present the SVM as well as the methods that have been derived from it to solve regression (the support vector regression, SVR) and density estimation (one-class SVM) tasks.

13.3 Support vector machines and kernel methods

The learning task of finding a function f by minimizing the risk functional of Equation (13.1) is intractable as $P(\mathbf{x}, y)$ is unknown but only data samples are given. In practice, one can compute and minimize the empirical risk instead:

$$R_{emp} = \frac{1}{N}\sum_{i=1}^{N} L(y_i, f(\mathbf{x}_i)). \tag{13.2}$$

A way to avoid overfitting when minimizing empirical risk is to make a smart choice of **F** and add a regularizer to (13.2). An interesting framework is provided by the use of *kernels* which are semi-positive definite functions $k(.,.) : \mathcal{X} \times \mathcal{X} \to \mathbb{R}^1$ which satisfy Mercer conditions (Mercer 1909). Such functions act as a dot product $k(\mathbf{x}, \mathbf{x}') = \langle k(., \mathbf{x}), k(., \mathbf{x}') \rangle$ and form a linear space $\mathcal{H}$ of functions $f(\mathbf{x}) = \langle f(.), k(\mathbf{x}, .) \rangle_{\mathcal{H}}$ known as a Reproducing Kernel Hilbert Space (RKHS). A norm in RKHS, $\|f\|^2_{\mathcal{H}}$ can be used as a regularizer when minimizing (13.2). The key benefit from this choice for learning problems is the fact that a large class of optimization problems involving a variety of loss functions L with RKHS regularizers have solutions that can be expressed as kernel expansions in terms of the training data (Schölkopf *et al.* 2001a):

$$f(.) = \sum_{i=1}^{N} \alpha_i k(., \mathbf{x}_i), \quad \alpha_i \in \mathbb{R}^1. \tag{13.3}$$

Kernel learning provides several important advantages. First, a regularized empirical risk minimization in a space of functions **F** is reduced to a finite-dimensional optimization problem for coefficients $\{\alpha_i\}_{i=1,\dots,N}$. Usually, it is a Linear or Quadratic Programming problem with equality and inequality constraints for which efficient numerical solvers are available. Secondly, any data processing algorithm which includes data samples only in the form of dot products can be transformed into a nonlinear kernel version by simply substituting the dot products with kernel functions. For vector spaces, kernels are usually taken as simple functional forms such as Gaussian radial basis functions $k(\mathbf{x}, \mathbf{x}') = exp(-\delta(\|\mathbf{x} - \mathbf{x}'\|^2))$, thus reproducing smooth patterns in the data. Such choice simplifies the interpretation of a resulting predictive model built according to Equation (13.3), as it becomes a linear combination of Gaussians centered at training data samples.

To benefit from the described framework, we briefly introduce basic machine learning methods based on kernels. All the kernel models presented provide a prediction function of the type described by Equation (13.3), and, depending on the task, the weights $\{\alpha_i\}_{i=1,\dots,N}$ are found by solving a different optimization problem. A common feature to these methods is that optimization problems are convex and thus have unique solutions as opposed to local minima issues in neural networks training (Haykin 2009). Then, due to the specific cost functions used most weights take zero values. The samples with weights $\alpha_i > 0$ are called *support vectors* giving a name to the family of support vector learning algorithms. This property of selecting the most important samples from the data is a key idea that we will explore in building active learning strategies for monitoring network optimization in Section 13.5.

13.3.1 Classification: SVMs

The SVM is a machine learning classification approach derived from the large margin classifier of Statistical Learning Theory (Vapnik 1998). It aims to deal

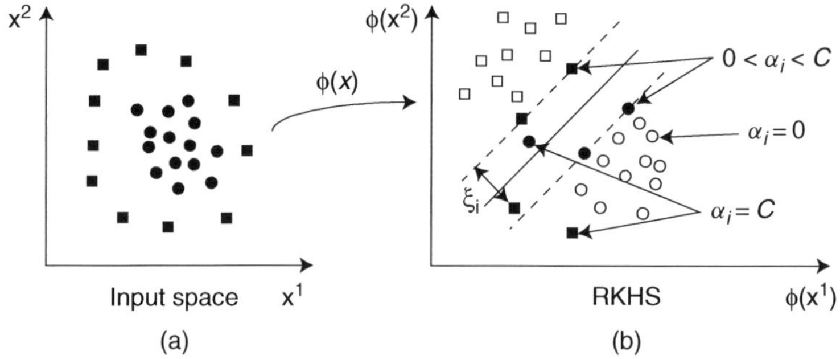

Figure 13.1 Principles of SVM. The kernel applies an implicit mapping from the input space (a) to the RKHS (b) where the problem becomes linearly separable. High generalization skill is achieved by maximizing the margin between the two classes which is completely defined by the set of support vectors [filled symbols in (b)].

with data of high dimensionality by approaching nonlinear binary classification problems in a robust and nonparametric way using the kernel framework presented above.

In classification, we deal with the training data $\{\mathbf{x}_i, y_i\}_{i=1}^N$, where $\mathbf{x}$ are the input features and $y \in \{+1, -1\}$ are the binary labels assigned to each sample (squares and circles in Figure 13.1). SVM uses the cost function

$$L(\mathbf{x}_i, y_i, f) = \max(0, 1 - y_i f(\mathbf{x}_i)), \tag{13.4}$$

which is simply a misclassification penalty, and the $\|f(.)\|_H^2$ RKHS regularizer. The final classification model of an SVM is a kernel expansion:

$$f(\mathbf{x}) = \sum_{i=1}^{N} y_i \alpha_i k(\mathbf{x}, \mathbf{x}_i) + b, \tag{13.5}$$

and a final decision is taken according to the sign of $f(\mathbf{x})$.

The weights in the kernel expansion (13.5) are found by solving a Quadratic Programming problem:

$$\max_{\boldsymbol{\alpha}} \sum_{i=1}^{N} \alpha_i - \frac{1}{2} \sum_{i=1}^{N} \sum_{j=1}^{N} y_i y_j \alpha_i \alpha_j k(\mathbf{x}_i, \mathbf{x}_j), \tag{13.6}$$

$$\text{subject to} \begin{cases} \sum_{i=1}^{N} y_i \alpha_i = 0, \\ 0 \leq \alpha_i \leq C \text{ for } i = 1, \ldots, N\,. \end{cases}$$

for which many well-developed numerical solvers exist (Chang and Lin 2001; Collobert and Bengio 2001).

13.3.2 Density estimation: One-class SVM

One-class SVM is an unsupervised kernel-based method which is used for estimating the support of the probability density distribution characterizing a set of observations in a d-dimensional space. The model tries to separate with maximum margin the center of the distribution (the support) from the observations which are atypical, noisy or can be considered as outliers. Given a set of observations $\{\mathbf{x}_i\}_{i=1}^N \subseteq \mathcal{X}$ drawn from some distribution $P(\mathbf{x})$, one wants to find a function f whose value is below a threshold only for a given portion of observations. This can be formulated in terms of risk functional

$$\min_{\mathcal{H}} \sum_{i=1}^{N} \max(0, \rho - f(\mathbf{x}_i)) + \lambda\Omega(f) \tag{13.7}$$

where ρ is an offset and $\Omega(f)$ is a regularization functional controlled by the parameter λ.

The use of RKHS regularizers results in a family of one-class support vector estimators $f(\mathbf{x}) = \sum_{i=1}^{N} \alpha_i k(\mathbf{x}, \mathbf{x}_i)$ (Tax and Duin 2000; Schölkopf *et al.* 2001a). Finding $\boldsymbol{\alpha}$ involves solving the convex Quadratic Programming problem:

$$\min_{\boldsymbol{\alpha}} \frac{1}{2} \sum_{i=1}^{N} \sum_{j=1}^{N} \alpha_i \alpha_j k(\mathbf{x}_i, \mathbf{x}_j) \tag{13.8}$$

$$\text{subject to} \begin{cases} \sum_{i=1}^{N} \alpha_i = 1 \\ 0 \leq \alpha_i \leq \frac{1}{\nu N} \text{ for } i = 1, \ldots, N \, . \end{cases}$$

where ν controls the fraction of support vectors (those samples with $\alpha_i \neq 0$) and outliers (Figure 13.2).

It is noticeable that this nonparametric nonlinear technique is focused on reproducing the low density regions (Smola *et al.* 2009). Hence its applications principally regard the detection of novelties and outliers, or rare events represented as points in the high-dimensional space of contributing factors. To obtain a good density estimate for $\mathbf{x}$ one would like to solve

$$\min_{\mathcal{H}} \sum_{i=1}^{N} g[f] - f(\mathbf{x}_i) + \lambda\Omega(f) \tag{13.9}$$

for some regularization functional $\Omega(f)$ and for a given threshold ρ declare those samples novel for which $f(\mathbf{x}) < \rho$ (Smola *et al.* 2009). This strategy is justified

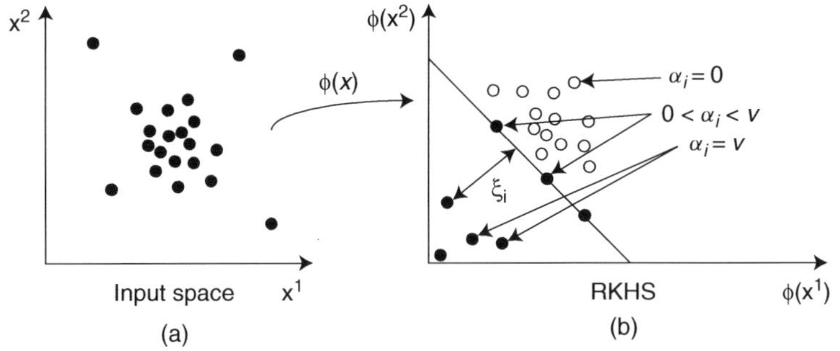

Figure 13.2 One-class SVM principle. The idea is to find a hyperplane of maximum margin between a low number of outliers (ν parameter) and the rest of the dataset (the support).

as it is actually a search for low density regions of the probability density

$$\begin{aligned} p(\mathbf{x}|f) &= \exp(f(\mathbf{x}) - g[f]), \\ g[f] &= \log \int_{\mathbf{x}\subset\mathcal{X}} \exp(f[\mathbf{x}]). \end{aligned} \tag{13.10}$$

A direct minimization of Equation (13.9) is intractable due to unknown $g[f]$; moreover, for novelty detection one is interested in regions where $f(\mathbf{x})$ is low. A workaround to this problem consists in substituting $g[f] - f(\mathbf{x}_i)$ in (13.9) with thresholded likelihood ratio

$$\begin{aligned} g[f] - f(\mathbf{x}_i) = -\log p(\mathbf{x}|f) \sim \\ \max(0, \log(\tfrac{e^{\rho - g[f]}}{p(\mathbf{x}|f)}) = \max(0, \rho - f(\mathbf{x})), \end{aligned} \tag{13.11}$$

directly providing (13.7) which is a useful approximation for solving (13.9) for a given family of RKHS $\mathcal{H}$.

13.3.3 Regression: Kernel ridge regression

Kernel ridge regression (KRR) is probably the simplest kernel method and is a straightforward extension of traditional ridge regression.[2] It aims to find a nonparametric regression $f(\mathbf{x})$ from a training set of pairs, $\{(\mathbf{x}_i, y_i)\}_{i=1}^N$ with $y_i \in \mathbb{R}^1$. KRR uses an RKHS-regularized least square approach with quadratic loss $L(\mathbf{x}_i, y_i, f) = (y_i - f(\mathbf{x}_i))^2$, hence a regression function takes the form of kernel expansion (13.3). For a pre-computed $N \times N$ kernel matrix $\mathbf{K}$, the prediction for any $\mathbf{x}$ can be obtained from the following closed form expression:

$$f(\mathbf{x}) = \mathbf{y}^T(\mathbf{K} + \delta\mathbf{I})^{-1}\mathbf{k}_{\mathbf{x}} \tag{13.12}$$

[2] Given a design matrix $\mathbf{X}$ and a vector of responses $\mathbf{y}$, ridge regression solves for coefficients α of linear regression $\mathbf{y} = \mathbf{X}\alpha$ by adding a ridge regularization term as $\alpha = (\mathbf{X}^T\mathbf{X} + \delta\mathbf{I})^{-1}\mathbf{X}^T\mathbf{y}$. For a Gaussian noise model $\mathbf{y} = \mathbf{X}\alpha + \varepsilon$, $\varepsilon \sim N(0, \mathbf{I})$ this approach results in a biased estimation which is however useful to tackle multi-collinearity (Hoerl and Kennard 1970).

where δ is a regularization parameter and $\mathbf{k_x}$ is vector column of $\mathbf{K}$. One can notice that the dot product term $\mathbf{X}^T\mathbf{X}$ of the coefficient estimator of linear ridge regression is replaced with kernel matrix $\mathbf{K}$. Note that an iterative method can be used in order to train the KRR model (Pozdnoukhov 2002) when matrix computations are not feasible due to the large number of samples.

13.3.4 Regression: SVR

The support vector regression (Smola and Schölkopf 2004) model is a regression estimation technique based on the family of so-called ε-insensitive loss functions (Huber 1964). The linear ε-insensitive loss is defined as

$$L(\mathbf{x}_i, y_i, f) = \begin{cases} |y_i - f(\mathbf{x}_i)| - \varepsilon & \text{if } |y_i - f(\mathbf{x}_i)| > \varepsilon \\ 0 & \text{otherwise.} \end{cases} \tag{13.13}$$

Again, using a RKHS regularizer $\|f(.)\|^2_H$ when minimizing the empirical risk, the prediction function takes form of a kernel regression:

$$f(\mathbf{x}) = \sum_{i=1}^{N}(\alpha_i^* - \alpha_i)k(\mathbf{x}, \mathbf{x}_i) + b, \tag{13.14}$$

with a constant offset b. The two types of weights (α_i^* and α_i) are derived by solving the following optimization problem:

$$\begin{aligned} \max_{\boldsymbol{\alpha},\boldsymbol{\alpha}^*} \quad & -\frac{1}{2}\sum_{i=1}^{N}\sum_{j=1}^{N}(\alpha_i^* - \alpha_i)(\alpha_j^* - \alpha_j)k(\mathbf{x}_i, \mathbf{x}_j) \\ & -\varepsilon\sum_{i=1}^{N}(\alpha_i^* + \alpha_i) + \sum_{i=1}^{N} y_i(\alpha_i^* - \alpha_i), \end{aligned} \tag{13.15}$$

$$\text{subject to} \quad \begin{cases} \sum_{i=1}^{N}(\alpha_i^* - \alpha_i) = 0 \\ 0 \le \alpha_i^* \le C, \ 0 \le \alpha_i \le C, \text{ for } i = 1, \ldots, N. \end{cases}$$

Either α_i or α_i^* take zero value for each i (13.14) (Figure 13.3). The solution of problem (13.15) requires a Quadratic Programming solver as well and exhibits similar properties to the SVM, that is, uniqueness of the solution and sparsity.

13.4 Active learning

The promise of active learning is the following: when the examples to be used for training are selected properly, the data requirements for many problems decrease drastically. In some cases, even the computational requirements decrease, and

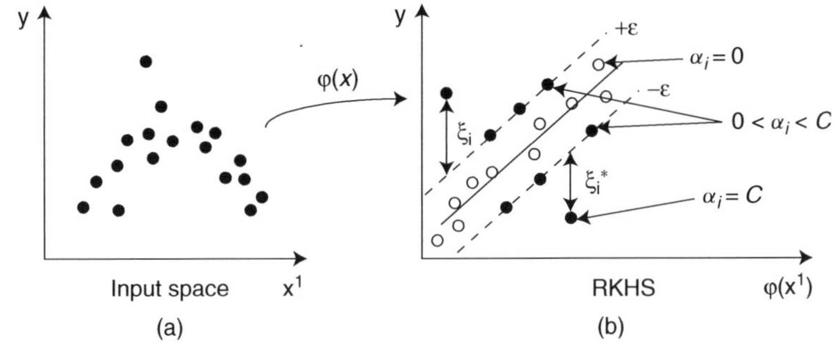

Figure 13.3 SVR principles. The data in the input space are implicitly mapped to the RKHS by means of kernels where a linear solution can be found. The noisy data inside the ϵ-tube (empty circles) do not become support vectors and are not involved in defining the regression function.

some NP-complete learning problems become polynomial in computation time. A wide range of models can be adapted to perform active learning: artificial neural networks, mixture of Gaussians, locally weighted regression, Gaussian processes, general regression neural networks, etc. For reviews, see Settles (2010) (theoretical) or Tuia *et al.* (2011c) (applications in remote sensing).

13.4.1 A general framework

In general, the main goal of active learning is to *incrementally* build a predictive model for the observed environment. That is, the goal is to select the new data which will minimize the generalization error in the most efficient way (Cohn 1996; Schohn and Cohn 2000). Formally, active learning studies the closed-loop phenomenon of a learner selecting smartly l samples $\{\mathbf{x}_k\}_{k=1}^{l}$, whose corresponding outputs are then discovered by an oracle (a user or a monitoring station). Once the outputs are known (selected samples are now composed of a set of couples $S^\epsilon = \{\mathbf{x}_k, y_k\}_{k=1}^{l}$), S^ϵ is added to the current training set (for a given iteration ϵ, the training set $X^{\epsilon+1} = X^\epsilon \cup S^\epsilon$). Many MLAs can provide an estimation of prediction uncertainties, which can be used for model-based network optimization. This approach is called uncertainty sampling.

This type of methodology is motivated by three reasons. First is the cost related to the obtention of new labeled training examples: very often, sampling requires considerable man-power and equipment that can be costly. For this reason, incorporating a large number of examples would require considerable resources. Secondly, increasing excessively the training set size grows the computational time of models, which can be a problem for some MLAs (typically SVM cost grows quadratically with the number of training examples). Compact training sets are thus desirable. Thirdly, the traditional optimal experimental design is usually considered as a 'one-shot' process, while in machine learning the active learning approaches are usually incremental by construction (see

Table 13.1 Algorithm 1. General active learning algorithm.

Inputs
- Initial training set $X^\epsilon = \{\mathbf{x}_i, y_i\}_{i=1}^{N} (X \in \mathcal{X}, \epsilon = 1)$.
- Pool of candidates $U^\epsilon = \{\mathbf{x}_j\}_{j=N+1}^{N+q} (U \in \mathcal{X}, \epsilon = 1)$.
- Number of samples l to add at each iteration (defining the batch of selected samples S).

repeat
  Train a model with current training set X^ϵ.
  for each candidate in U^ϵ **do**
    Evaluate a user-defined *criterion*
  end for
  Rank the candidates in U^ϵ according to the score of the criterion
  Select the l most interesting samples. $S^\epsilon = \{\mathbf{x}_k\}_{k=1}^{l}$
  The oracle assigns a label to the selected samples. $S^\epsilon = \{\mathbf{x}_k, y_k\}_{k=1}^{l}$
  Add the batch to the training set $X^{\epsilon+1} = X^\epsilon \cup S^\epsilon$.
  Remove the batch from the pool of candidates $U^{\epsilon+1} = U^\epsilon \backslash S^\epsilon$
  $\epsilon = \epsilon + 1$
until a stopping criterion is met.

Table 13.1 for a general description of active learning strategies): by proceeding incrementally, the model is allowed to adapt to the input given by the oracle and to change behavior in case a new unexpected situation occurs after an active learning cycle.

13.4.2 First steps in active learning: Reducing output variance

Being nonlinear universal modeling tools, MLAs have adapted many ideas from the theory of optimal experimental design (OED) and model-based design (MBD) developed in the statistical literature. For example, the first models of active learning were highly influenced by the books of Fedorov (1972); and Fedorov and Hackl (1997). The first steps in active learning algorithms were undertaken in the field of artificial neural networks (multilayer perceptron, MLP): the first ideas of active learning dealt with the selection of training data optimizing MLP weights (McKay 1992; Cohn 1996; Cohn *et al.* 1996) or minimizing the estimated variance of a model as a covariance trace maximization problem (Yu *et al.* 2006).

Most of the research in active learning assumed that the bias is negligible and only the variance should be minimized during the optimization. An approach to active learning based on minimizing a statistical bias for locally weighted regression model was elaborated in Cohn (1995). Several techniques were used to estimate the bias in local linear models: fitting higher order models and measuring the difference, bootstrapping of the residuals and cross-validation. Both the bootstrap and cross-validation methods produced fairly accurate bias estimates. An interesting integrated approach, called ensemble active learning, combining active learning and model selection at the same time was proposed in Sugiyama and Rubens (2008). An application of Gaussian processes (GPs) to active learning was studied in Sambu Seo Wallat *et al.* (2000), Krause *et al.* (2008). GPs are well known kriging-like models extended to high dimensional spaces (Rasmussen and Williams 2006).

13.4.3 Exploration–exploitation strategies: Towards mixed approaches

In machine learning there is a clear distinction between input space (space of independent variables, or feature space) and output space (dependent variables). As the input feature space is usually high dimensional (for environmental problems it can be more than 10d, and for remote sensing images more than 100d), an optimization in the input space can be considered as an exploration. Some methods are well adapted to the exploration, e.g., space filling algorithms (Müller *et al.* 2011; Pronzato and Müller 2012) or, in a more general setting, the manifold filling. Optimization during active learning in an output space corresponds to the exploitation. Exploitation is usually based on modeling of learner's confidence, uncertainties, and estimates of the generalization error. Hybrid exploration–exploitation methodology is possible and can be quite an efficient strategy. The strategy of exploration–exploitation using different criteria and modeling techniques, including machine learning algorithms, is being developed in surrogate modeling and computer-aided engineering (Crombecq *et al.* 2009; Gorrisen *et al.* 2010). Some of the ideas were first to explore the space using Voronoi polygons, i.e., to split the largest polygon to make sampling more homogeneous and then to exploit the modeling solution by considering the deviations from local linear models. Some new strategies for active learning based on hierarchical exploration of the space were developed in Castro *et al.* (2005), Singh *et al.* (2006) and Castro (2007).

13.5 Active learning with SVMs

Large margin methods such as SVM and SVR are natural candidates for active learning strategies. First, they are sparse (they rely only on a part of the available training data) and, secondly, the support vectors (the important samples), are clearly defined and identifiable by their position with respect to the current margin. This fact has been exploited in the first SVM-based active learning schemes proposed in 2000 (Campbell *et al.* 2000; Schohn and Cohn 2000), named the *margin sampling*.

13.5.1 Margin sampling

The evaluation of the distance to the hyperplane given in Equation (13.5) is the base ingredient of most large margin heuristics. These heuristics consider that a sample away from the decision boundary (with a high value of the decision function) has a high confidence about its class assignment and is thus not interesting for future sampling, since it will not become a support vector if added to the training set. This principle, called Margin Sampling (MS), is summarized in Figure 13.4.

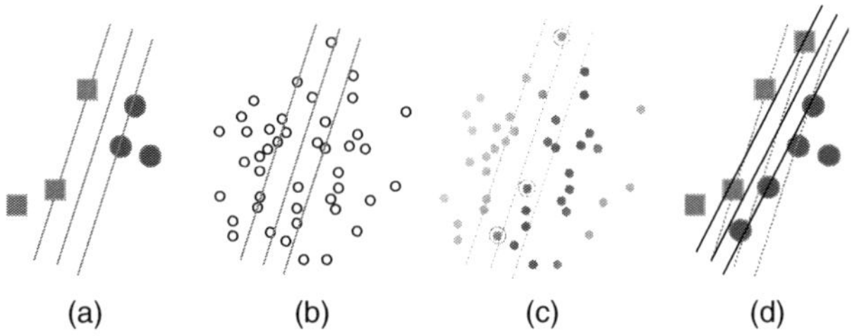

Figure 13.4 Margin sampling: (a) given a training set and a suboptimal model, (b) MS ranks the candidates points according to a criterion (c) inversely proportional to the distance to the margin. (d) Once added, these samples become support vectors and make the SVM hyperplane change.

Since SVMs rely on a sparse representation of the data, large margin based heuristics aim at finding the pixels in U^{ϵ} that are most likely to receive a nonzero α_i weight if added to the current training set X^{ϵ}. The points more likely to become support vectors are the ones lying within the margin of the current model (Tong and Koller 2001). Recent modifications of this idea can be found in Zomer *et al.* (2004), Cheng and Shih (2007) and Pasolli *et al.* (2011). They basically aim at minimizing the risk of selecting points that will not become support vectors.

This intuitive idea has been successfully applied in spatial sample design in applications such as soil mapping (Pozdnoukhov and Kanevski 2006) and remote sensing data classification (Mitra *et al.* 2004; Tuia *et al.* 2009), where it has been proven that MS can select the locations where, after adding new monitoring stations, the model uncertainty will decrease.

Another popular criterion is to fit a sigmoid function to the SVM decision function to obtain a probabilistic output and then use this as an uncertainty criterion. In Luo *et al.* (2005), the authors use the difference between the posterior probability of the two most probable classes as criterion: if these probabilities are alike, the sample is not well handled by the current model and its inclusion in the training set will be beneficial for the model. This criterion has been used successfully in remote sensing classification in Li *et al.* (2010) and Tuia *et al.* (2011b).

When dealing with one-class classifiers such as the one-class SVM (Section 13.3.2), margin techniques can be useful, as they increase the probability of detecting samples belonging to the class of interest: in contrast to traditional classification, only these target samples are included in the training set, thus making active learning strategies very important for acquiring new informative samples. Active one-class SVMs have been proposed in Juszczak and Duin (2003), where the authors consider buffers around the Support Vector Domain Description sphere to search for new samples. It must be mentioned that the latter corresponds to the one-class SVM proposed by Schölkopf *et al.* (2001b) when using normalized data and isotropic Gaussian kernels. Moreover, experimental results show that exceeding the buffer towards the background region increases probability of fast expansion of the sphere, if there is a target region not yet discovered.

13.5.2 Diversity of batches of samples

When dealing with large sets of samples it becomes necessary to select batches of samples in a single iteration. Since the sets of candidate sample locations are often clustered, MS becomes suboptimal, since it will tend to select samples that are useful for the current model, but very similar to each other. Since no cross-information among the samples is considered in the MS, the questions of diversity in batches of samples have been recently considered: these methods constrain the uncertainty-based traditional active learning criteria with a criterion of diversity aimed at selecting samples that are different to each other. Brinker (2003) and Ferecatu and Boujemaa (2007) measure spectral angles among samples, while Tuia *et al.* (2009) add a constraint on proximity to current support vectors. Demir *et al.* (2011) use nonlinear clustering to cluster the SVM margin and select diverse samples in the batch. Finally, Volpi *et al.* (2012) use memory of the samples selected in previous iterations to avoid sampling similar examples along iterations.

Regarding one-class models, in Görnitz *et al.* (2009), diversity of samples is considered through graph regularization: by considering weighting relative to proximity to target and other unlabeled samples, independence among the samples selected is maximized, along with their informativeness.

13.5.3 Committees of models

Another, more flexible view on active learning is given by the use of committees of models. Relying on committees of weaker classifiers gives freedom in the choice of the model and helps designing criteria not captive of the margin definition. The use of query-by-committee (QBC) for active learning was proposed by Seung *et al.* (1992) and Freund *et al.* (1997), where maximal disagreement between a committee of models was used to assess the uncertainty in predicting the class membership in binary classification problems.

Later on, strategies based on ensemble classifiers were proposed and deeply tested by Abe and Mamitsuka (1998) and Melville and Mooney (2004). In Tuia *et al.* (2009), the model of Abe and Mamitsuka (1998) was extended to multiclass problems by using a score based on the empirical entropy of the predictions of the committee. These models build the committee by using partial training sets drawn from the available data, but considering the totality of the feature space.

Another possibility is to split the feature space in different subsets to obtain separate views on the problem (Muslea 2006). In this case the disagreement between the views, that are modalities based on subsets of the feature space as independent as possible, is used as a sampling criterion. In Di and Crawford (2011) multiview uncertainty is coupled to a manifold-based regularizer in order to increase consistency of the chosen sample with respect to its neighbors.

Committees can also be used to assess prediction variance and thus make intuitive strategies for regression: in Krogh and Vedelsby (1995), committees of neural networks are used to assess predictions ambiguity, while in Burbidge *et al.* (2007) an extensive study is performed to test the robustness of such an approach, by varying the noise and the degree of specification of the problem. In the following, we will use QBC strategies for SVM and SVR.

13.6 Case studies

In this section, we present three case studies concerning the optimization of networks monitoring environmental parameters. First, we show an exploratory analysis of potential sampling sites for spatial prediction of average monthly air temperature and precipitation in Northern Austria. Second is a case study analyzing the concentration of Cesium-137 (Cs-137) in the soils of Russia after the Chernobyl accident. The last example considers the suitability of sites for the construction of wind power plants. The aim of the study is to illustrate how active learning can be used to minimize the number of additional samples necessary to retrieve reliable prediction maps.

13.6.1 Austrian climatological data

As a first application, we considered the data from 37 climatological stations from the region of Oberösterreich[3] (Figure 13.5). We considered a series of four monthly temperatures (T) and precipitation totals (P), accounting for January, March, July and September. Monthly data were averaged over the period 1994–2009, when available. A committee of 50 SVR models using 90% of the stations data is trained, taking (X, Y) coordinates as inputs and climatic variables as independent outputs. The prediction variance of the committee on a regular grid is used as a ranking criterion for underlining the most uncertain areas.

13.6.1.1 Air temperature

Since a clear trend between the air temperature and altitude (Z) is observed, we first detrended the data by a linear regression $\hat{T} = aZ + b$ (simple linear atmospheric model). Then, we computed the residual between measured and predicted temperature $r = T - \hat{T}$ and used it as the output variable to predict. Figure 13.6 illustrates the SVR prediction maps for the four months considered, as well as the standard deviation of predictions and resulting uncertainty map, obtained by thresholding the latter at the 90th quantile of the distribution of the standard deviation.

[3] http://www.zamg.ac.at.

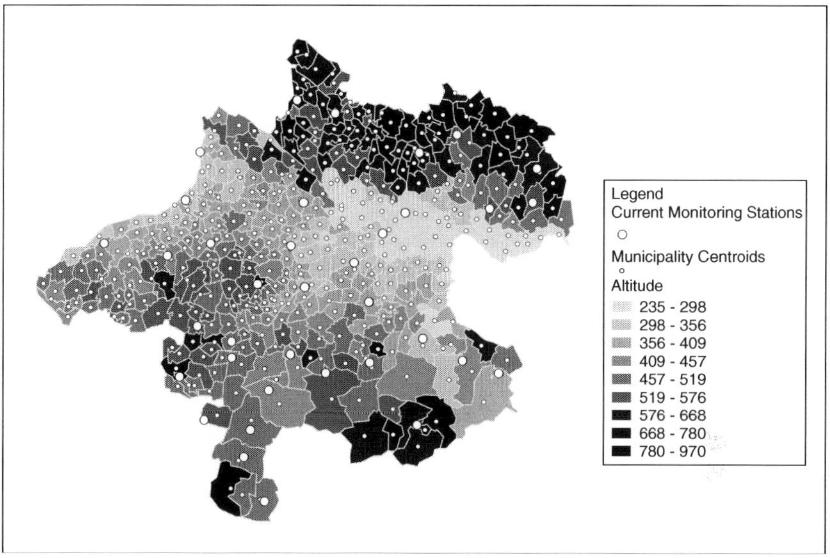

Figure 13.5 Network of 37 weather stations (large circles) and the elevation at the centroids (small dots) of municipalities in Oberösterreich region, Austria.

The maps of average residual temperatures (left column in Figure 13.6) represent how much the measured temperature deviates from the ideal situation of linear decrease of temperature with height, which is a reasonable assumption for the atmosphere at the temporal scale of analysis considered. However, some discrepancies can be noticed between different months. The spatial structure of the months January and September is clearer than that of March and June.

The map of standard deviations (middle column in Figure 13.6) summarizes the regions of uncertainty which are usually located at high elevations, i.e., where the meteorological network is sparse and where the linear atmospheric model is not well fitted due to the imbalance of low and high elevated weather stations.

Eventually, to obtain temperature maps from residuals, it suffices to take a digital elevation model, to compute $\hat{T} = aZ + b$ for every grid cell and to add it to the map of interpolated residuals.

13.6.1.2 Precipitation

Since there is only a slight trend between altitude and precipitation, raw precipitation was considered for predictions. Figure 13.7 illustrates the prediction map for precipitation, as well as the standard deviations of predictions for the 50 SVR models and the thresholding at the 90th quantile, analogously to the temperature study above.

Precipitation patterns are in general smoother than temperature ones (left column of Figure 13.7). There is a general increase in precipitation under the

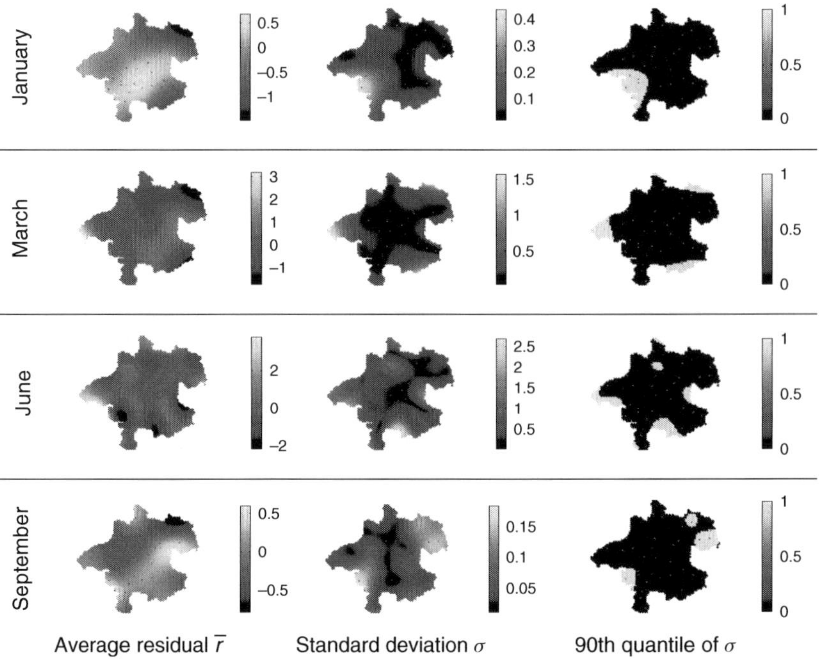

Figure 13.6 Predictions by the committee of SVR models for January, March, June and September residual average temperatures ($r = T - \hat{T}$). Left column: mean of the estimated average monthly temperature residuals $\bar{r}$; middle column: standard deviation of the 50 SVR predictions; right column: 90th quantile of the distribution of standard deviation, corresponding to uncertain areas for each month. (Please see plate section for color version of the figure.)

form of a trend when going from north to south which is due to the approaching of the Alpine chain. No clear seasonality of precipitation can be detected from the patterns excepting the higher rainfall rates in summer months. The map of uncertainty (middle column of Figure 13.7) reveals a higher uncertainty in March and January, probably caused by the higher spatial variability of rainfall which is detected by the committee of SVR models.

13.6.1.3 Where to add measurements

Figure 13.8 illustrates a stack of the four monthly uncertainty maps for temperature and precipitation estimation. Both correspond to a sum of the maps illustrated in the right columns of Figures 13.6 and 13.7. Both maps show potential regions for planning new measurement sites and share similar places, especially in the north-east. Additionally, new measurements would be beneficial for temperature estimation in the west of the region of study, while precipitation would benefit of new samples in the south. The different locations retrieved show the interest

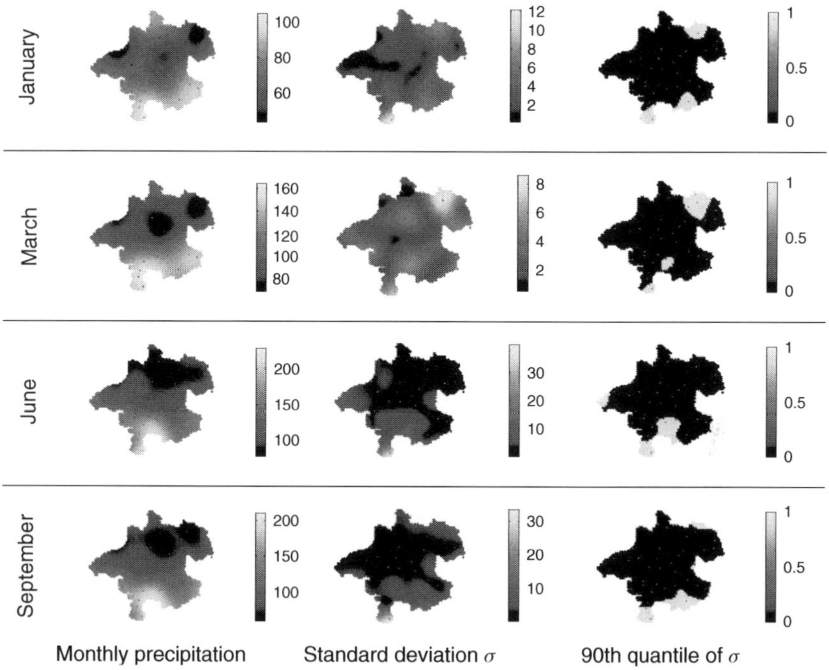

Figure 13.7 Predictions by the committee of SVR models for January, March, June and September precipitation. Left column: mean monthly precipitation of the 50 SVR predictions; middle column: standard deviation of the 50 SVR predictions; right column: 90th quantile of the distribution of standard deviation, corresponding to uncertain areas for each month.

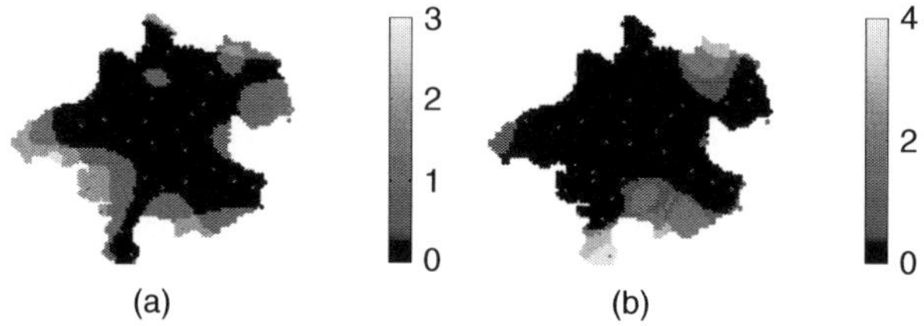

Figure 13.8 Map of uncertain areas on the multitemporal series for (a) temperature and (b) precipitation.

of using active learning for improving a specific desired prediction output, that is monthly temperature and precipitation. Adding samples in these regions will reduce prediction variance for the SVR models and thus converge to more accurate prediction maps. However, it is not possible to test this hypothesis using this dataset, as no measures were available at the potential sites. In order to test the hypothesis on real candidate sites, in the next case study we present the modeling of radioactivity in Russia.

13.6.2 Cesium-137 concentration after Chernobyl

The second study considers Cs-137 concentrations measured in soils after the Chernobyl accident in 1986 (Kanevski and Maignan 2004; Kanevski *et al.* 2009). Figure 13.9 illustrates the data available, consisting of 684 samples.

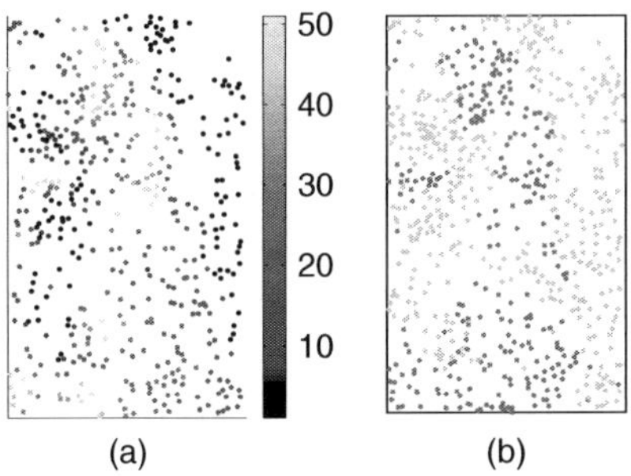

Figure 13.9 Chernobyl Cs-137 dataset. (a) Cs-137 concentrations (Ci/km^2). (b) Areas having Cs-137 concentrations higher than 15 Ci/km^2 (dark gray dots).

13.6.2.1 Experimental set-up

The principal aim is to map the soil contamination using as few samples as possible. Starting from the 100 measurements illustrated in Figure 13.10(a), we would like to improve the quality of prediction by adding new measurements using an active learning strategy. In this case study, the selection is performed among 384 additional candidate samples. For the sake of illustration, we already

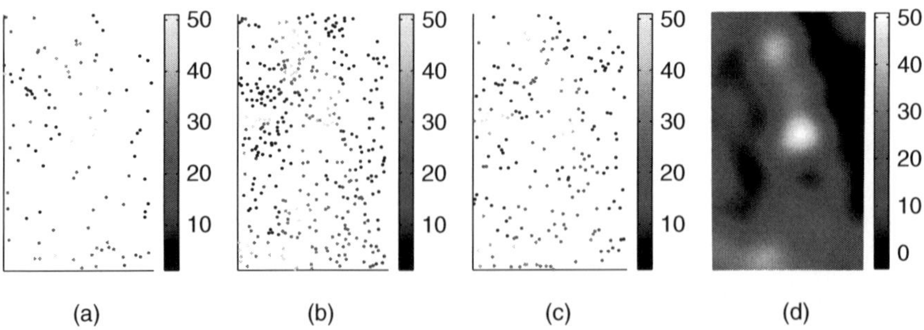

Figure 13.10 Chernobyl Cs-137 dataset. (a) 100 training samples. (b) 384 candidates sites. (c) 200 test sites. (d) Map obtained by using the 100 initial training samples. All map units are Ci/km^2.

possess the concentration at these 384 locations [Figure 13.10(b)]: in a real application these data should be acquired on the field. Quality of the model with the increased training set is assessed on a 200 points independent test set, as shown in Figure 13.10(c). We sampled 100 additional measurements from the set of candidates in 20 iterations of 5 samples each. SVR was used to compute the mapping of Cs-137 as a function of (X, Y) spatial coordinates.

Figure 13.10(d) illustrates the best prediction by SVR when using the 100 initial samples. Note that the highly contaminated area in the middle left part of the plot is hidden to the model, since no training samples are available in that region.

Two approaches are compared: the first ('Random' hereafter) picks the new measurement sites randomly in the region, while the second ('Active') uses a QBC active learning approach. We train 50 SVR models with 90% of the available training data and then select the sites showing maximal variance of the predictions. This way, sampling is focused on the areas where the current models are the most uncertain.

13.6.2.2 Active learning results

Results are shown in Figure 13.11: active sampling, i.e., sampling where prediction variance is maximal, efficiently reduces the prediction error, both globally [Figure 13.11(a)–(c)] and in the areas at risk [where Cs-137 concentration is higher than 15 Ci/km^2; Figure 13.11(d) and 13.11(e)]. The decrease in error when using active learning is much faster than the one obtained by random sampling.

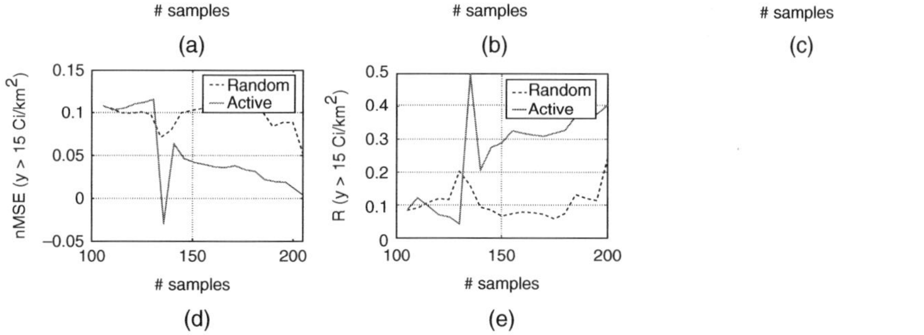

Figure 13.11 *Numerical results for Cs-137 predictions: (a)–(c) using the whole test set; (d) and (e) considering only test samples with Cs-137 concentration higher than* 15 *Ci/km*^{2}.

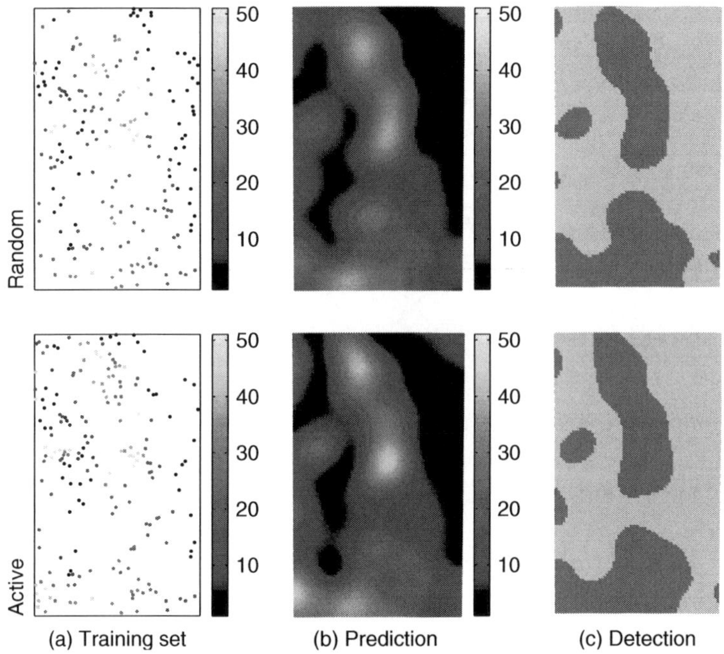

Figure 13.12 Numerical results for Cs-137 predictions: (a) training set constructed by Random and active sampling; (b) the corresponding prediction maps (Ci/km^2); (c) detection maps for areas showing Cs-137 concentration higher than 15 *Ci/km^2.*

This is mainly due to the focused sampling reducing the uncertainty of the model. The error is measured by three indicators:

- the normalized mean squared error, nMSE $= \log_{10}\left[\sqrt{\frac{1}{N\sigma^2}\sum_{i=1}^{N}(y_i - y_i^*)^2}\right]$, where y is the true value, y^* is the predicted value and σ^2 is the estimated variance of the data,
- the correlation coefficient R between true and predicted values,
- the Cohen's Kappa coefficient, estimating the detection of the area for which Cs-137 concentration is higher than 15 Ci/km^2.

Looking at the final prediction maps of Figure 13.12 and respective training sets, it is clear that active learning permits the highly contaminated area in the left part of the region to be found and samples many measurements in that region. It is not the case for the Random strategy, where poor attention is paid to this hidden contaminated area which is only detected after several iterations.

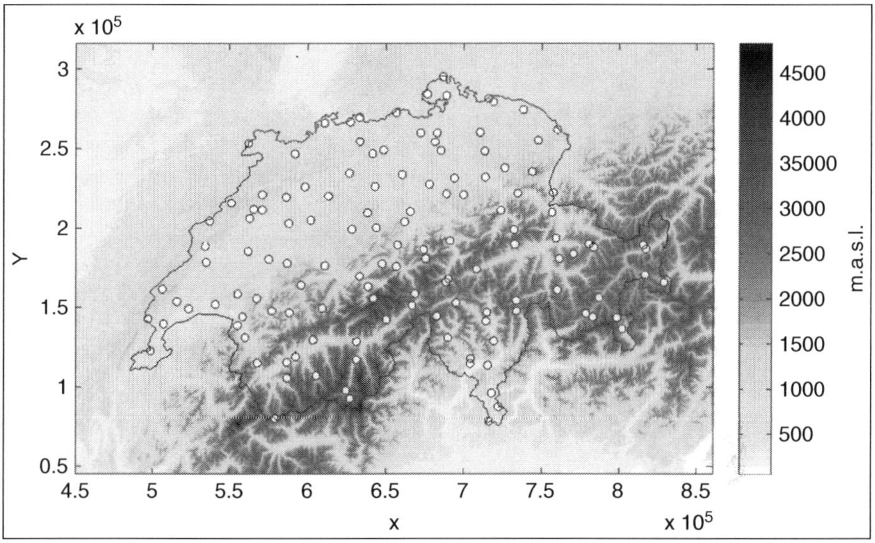

Figure 13.13 Sensor network used for wind speed measurements in Switzerland overlaid on the digital elevation model RIMINI.

13.6.3 Wind power plants sites evaluation

This case study deals with renewable energy resources in Switzerland. The Swiss Law (Bylaw on Energy, Ord 730.1)[4] states that wind power plants must be constructed in sites showing an average wind speed exceeding 4.5 m/s at a height of 50 m above the ground, with an estimated roughness length of 0.1 m. Moreover, wind plants must keep a productivity of 100–150%. Keeping these facts in mind, the planning of new wind plants must monitor wind speed over time to ensure enough speed and wind constancy. Therefore, maps of average winds are useful documents when deciding on the suitability of a site for the construction of a plant. In this optic, prediction maps must be as accurate as possible.

For this study, we consider time series of monthly wind speeds during the period 2003–2008 measured by the Swiss monitoring network illustrated in Figure 13.13. Assuming a new measuring station can be added to the sensor network of Figure 13.13, the new station could be located as a function of the objective that is being optimized, i.e. the fulfillment of the 4.5 m/s requirement for new plants. See also Tuia *et al.* (2011a) for an extended version of this case study.

The problem is cast as the classification problem of assessing (i) the class assignment according to the location and topographic conditions of a site (suitable/not suitable) and (ii) the confidence of such assignment on a multitemporal

[4] Available at www.admin.ch/ch/f/rs/c730_01.html.

series reporting the monthly average wind speeds. To do so, an active learner based on SVM classifiers trained to discriminate areas with mean monthly wind speed higher than 4.5 m/s is used.

First, classification maps of wind speed are computed for each month in the period 2003–2008 and then they are averaged to account for the variability of the limit suitable/not suitable between months. Using this information, the areas suitable for wind power plants and the confidence of such assignments are extracted. Using this confidence, we will extract the topographic configurations related to maximal uncertainty and define regions where a new measuring station will bring relevant information to solve the problem.

13.6.3.1 Data

The Swiss Digital Elevation Model (DEM) underlying Figure 13.13 was employed to extract 13 informative features at a resolution of (250×250) m^2 (see Foresti *et al.* (2011) and Robert *et al.* (2012) for details):

- Spatial coordinates (X, Y) (features 1 and 2);
- Altitude (Z) (feature 3);
- Difference of Gaussians (features 4–6). By subtracting two smoothed DEM surfaces obtained with different smoothing bandwidths, the ridges and canyons of different characteristic length scales are highlighted. The resulting set of features describes terrain convexity at three different spatial scales (small, medium and large);
- Slope (features 7–9). The norm of the terrain gradient, which is proportional to slope, is computed on smoothed DEM surfaces at three spatial scales as well. Terrain slope is expected to explain the lower wind speeds in the flanks of inner Alpine valleys;
- Directional derivatives (features 10–13). These features highlight natural obstacles represented by relief, that break winds by being perpendicular to its direction. They could also describe patterns of wind shadowing with respect to the predominant mesoscale winds. In summer they could potentially explain some thermal effects due to sun exposure. Zero values occur in flat regions (along North-South and East-West directions).

Topographic features are known both on the DEM grid and at the location of weather stations where they are stacked to geographical coordinates (X, Y, Z).

All stations data are used to classify the monthly wind speed $\bar{s}_{50}$. The positive class is given when $\bar{s}_{50} \geq 4.5$ m/s while the negative when $\bar{s}_{50} < 4.5$. This value is given by an average model for Switzerland, that extrapolates the speed at 50 m by using the speed at 10 m (height above the ground of the stations) using the following wind shear model:

$$\bar{s}_{50} = \bar{s}_{10} * \frac{\ln(\frac{50}{0.1})}{\ln(\frac{10}{0.1})} = \bar{s}_{10} * 1.3495 \tag{13.16}$$

where 0.1 is the average roughness length in Switzerland (see Swiss Bylaw on Energy).

13.6.3.2 SVM classification for detection of suitable areas

Figure 13.14 shows nine among the 72 derived classification maps for the months of January, May and September and the years 2003, 2005 and 2008. White areas are those with average wind speed fulfilling the Swiss law requirements for wind power plants at 50 m. The patterns retrieved are similar from year to year, showing the consistency of winds in Switzerland. However, September 2008 shows stronger winds than the previous years. This also shows that decision making with single months may lead to wrong conclusions. For these reasons, average classifications through the 6-year period are considered.

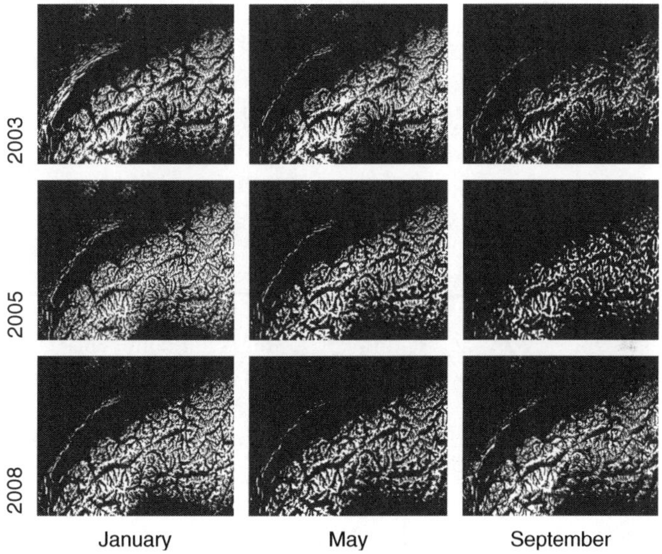

Figure 13.14 Example of single classifications of wind speed in three months in different years. White areas fulfill the requirements for wind power plants.

Figure 13.15 shows the average classification result for the 2003–2008 period. Areas in bright tones fulfill almost always the requirements in different months and years, while areas in black never fulfill them. This map brings multitemporal information about persistence of winds at the necessary speed for correct functioning of wind power plants and can be a useful document for renewable resources assessment.

13.6.3.3 Active learning to find useful new monitoring station locations

The map in Figure 13.15 does not provide information about the confidence of the suitability assignments. In order to improve this map, new measurement

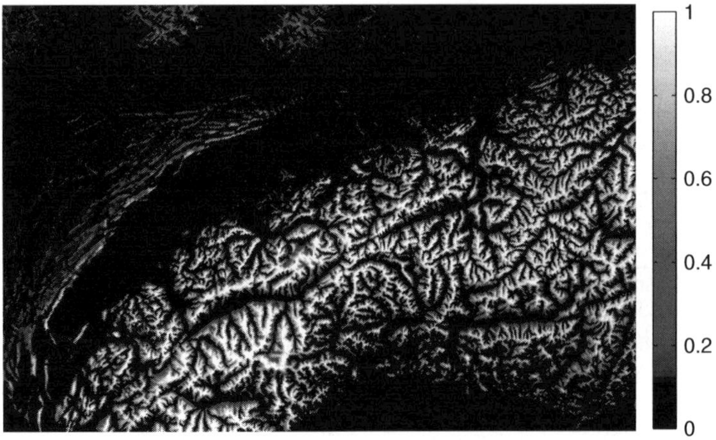

Figure 13.15 Average classification for the 2003–2008 period. Colors are proportional to the fraction of months for which $s_{50} \geq 4.5$ m/s.

stations could be added to represent the topographic conditions where the spatial classification of wind speed is the most uncertain.

The SVM models, whose predictions are represented in Figures 13.14 and 13.15, also provide information about the confidence of the class assignment. The decision function of the SVM–see Equation (13.5) –can be used to assess the level of confidence of the assignment. Using such confidence, we can detect the topographic situations which are under represented by the current sensor network for the specific task of wind speed modeling.

We will focus on uncertain areas with respect to our specific task, which is the modeling of the boundary defining the areas where winds are close to 4.5 m/s, that is within the margin of SVM. An increase in precision of the definition of these areas can have enormous impacts on decision-making processes.

Figure 13.16 illustrates the confidence of the class assignments over the 72 months, measured as the average decision function. The areas in dark tones are those of reduced confidence in the class assignment, i.e., the areas where the mean speed is often very close to 4.5 m/s.

With this knowledge, the specific topographic conditions of the uncertain areas–dark tones in Figure 13.16–can be analyzed. However, due to the high complexity of the features at these uncertain locations, it is necessary to extract general trends from the data. To do so, principal component analysis is applied to the data within the SVM margin ($|f(x)| < 1$) to analyze the behavior of the features in the maximal variance directions. A maximal variance direction can be interpreted as a linear combination of terrain features explaining the topographic conditions representative of the SVM margin shown in Figure 13.16. Being a combination of terrain features, it accounts for many of them at the same time, providing a useful visualization tool for uncertain topographical conditions.

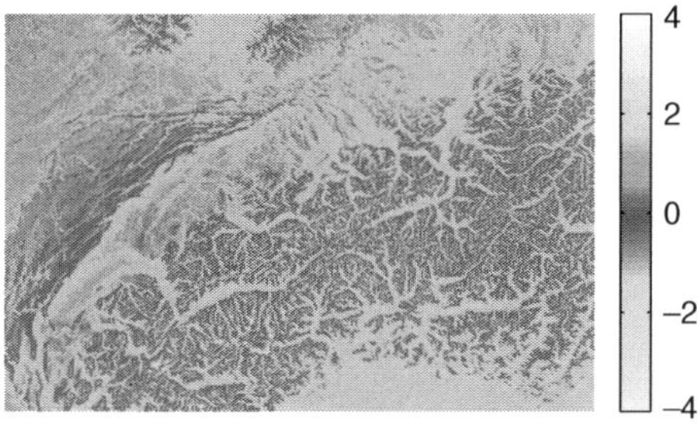

Figure 13.16 Average SVM decision function for the 2003–2008 period. Black: $\overline{|f(x)|} < 1$ = *uncertain prediction areas; white* $\overline{|f(x)|} > 1$ *certain prediction areas (for both the positive and negative class).*

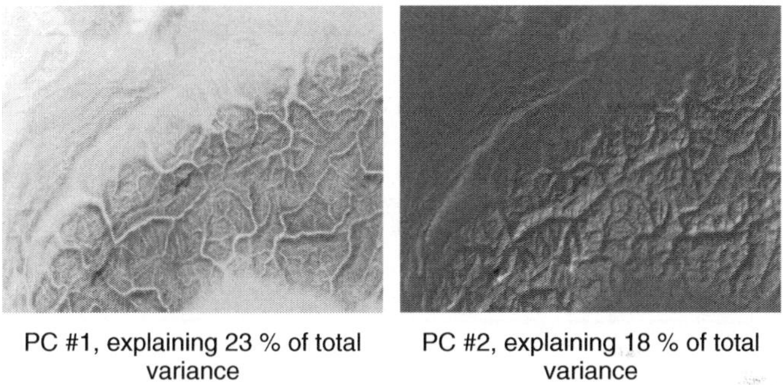

Figure 13.17 The first two principal components of the space related to uncertain samples: (a) PC #1; (b) PC #2.

Figure 13.17 illustrates the first two principal components related to maximal variance (components #1 and #2). We can observe spatial trends and, especially for component #1, precise localization of the data related to maximal scores [dark tones in Figure. 13.17(a)]. These locations could host the new stations for wind speed measurement. However, more generally, these extracted features provide information about the topographic conditions related to the uncertainty of the SVM:

- PC #1. Distribution of PC #1 has a single tail [Figure 13.18(a)]. Data forming this tail [Figure 13.18(b)] are pixels of high altitude and strong slope (see Section 13.6.3.1). Moreover, the directional derivatives of these samples are low.
- PC #2. Distribution of PC #2 has two tails [Figure 13.19(a)]. In the lower tail [Figure 13.19(b)] locations with strong slope are found, while the upper tail [Figure 13.19(c)] is mainly represented by samples with strong directional derivatives.

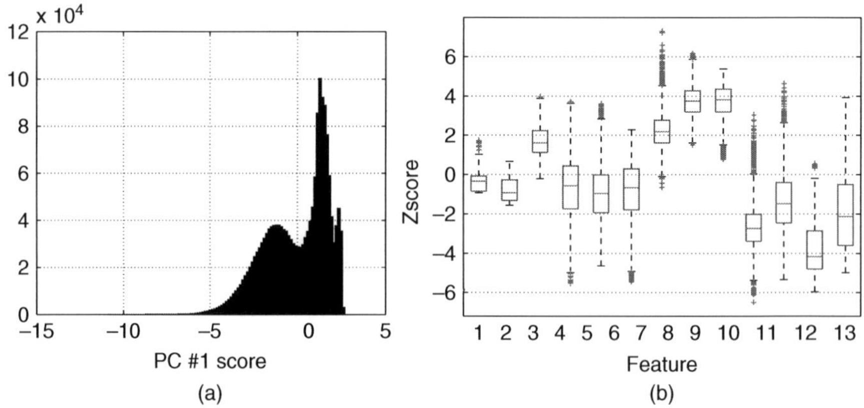

Figure 13.18 (a) Histogram of PC #1 and (b) boxplots of feature values constructed with data selected from the left tail of the histogram.

Such analyses confirm the well-known under representation of Alpine slopes by the MeteoSwiss network and the uncertainty of renewable energy assessment in complex Alpine environments.

13.7 Conclusions

Kernel-based active learning strategies were studied for the optimization of environmental monitoring networks. The chapter first introduced the basic machine learning algorithms originated in the statistical learning theory of Vapnik (1998). Such framework allowed the design of efficient nonparametric algorithms for data classification, regression and estimation of the support of probability density functions in high-dimensional spaces.

Kernel methods endow nice properties which can be adapted for active learning purposes and guide the design and re-design of monitoring networks. They represent nowadays a valid alternative to what is traditionally done in the field of geostatistics, which is often limited by constraints about the types of data that can be considered (spatially distributed and low dimensional). Kernel methods are more flexible in modeling data of variable dimensionality presenting nonlinear relationships and different types of noise. Moreover, they share a solid common framework that allows the base methodology to be adapted to respond to the different problems encountered in environmental modeling.

The SVMs lead to sparse solutions and only the most representative samples, the support vectors, are combined to compute the predictions. Support vectors give rise to a class of active learning approaches exploiting their respective weights as a criterion : the region within the margin is the one related to highest uncertainty and samples found in that region are those with maximal chance of becoming support vectors. By focusing the sampling within the margin, it is possible to achieve high accuracies with a lower number of training samples which

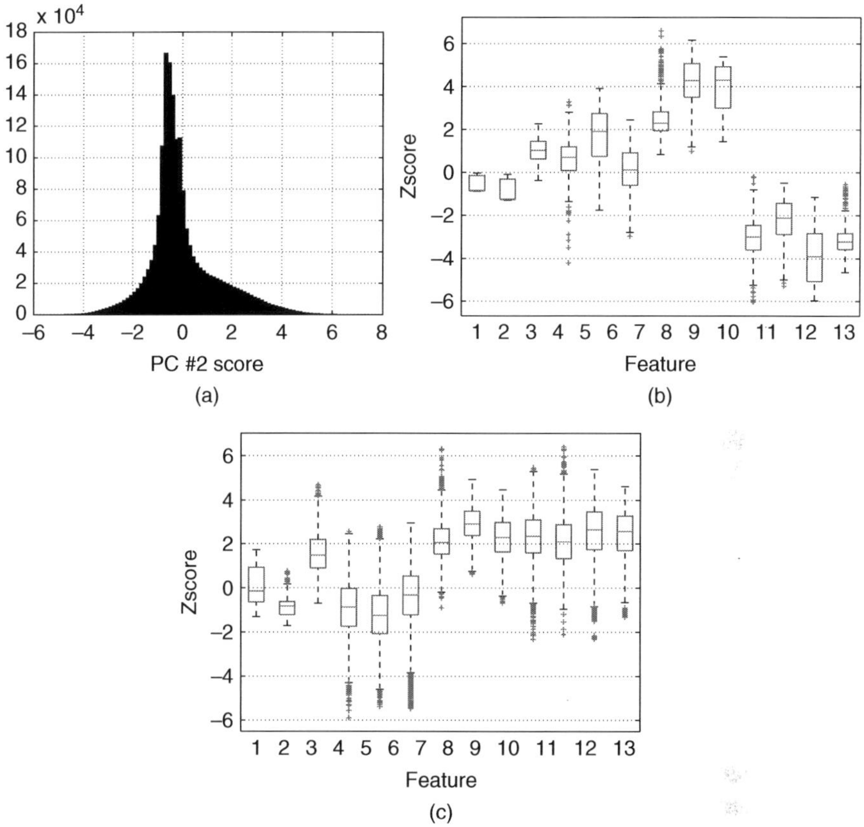

Figure 13.19 (a) Histogram of PC #2 and boxplots of feature values constructed with data selected from (b) the left tail and (c) the right tail of the histogram.

is especially targeted to discriminate the classes accurately. However, kernel-based active learning allows detecting the stations which are most important to reproduce the smooth pattern. In order to reproduce the histogram and variogram of the training data, further extensions or combination of stochastic simulations with kernel methods could be imagined.

While margin sampling is becoming the state-of-the-art in several fields, this chapter also explored the use of a committee of machines to characterize the effect of sampling uncertainty to predict the uncertainty. By training a committee of models, each of them with reduced training sets, it was possible to evaluate the prediction uncertainty at a given spatial location. Such a measure was also used to guide the selection of candidate samples to extend the monitoring network in an intelligent way which helps to evaluate the modeling uncertainty of the environmental phenomenon at a given location.

It must be mentioned that monitoring network optimization with active learning strategies is problem- and task-oriented. For example, an optimal monitoring network especially designed to the classification of suitable locations for wind resource exploitation would not necessarily be optimal for evaluating extreme wind statistics or to be used as input in numerical weather prediction models. In general, when designing a 'multi-objective' monitoring network, an objective criterion summarizing the conflicting application needs should be considered. In this sense, the preliminary analysis in the other case study, the temperature and precipitation fields in Austria, allowed the uncertainty related to different environmental processes to be pointed out, which, as a consequence, requires distinct networks for adequate monitoring. Another issue considered was the detection of areas polluted by Cs-137 following the Chernobyl fallout. The committee-based active learning methods gave a good basis for the selection of new samples in the regions characterized by high levels of Cs concentration, thus requiring less samples than traditional random sampling strategies to obtain accurate results. The last experimental study in this chapter was about discovering the uncertainty related to the suitability for windfarm construction, modeled by taking into account wind speed observation stations and high-resolution topographic data. This case study revealed that, even though a monitoring network can be judged as representative in the geographical space, it can become highly unrepresentative and clustered if additional dimensions are considered to characterize external factors such as the relief features. Monitoring network design also needs to account for such higher-dimensional environmental data, which goes beyond the possibilities of classical geostatistics.

Active learning, monitoring network optimization and sampling issues are nowadays becoming a central part of geostatistical, machine learning and applied research. On the one hand, new efficient algorithms are being developed to minimize the number of samples needed to characterize a particular phenomenon. On the other hand, applied research looks for a compromise solution when designing monitoring networks since they are more and more often conceived for measuring several environmental parameters at a time. Future studies have to be targeted towards the development of multi-objective monitoring network design algorithms to measure complex, nonlinear and high-dimensional environmental processes that vary in space and time.

Acknowledgments

The research was partly funded by the Swiss National Science Foundation projects 200020-121835/1, 200021-126505, PBLAP2-127713/1 and PZ00P2-136827. A.P. acknowledges the support of Science Foundation Ireland under the National Development Plan, particularly through Stokes Award and Strategic Research Cluster grant (07/SRC/I1168).

References

Abe N and Mamitsuka H 1998 Query learning strategies using boosting and bagging. *Int. Conf. on Machine Learning (ICML)*, pp. 1–9. Morgan Kaufmann.

Bakir G, Hofmann T, Schölkopf B, Smola AJ and Vishwanathan S 2007 *Predicting Structured Data*. MIT Press.

Bishop CM 2006 *Pattern Recognition and Machine Learning*. Springer.

Brinker K 2003 Incorporating diversity in active learning with support vector machines. *Proc. of the 20th Int. Conf. on Machine Learning (ICML)*, pp. 59–66. AAAI Press.

Burbidge R, Rowland JJ and King RD 2007 Active learning for regression based on query by committee. *Proc. IDEAL*, pp. 209–218. Springer.

Camastra F and Vinciarelli A 2007 *Machine Learning for Audio, Image and Video Analysis: Theory and Applications*. Springer.

Campbell C, Cristianini N and Smola A 2000 Query learning with large margin classifiers. *Proc. of the 17th Int. Conf. on Machine Learning (ICML)*, pp. 111 – 118. Morgan Kaufmann.

Camps-Valls G and Bruzzone L 2009 *Kernel Methods for Remote Sensing Data Analysis*. John Wiley & Sons, Ltd.

Castro R 2007 *Active Learning and Adaptive Sampling for Non-Parametric Inference*. PhD thesis, Rice University.

Castro R, Willett R and Nowak R 2005 Faster rates in regression via active learning. Technical Report ECE-05-3, University of Wisconsin-Madison.

Chang CC and Lin CJ 2001 Libsvm: a library for support vector machines. Technical Report. Software available at http://www.csie.ntu.edu.tw/~cjlin/libsvm.

Cheng S and Shih FY 2007 An improved incremental training algorithm for support vector machines using active query. *Pattern Recognition* **40**, 964 – 971.

Cherkassky V and Mulier FM 2007 *Learning from Data: Concepts, Theory and Methods*, 2nd edn Wiley-IEEE Press.

Cohn DA 1996 Neural network exploration using optimal experiment design. *Neural Networks* **9**, 1071–1083.

Cohn DA, Gharamani Z and Jordan MI 1996 Active learning with statistical models. *Journal of Artificial Intelligence Research* **4**, 129–145.

Cohn DA 1995 Minimizing statistical bias with queries. Technical Report, A.I. Memo, No. 1552.

Collobert R and Bengio S 2001 SVMTorch: support vector machines for large-scale regression problems. *Journal of Machine Learning Research* **1**, 143–160.

Crombecq K, Couckuyt I, Laermans E and Dhaene T 2009 A novel hybrid active learning strategy for nonlinear regression. *Proc. of the 18th Annual Belgian-Dutch Conference on Machine Learning*, pp. 109–110.

de Gruijter J, Brus D, Bierkens M and Knotters M 2006 *Sampling for Natural Resource Monitoring*. Springer.

Demir B, Persello C and Bruzzone L 2011 Batch mode active learning methods for the interactive classification of remote sensing images. *IEEE Transactions ôn Geosciences and Remote Sensing* **49**(3), 1014–1032.

Di W and Crawford MM 2011 Active learning via multi-view and local proximity co-regularization for hyperspectral image classification. *IEEE Journal of Selected Topics in Signal Processing* **5**(3), 618–628.

Dubois G 2005 *Automatic Mapping Algorithms for Routine and Emergency Monitoring Data. European Commission, IRC Ispra, EUR21595EN*. Springer.

Fedorov VV 1972 *Theory of Optimal Experiments*. Academic Press.

Fedorov VV and Hackl P 1997 *Model-Oriented Design of Experiments* number 125 in *Lecture Notes in Statistics*. Springer.

Ferecatu M and Boujemaa N 2007 Interactive remote sensing image retrieval using active relevance feedback. *IEEE* Transactions on Geosciences and Remote Sensing **45**(4), 818–826.

Foresti L, Tuia D, Kanevski M and Pozdnoukhov A 2011 Learning wind fields with multiple kernels. *Stochastic Environmental Research and Risk Assessment* **25**, 51–66.

Freund Y, Seung HS, Shamir E and Tishby N 1997 Selective sampling using the query by committee algorithm. *Machine Learning* **28**, 133 – 168.

Görnitz N, Kloft M and Brefeld U 2009 Active and semi-supervised data domain description. *Eur. Conf. on Machine Learning (ECML)*, pp. 407–422. Springer.

Gorrisen D, Couckuyt I, Demeester P, Dhaene T, and Crombecq K 2010 A surrogate and adaptive sampling toolbox for computer based design. *Journal of Machine Learning Research* **11**, 2051–2055.

Hastie T, Tibshirani R and Friedman J 2009 *The Elements of Statistical Learning: Data Mining, Inference, and Prediction*, 2nd edn. Springer.

Haykin S 2009 *Neural Networks and Learning Machine*, 3rd edn. Prentice Hall.

Heuvelinka GB, Jianga Z, Bruina SD and Twenhöfelb CJ 2010 Optimization of mobile radioactivity monitoring networks. *International Journal of Geographical Information Science* **24**(3), 365–382.

Hoerl AE and Kennard RW 1970 Ridge regression: biased estimation for nonorthogonal problems. *Technometrics* **12**(1), 55–67.

Huber PJ 1964 Robust estimation of a location parameter. *The Annals of Mathematical Statistics* **35**(1), 73–101.

Jiang, B, Zhang X and Cai T 2008 Estimating the confidence interval for prediction errors of support vector machine classifiers. *Journal of Machine Learning Research* **9**, 521–540.

Juszczak PJ and Duin RPW 2003 Uncertainty sampling methods for one-class classifiers. *Int. Conf. on Machine Learning (ICML)*, pp. 81–88.

Kanevski M and Maignan M 2004 *Analysis and Modelling of Spatial Environmental Data*. EPFL Press.

Kanevski M, Pozdnoukhov A and Timonin V 2009 *Machine Learning for Spatial Environmental Data. Theory, Applications and Software*. EPFL Press.

Krause A, Singh A and Guestrin C 2008 Near-optimal sensor placements in gaussian processes: theory, efficient algorithms and empirical studies. *Journal of Machine Learning Research* **9**, 235–284.

Krogh A and Vedelsby J 1995 Neural networks ensembles, cross validation and active learning. *Advances in Neural Information Processing Systems (NIPS)* **7**, 231–238.

Le N and Zidek J 2006 *Statistical Analysis of Environmental Space-Time Processes*. Springer.

Li J, Bioucas-Dias M and Plaza A 2010 Supervised hyperspectral image segmentation using active learning. *Proc. of the 2nd Workshop on Hyperspectral Image and Signal Processing (WHISPERS)*.

Liitiainen E, Verleysen M, Corona F and Lendasse A 2009 Residual variance estimation in machine learning. *Neurocomputing* **72**, 3692–3703.

Lovejoy S, Schertzer D and Ladoy P 1986 Fractal characterization of inhomogeneous geophysical measuring networks. *Nature* **319**, 43–44.

Luo T, Kramer K, Golgof DB, Hall LO, Samson S, Remsen A and Hopkins T 2005 Active learning to recognize multiple types of plankton. *Journal of Machine Learning Research* **6**, 589–613.

McKay D 1992 Information-based objective functions for active data selection. *Neural Computation* **4**, 590–604.

Melville P and Mooney RJ 2004 Diverse ensembles for active learning. *Proc. of the Int. Conf. on Machine Learning (ICML)* pp. 74–82. ACM Press.

Mercer J 1909 Functions of positive and negative type, and their connection with the theory of integral equations. *Philosophical Transactions of the Royal Society of London, Series* A, **209**(1), 415–446.

Mitra P, Uma Shankar B and Pal S 2004 Segmentation of multispectral remote sensing images using active support vector machines. *Pattern Recognition Letters* **25**(9), 1067–1074.

Müller WG, Pronzato L and Waldl H 2011 Beyond space-filling: an illustrative case. Technical Report 2011-53, Institut für Angewandte Statistik (IFAS).

Müller WG 2007 *Collecting Spatial Data. Optimum Design of Experiments for Random Fields*. Physica Verlag.

Muslea I, Minton S and Knoblock CA 2006 Active learning with multiple views. *Journal of Artificial Intelligence Research* **27**, 203–233.

Pasolli E, Melgani F and Bazi Y 2011 SVM active learning through significance space construction. *IEEE Geoscience and Remote Sensing Letters* **8**(3), 431–435.

Pozdnoukhov A 2002 The analysis of kernel ridge regression learning algorithm. Idiap-RR 54-2002, IDIAP, Martigny, Switzerland.

Pozdnoukhov A and Kanevski M 2006 Monitoring network optimisation for spatial data classification using support vector machines. *International Journal of Environment and Pollution* **28**(3-4), 465–484.

Pronzato L and Müller WG 2012 Design of computer experiments: space filling and beyond. *Statistics and Computing* **22**(3), 681–701.

Rasmussen CE and Williams CKI 2006 *Gaussian Processes for Machine Learning*. The MIT Press.

Sambu Seo Wallat M, Graepel T and Obermayer K 2000 Gaussian process regression: active data selection and test point rejection. *Proc. of the Int. Joint Conf. on Neural Networks*, vol. 3, pp. 241–246.

Schohn G and Cohn D 2000 Less is more: active learning with support vector machines. *Proc. of the 17th Int. Conf. on Machine Learning (ICML)*, pp. 839–846. Morgan Kaufmann.

Schölkopf B, Herbrich R and Smola AJ 2001a A generalized representer theorem. *LNCS, Computational Learning Theory* **2111/2001**, 416–426.

Schölkopf B, Platt JC, Shawe Taylor JC, Smola AJ and Williamson RC 2001b Estimating the support of a high-dimensional distribution. *Neural Computation* **13**(7), 1443–1471.

Settles B 2010 Active learning literature survey. Technical Report 1648, Computer Sciences, University of Wisconsin-Madison.

Seung HS, Opper M and Sompolinsky H 1992 Query by committee. *Proc. of the Annual Workshop on Computational Learning Theory*, pp. 287–294.

Shawe-Taylor J and Cristianini N 2004 *Kernel Methods for Pattern Analysis*. Cambridge University Press.

Singh A, Nowak R and Ramanathan P 2006 Active learning for adaptive mobile sensing networks. *Proc. of the 5th Int. Conf. on Information Processing in Sensor Networks*, pp. 60–68.

Smola AJ, Song L and Teo CH 2009 Relative novelty detection. *AISTATS: Artificial Intelligence and Statistics, JMLR WP 5* **5**, 536–543.

Smola AJ and Schölkopf B 2004 A tutorial on support vector regression. *Statistics and Computing* **14**(3), 199–222.

Sugiyama M and Rubens NA 2008 Batch ensemble approach to active learning with model selection. *Neural Networks* **21**, 1278–1286.

Tax DMJ and Duin RPW 2000 Data description in subspaces. *15th Int. Conf. on Pattern Recognition*, vol. 2, pp. 672–675.

Tong S and Koller D 2001 Support vector machine active learning with applications to text classification. *Journal of Machine Learning Research* **2**, 45 – 66.

Tsui APM and Jones AJ 2002 The construction of smooth models using irregular embeddings determined by a gamma test analysis. *Neural Computing and Applications* **10**(4), 318–329.

Tuia D and Kanevski M 2008 *Advanced Mapping of Environmental Data*, pp. 19–46. John Wiley & Sons, Ltd.

Tuia D, Joost S and Pozdnoukhov A 2011a Active multiple kernel learning of wind power resources *Machine Learning for Sustainability*. NIPS Workshop, Granada, Spain.

Tuia D, Pasolli E and Emery WJ 2011b Using active learning to adapt remote sensing image classifiers. *Remote Sensing of Environment* **115**(9), 2232–2242.

Tuia D, Ratle F, Pacifici F, Kanevski M and Emery WJ 2009 Active learning methods for remote sensing image classification. *IEEE Transactions on Geoscience and Remote Sensing* **47**(7), 2218–2232.

Tuia D, Volpi M, Copa L, Kanevski M and Muñoz-Marí J 2011c A survey of active learning algorithms for supervised remote sensing image classification. *IEEE Journal of Selected Topics in Signal Processing* **5**(3), 606–617.

Vapnik VN 1998 *Statistical Learning Theory*. Wiley-Interscience.

Volpi M, Tuia D and Kanevski M 2012 Memory-based cluster sampling for remote sensing image classification. *IEEE Transactions on Geoscience and Remote Sensing* in press.

Yu K, Bi J and Tresp V 2006 Active learning via transductive experimental design. *Proc 23rd Int. Conf. on Machine Learning (ICML)*, pp. 1081–1088. ACM Press.

Zomer S, Sànchez MN, Brereton RG and Pérez-Pavón J 2004 Active learning support vector machines for optimal sample selection in classification. *Journal of Chemometrics* **18**(6), 294–305.

14

Stationary sampling designs based on plume simulations

Kristina B. Helle and Edzer Pebesma
Institute for Geoinformatics (IFGI), University of Muenster, Germany

14.1 Introduction

Monitoring of pollution in air or water can be improved by considering the physical laws underlying the dispersion process of such substances. Pollutants usually spread from point sources in the shape of a plume. The randomness of these fields is mainly due to the uncertainty of flow parameters or weather forecasts and to the limited information about the source characteristics. Simulations of plumes are a feasible way to study these fields for sampling design optimisation. They give a detailed estimation of what sensors could measure and how good the conclusions are that we draw from such measurements. Although these phenomena are dynamic, we focus on optimising networks with stationary sensors, meaning that the sensors do not move during plume passage. Optimising sampling designs with the help of plume simulations is quite common for groundwater monitoring. It is a bit less common for the monitoring of air pollution because winds change faster than ground water fluxes and therefore the dispersion parameters of the fields are less well known in most applications. As for groundwater monitoring there are already good overviews available (Loaiciga *et al.* 1992; Montas *et al.* 2000; McPhee and Yeh 2005); we mainly focus on plume monitoring in the atmosphere.

Spatio-temporal design: Advances in efficient data acquisition, First Edition.
Edited by Jorge Mateu and Werner G. Müller.

We illustrate plume simulation, choice of a cost function, and optimisation by searching for the optimal sampling designs to detect radioactive plumes that could threaten Upper Austria. Our main concern is monitoring of pollution that spreads from a few potential sources by single plumes like poisonous or radioactive gases or aerosols. The random fields of such plumes are the concentrations or dose rates in the space around the source in the time after the beginning of the release. Uncertainty in these fields is due to the incomplete knowledge of the boundary conditions of the dispersion like the wind fields and sometimes also due to uncertainty about the source. Probabilities of these conditions are very complex as is their propagation through the dispersion. It would be very hard to model the resulting probabilities of the pollutant concentrations completely. However, by adopting a dispersion model we can generate realisations to test proposed sampling designs. In Section 14.2 we give an overview over sampling design studies that used simulations of atmospheric plumes. In addition we explicate our choices for the simulation of radioactive plumes with the RIMPUFF dispersion model (Mikkelsen *et al.* 1984).

Sampling designs are hardly ever optimal in a general sense but they can be optimised for a certain purpose. Cost functions quantify how good sampling designs are. They can be computed from the simulations. In Section 14.3 we describe for which aims the use of simulations is recommended. Computational issues are touched upon as we show how we could summarise the simulations before starting the optimisation for our cost function, the fraction of undetected plumes.

Greedy search, spatial simulated annealing, and genetic algorithms are commonly used to optimise spatial sampling designs. In Section 14.4 we explain how we adapt them to our use case. Section 14.4.6 gives an overview of how we use all three algorithms to optimise the sampling design for Upper Austria and compare their performance and the different resulting sampling designs.

The properties of the simulated plumes and the results of the optimisations are discussed in Section 14.5. The optimisations and their evaluation are carried out with the statistical software R (R Development Core Team 2011).

Finally we draw conclusions about the optimal sampling design and about the suitability of the different optimisation algorithms in our use case in Section 14.6.

14.2 Plumes: From random fields to simulations

The concentrations or dose rates of a pollutant after a single release can be modelled as a random field in the area of interest – that often, but not necessarily, contains the source – during a time interval that usually starts with the beginning of the release and ends after plume passage. The uncertainty of this field can be due to uncertainties about the source or due to the incomplete information about the conditions of the spread. These are propagated by the physical laws that control dispersion to result in a very complex multivariate distribution that hardly can be described completely by a parametric model. A way to approach this distribution is to generate realisations by simulation with a dispersion model. Such models are for example introduced by Seinfeld and Pandis (2006). The realisations are usually given as concentrations or doses of single plumes on points on a grid at time steps or as averages over spatial grid cells and time intervals; sometimes also time-integrated values are used. The space for dispersion modelling is three dimensional in most cases. For the monitoring of air pollution we are usually only interested in the values at the ground or at the typical height of sensors. Also for groundwater monitoring the sampling designs are often considered as two dimensional and averages over depth are taken.

We base our sampling design on simulations calculated with the Lagrangian atmospheric puff diffusion model RIMPUFF (Mikkelsen *et al.* 1984). It specifies plumes from puff releases with Gaussian concentration profiles. Plume growth is modelled due to Carruthers *et al.* (1992) close to the source and due to Mikkelsen *et al.* (2007) farther away. It takes into account wet and dry deposition according to Gering and Müller (1999). Gamma dose rates are derived by numerical integration over the plume volume and over radiation from deposits (Thykier-Nielsen *et al.* 1995).

We assume that the simulated plumes reflect the true world, although it is known that dispersion models are simplifications and that they may suffer from low resolution and errors in the underlying weather data. Dependent on the research questions, there are different ways to model the uncertainty in the model input, the source, and the dispersion parameters. Comparing simulations on an European scale to measurements from the Chernobyl accident and the ETEX test releases revealed that doses can differ up to a factor of 2–3 (Brandt *et al.* 2000). For stable winds, simulations of RIMPUFF and measurements of low doses of krypton-85 (Kr-85) within 80 km from the La Hague plant showed a similar relationship (Connan *et al.* 2011). For short range up to 5 km, more controlled experiments were possible, showing that about half of the values differed by less than factor 2 and that the model was very sensitive to the meteorological and land use parameters (Dyer and Astrup 2010).

Heuvelink *et al.* (2010) investigated the uncertainty of weather predictions as they wanted to optimise sensor locations for an outbreak in the immediate future based on the weather information that would be available in such cases. They assumed that the uncertainty of parameters would cause an additive, spatially correlated and stationary error. In applications for groundwater monitoring and remediation the dispersion parameters like conductivity and hydraulic head are usually known with less uncertainty than wind fields. Meyer *et al.* (1994) identified conductivity as the most uncertain parameter and tested its influence. In most studies, sampling designs are optimised for plumes that could occur under all weather conditions that are likely at the source. Usually they are based on weather data from one year. Serón *et al.* (1993) modelled daily averages of pollutants that were released routinely. The weather data for the simulations were sampled from statistics of wind directions and velocities and the related stability classes. In contrast, Abida *et al.* (2008) and Melles *et al.* (2011) used sample sequences from full weather data to preserve the dependencies among the weather parameters to model accidental plumes. They used wind fields with several horizontal layers or data from one meteorological station, respectively. We also use sample sequences of full weather data: wind fields and rain for whole Europe are based on hourly reanalysis data (Kalnay *et al.* 1996; NOAA 2011), downscaled by WRF (2011) to 40 km resolution and hourly values for the whole year 2007. With these data 292 plumes from each source are simulated, a new one starting every 30 h.

The source can be described as the amount of pollutant released at any point in space and time during the outbreak. In most applications, the location of the source is known. However, it can also be taken from a distribution as in Melles *et al.* (2011) where towns and transportation routes were sampled with different probabilities. It can even be the main aim of the sampling design to find the source location(s) as in Mahar and Datta (1997). As there are no commercial nuclear power plants in Austria, we consider power plants in the neighbouring countries as the potential sources of radioactivity. From the multitude of plants – that can be found at EURDEP (European Comission Joint Research Centre 2011) – we choose 13 that are close to Upper Austria or represent regions with several nuclear facilities: Bugey for south eastern France, and Cattenom for northern France and Belgium (Figure 14.1).

Figure 14.1 Study area (Upper Austria: grey) and the 13 sources of radioactive plumes. Boundaries from GADM (2012).

The amount of released pollutant and how it develops in time can be a major source of uncertainty. It may be the main aim of monitoring to reconstruct it, or at least the total released mass or radiation, by data assimilation, as in Montas *et al.* (2000), Reed *et al.* (2000), Wu *et al.* (2006), and Twenhöfel *et al.* (2007). For radioactive outbreaks, detailed source terms for different kinds of plants and accidents can be found in expert systems such as RODOS (Ehrhardt and Weis 2000; RODOS 2011). For feasibility we choose only 6 nuclides that represent typical decay and deposition properties of the main nuclide types: noble gases, aerosols, and iodine (Table 14.1).

Table 14.1 Source term: the nuclides are released by puffs every 30 min at 50 m above ground.

	1st hour	2nd–12th hour
Heat (MW)	0	0
Te-132 (Bq/s)	1.0E+15	1.0E+13
I-131 (Bq/s)	1.0E+15	1.0E+13
I-132 (Bq/s)	1.0E+15	1.0E+13
Xe-133 (Bq/s)	1.0E+15	1.0E+13
Xe-135 (Bq/s)	1.0E+15	1.0E+13
Cs-137 (Bq/s)	1.0E+15	1.0E+11

The extent of the simulations is from several metres for hydrological models, to a few kilometres for atmospheric models for well known sources, up to hundreds or thousands of kilometres. The extent of the simulations should cover the area where one wants to know the plume, not only that of potential sensor locations. Also the spatial resolution has to fit the purpose: the optimisation determines sensor locations only up to the grid cells. Abida *et al.* (2008) noticed that the resolution they had chosen did not support optimisation of an arbitrary number of many sensors, a sensor in each fifth cell was the limit. The temporal resolution should fit the properties of the modelled sensors or of legal thresholds to be compared with the 'measurements'.

We only consider the hourly averages of the sum of groundshine and cloudshine of all nuclides at 1 m above ground. The spatial resolution is 4 km, the simulations cover the whole of Upper Austria and the source as well as a wide corridor in between. To compute the cost function and for optimisation we only use the 736 cells within Upper Austria. The simulation is run for 7 days, assuming that the plume has passed the region by that time. Although the spatial resolution is limited by computational resources, choosing a finer resolution would probably not result in substantially different sets of sensors because of the limited resolution of the weather data. Furthermore, finding good locations for placing stationary sensors on a finer scale is constrained by local conditions, for example the presence of a flat, big lawn. Hourly averages fit a typical reporting system of the sensors we want to model. So for each plume $p \in P$, for all cell(centre)s in the grid $x \in X$, and for all hours during plume passage $t \in T_p$ we compute the average dose rate $r_p(x, t)$.

14.3 Cost functions

We use simulated plumes to optimise sampling designs because this tells us which information is encompassed by the sensors and what passes unnoticed. We can compare that to the information we actually need for our purpose. An intuitive way to evaluate sampling designs is to use the values of the simulations at the potential sensor locations and evaluate them like measurements. To rate the value of the sampling design, the derived information is compared with the information from the whole simulation. To optimise sensor locations we need to quantify this fitness for use, and we do this by cost functions.

The most basic aim of plume monitoring is *detection*, which also is the aim in our use case. Beyond this, one likes to know more details about the plumes, mainly given as *maps*. A different perspective is to quantify the loss caused by the limits of our knowledge (James and Gorelick 1994).

14.3.1 Detecting plumes

A very simple cost function is the fraction of plumes that are not detected by the sampling design. We use this criterion for optimisation. We assume a plume to be detected if for at least 1 hour and 1 sensor the dose rate of the plume exceeds the detection limit of 100 nSv/h. This threshold is above the detection level of most devices – which often is 10 nSv/h – as smaller changes are hard to distinguish from natural background variation and therefore alert levels are often in the range of 100 nSv/h above the average background. We can preprocess the raw data in advance to reduce the computation needed during optimisation and also the amount of data we have to handle. For each proposed sampling design $S = \{x_1, \ldots, x_n\} \subset X$ we want to know which plumes it can detect. To compute this, we only need to know for each location $x \in X$ and each plume $p \in P$ whether it can ever be detected there:

$$I_p(x) = \begin{cases} 1 & : \text{ if } \forall\, t \in T_p : r_p(x,t) \leq 100\,\text{nSv/h (not detected)} \\ 0 & : \text{ else (detected).} \end{cases}$$

Then for each sampling design S we only need to compute $I_p(S) = \prod_{x_j \in S} I_p(x_j)$, which is 1 for undetected plumes and 0 else, to get the fraction of undetected plumes

$$c(S) = \frac{1}{|P|} \sum_{p \in P} I_p(S).$$

This can be modified to focus on plumes with high impact or on early detection as Melles *et al.* (2011) did. They rewarded early detection or the detection with multiple sensors by counting how many sensors detected the plume in each time step and assigning larger weights to undesired cases. The focus can also lie on locations where exceedances of legal thresholds are likely to be detected; Serón *et al.* (1993) assigned different weights to values in different intervals, rewarding measurements of the high ones. For groundwater remediation it is not only important to detect most of the plumes, but also to catch them when they are still small; for example Meyer *et al.* (1994) tried to achieve these conflicting aims.

14.3.2 Mapping and characterising plumes

For many applications, maps of the developing or past plumes, based on the data available from the sensors, are desired. Various interpolation algorithms can be used to provide estimates between the measurements. Plumes are difficult to interpolate: at the source they change abruptly whereas farther away they are rather smooth, and the distribution of the values is usually skewed and may contain many zeros.

Several authors compared different interpolation methods and related optimal sampling designs. It seems more important to restrict interpolation to small neighbourhoods than to use complex interpolation methods. Abida *et al.* (2008) mapped radioactive plumes over France and tried to minimize the Hölder norms with power $p \in \{0.5, 1, 2\}$. For $p = 2$ the sensors were clustered around the sources whereas the optimum for $p = 0.5$ was closer to equal spread. For the resulting optimal sampling designs it did not make a big difference if nearest neighbour interpolation or kriging was used. The final aim of Reed *et al.* (2000) was not the map but an estimate of the mass of the plume. They used both ordinary kriging and inverse distance weighted interpolation (IDW). They found it most effective to use IDW in the first phase of the optimisation because it was much faster, and to use kriging for the final adjustment. For both interpolation methods they chose to use only neighbouring values within a certain range and added a high penalty to the cost function if in a location no interpolation was possible. Wu *et al.* (2006) applied ordinary kriging and also used such a penalty. Heuvelink *et al.* (2010) considered the differences between 'true' and simulated plume in a setting where they were relatively similar, so the residuals were less extreme and could be interpolated using simple kriging.

There is a variety of other plume properties one would like to infer from the measurements. Often the aim is calibration of the model parameters, which can be a very complex problem and require data assimilation. In groundwater monitoring, often the mass and the first two moments of a plume are of interest (Montas *et al.* 2000; Reed *et al.* 2000; Wu *et al.* 2006). Mahar and Datta (1997) focussed on source identification. Schaetzen *et al.* (2000) calibrated the roughness parameter of pipes combining Shannon entropy and a sensitivity function. For the NPKPUFF model, up to 8 parameters were calibrated by Twenhöfel *et al.* (2007). As cost function they used a linear combination of 10 measures, each comparing the 'measurements' and the simulations of the plumes at the sensor locations.

14.3.3 Combined cost functions

Often sensors should fulfil several purposes. This can be achieved by combining the respective cost functions; linear combinations are most common. By varying the weights Meyer *et al.* (1994) explored the trade-off between two purposes and how the optimal sampling designs changed accordingly. Cieniawski *et al.* (1995) used a genetic algorithm and sorted the sampling designs by Pareto optimality to find the best points of the trade-off curve. Schaetzen *et al.* (2000) suggested to normalise and combine two cost functions by

$$c_{1,2}(x) = \sqrt{\sum_{i=1}^{2} \left(\frac{c_i(x) - \max_{y \in X}\{c_i(y)\}}{\min_{y \in X}\{c_i(y)\} - \max_{y \in X}\{c_i(y)\}} \right)}.$$

Whereas Serón *et al.* (1993) used $c_1(x)c_2(x)^b$ for a cost function c_1 related to the area covered by a sensor, and c_2 related to its ability to detect high values.

14.4 Optimisation

In general, sampling design optimisation based on simulations is a discrete combinatorial problem: to find the best N places within $Y \subset X$. In addition, the number of sensors N to fulfil a required aim may not be known in advance. A complete search is usually not feasible as $M = |Y|$ possible locations for N sensors allow $\binom{M}{N}$ different sampling designs. Therefore, heuristics that do not search all possibilities exhaustively are usually used. Lee and Ellis (1996) compared several of the common methods to minimise the average kriging variance. We apply spatial simulated annealing, a genetic algorithm, and greedy one by one search to improve plume detection and compare the results.

14.4.1 Greedy search

Greedy algorithms are deterministic, taking the local optimum in each step. For sampling design optimisation they usually change sensors one by one. They can either be used to add sensors or to delete them. To add a sensor, they test for all candidate locations how cost would change if a sensor was added there; then they add a sensor in one of the best locations:

- add a sensor in the location $x \in X$ to the current sampling design S such that the cost is minimal

$$x = \arg\min_{x' \in X} c(S \cup \{x'\}).$$

This is repeated until the desired sample size or cost is reached. Deleting sensors follows a similar rule, it tests for all sensors in the current sampling design what happens if they were deleted and deletes the least important sensor:

- delete the sensor $s \in S$ of the current sampling design such that the cost is minimal

$$s = \arg\min_{s' \in S} c(S \setminus \{s'\}).$$

Usually these algorithms need few iterations, they just add or, respectively, delete sensors until a certain goal is reached. If there are many candidates, they may however need a lot of evaluations of the cost function for each iteration. It is usually assumed that the result is globally optimal, but the algorithm can get trapped in local optima as illustrated in Figure 14.2.

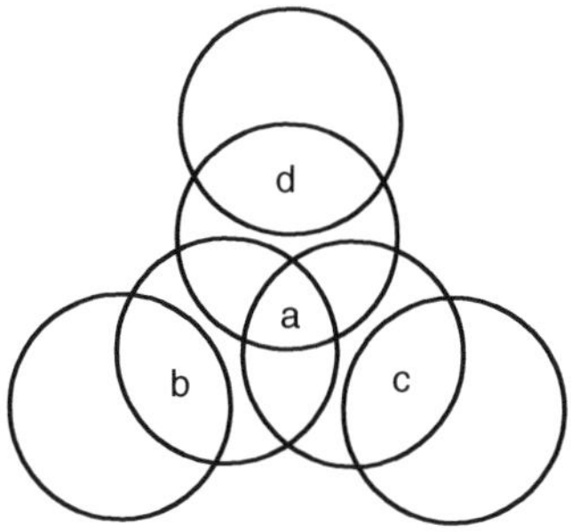

Figure 14.2 Example where greedy one by one addition does not find the optimum. The aim is at least 1 sensor in each circle. The first sensor is put in a, but the optimal 3 sensors are in b, c, d.

We use greedy addition of sensors, starting from no sensor. First we add sensors until they detect all plumes. Then we delete redundant sensors through until cost increases. We repeat these two phases until the sampling design is a repeat of an earlier iteration – and the algorithm would start to loop. Switching between both phases helps to overcome trapping in local optima. In addition we are interested in the optimal sampling designs with less sensors. Therefore we run the greedy algorithm again for all smaller sample sizes, stopping the adding phase at the desired sample size, then deleting, etc. The resulting sampling designs are referred to as greedy optimum below.

For adding a sensor, the cost function must be evaluated once for each grid cell without a sensor; for deleting a sensor it must be evaluated once for each sensor in the current sampling design. Our region has 736 cells and the sampling designs are small, this means adding a sensor needs more than 700 evaluations of the cost function whereas deleting one only needs a few.

14.4.1.1 Greedy algorithms tailored to specific cost functions

The following two examples are other typical applications of greedy search; in both cases the way to update the cost is specified. The aim of Serón *et al.* (1993) was to maximise the coverage of the sensors. It was based on the correlation between the values at all pairs of points. If correlation exceeded a threshold, it was assumed that it is sufficient to have a sensor at one of the locations. The optimisation algorithm maximised coverage with sensors where high values were likely; it always selected the location with the best performance given the already existing sensors. Then all potential sensor locations within the coverage of this sensor were excluded from further optimisation, and the next sensor was added in the same way.

The algorithm of Kim and Lee (2007) was based on a set of simulated realisations of a plume in groundwater. The aim was to delineate the plume, to know in as many locations as possible that the plume was there or that it was not. In each location $x \in X$, the fraction of plumes being non zero there $p(x)$ was determined. The cost function $E(f)$ was the spatial integral over

$$f(x) = \begin{cases} p(x) & p(x) \leq 0.5 \\ 1 - p(x) & p(x) > 0.5 \end{cases} .$$

When a sensor was placed in a location, all realisations that did not comply with the measurement there were deleted; this reduced the overall uncertainty. To find the best location, they were ranked according to the *expected value of sample information*:

$$\text{EVSI}(x) = p(x)E(f|x = 1) + (1 - p(x))E(f|x = 0)$$

where $E(f|x = 1)$ was the remaining global uncertainty if we knew that $x = 1$. James and Gorelick (1994) used a similar cost function based on a function f related to the cost of remediation.

Updating the cost for all candidates is usually expensive. Parallel changes where several sensors are changed at once can save a lot of effort. Parallelisation causes least impairment if there is little dependence between the sensors. Kim and Lee (2007) generalised the cost function above to take into account several parallel changes and minimised it with a genetic algorithm. When only 2 or 3 sensors were added in parallel, the resulting cost was similar to sequential addition; adding more sensors in parallel reduced the cost only half as much as sequential addition. The cost function of Séron *et al.* (1993) above is not at all suited for parallel addition of sensors. Neighbouring sensors usually have similar properties. They would be added in parallel, but their measurements would be redundant.

14.4.2 Spatial simulated annealing

Spatial simulated annealing is a random search algorithm that explicitly addresses spatial vicinity. The spatial version of this algorithm was developed by Van Groenigen (1997) for spatial sampling design optimisation from the general simulated annealing algorithm which mimics cooling of metals. In the first iterations, the sensors can change a lot, with low probability even to worse locations. As the process cools, changes become smaller and acceptance of worse results becomes less likely. This algorithm is widely used, especially for sampling designs for mapping. In most applications optimisation takes several hundreds or thousands of iterations. Heuvelink *et al.* (2010) needed 600 iterations to place 25 sensors among approximately 3000 locations, reducing the cost by 13% up to 45% depending on the noise in the data. Abida *et al.* (2008) ran about 11 000 iterations to find the optimal locations of 100 sensors from about

500 candidate locations. Both used the 'measurements' for mapping by kriging or inverse distance weighted interpolation. Spatial simulated annealing was also used to optimise sensor locations for plume detection by Melles *et al.* (2011). Using greedy one by one optimisation first and then refining the result by spatial simulated annealing was successfully applied by Delmelle and Goovaerts (2009).

Spatial simulated annealing optimises sampling designs of a fixed size. Usually the initial sensors are chosen randomly. Each iteration $i \in I$ consists of three steps:

(i) Randomly pick the sensors to be changed. Each has independently the same probability $q_0 q^i$ to be changed, where q_0 is the initial change rate and q is the cooling coefficient.

(ii) Randomly move these sensors to a grid point in their neighbourhood. The size of the neighbourhood decreases as i grows. We use a rectangle with $0.4(1 - \frac{i}{I})$ of the extent of Upper Austria around the previous sensor location.

(iii) Compute the cost of the new sampling design and compare it with the cost of the previous sampling design. If cost decreases, the change is always accepted. The rate to accept a sampling design with higher cost is again $q_0 q^i$.

The initial change rate $q_0 = 0.2$ as well as the cooling coefficient $q = \exp(-\frac{10}{I})$ and the decreasing neighbourhood size are taken from the function *ssaOptim* in the *R* package *intamapInteractive* (Pebesma *et al.* 2012). The random choice in (i) of how many sensors are changed, is different from other implementations of spatial simulated annealing. Usually a fixed number of sensors is changed – often 1, which would not be likely to give better results than greedy one by one changes. The other extreme of changing all sensors in each step seems to introduce too much randomness. There also exist other versions of steps (ii) and (iii), the acceptance of worse sampling designs may for example depend on how much worse they are. We run the optimisation for $I = 15000$ iterations to employ a similar effort as for the greedy search. We keep the best sampling design as it may be better than the final one.

14.4.3 Genetic algorithms

The basic idea of genetic algorithms is to optimise a set of dependent parameters together. It is a stochastic algorithm that tries to evolve the set of parameters towards the fittest solution like genes evolve in nature. One of the main developers of genetic algorithms was Holland (1975) who based them on a broad probabilistic framework. He proposed the main mechanisms of evolution: reproduction, crossover, and mutation.

In the case of our sampling designs the parameters are the $M = 736$ grid cells that may contain a sensor or not. The fitness of each sampling design is determined by an evaluation function C. Sampling designs may have different size, and bigger sampling designs are likely to detect more plumes. The original cost function c would prefer sampling designs of many sensors which are not optimal in terms of monetary cost. For a fairer relationship between the fraction of undetected plumes and sample size we use a logarithmic model

$$C(S) = c(S) + \alpha \ln(|S|),$$

where the factor $\alpha = 0.0003$ was adapted manually.

We use the genetic algorithm *rbga.bin* for 'binary chromosome' of the *R* package *genalg* (Willighagen 2005). The steps of the genetic algorithm are illustrated in Figure 14.3. In each iteration i the population $P(i)$ consists of 200 sampling designs.

- Initialise. The initial sampling designs are random; they have on average 15 sensors.
- For each iteration i:
  - (i) Evaluate. Determine the cost $C(S)$ for each sampling design $S \in P(i)$ of the population.
  - (ii) Reproduce. The best 20% of the sampling designs become part of the new population $P(i + 1)$.
  - (iii) Recombine (crossover). Sort the sampling designs according to their cost. Randomly draw pairs, selecting better ones with higher probability. For each pair, cut the tuples representing the sampling designs at a random location and exchange the parts. The two new sampling designs become part of the other 80% of the new population $P(i + 1)$.
  - (iv) Mutate. Randomly change all sampling designs of the new population. On average in 100 locations the current setting is replaced by a random choice. The random choice generates sampling designs of on average 15 sensors.
- Stop after 100 iterations.

This is a stochastic heuristic that can be adjusted by the parameters to search widely or to focus, like spatial simulated annealing. There are many variations:

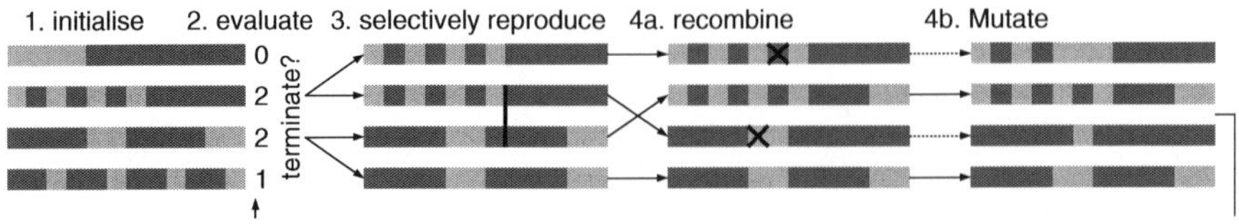

Figure 14.3 Iteration of a genetic algorithm, modified after Lucasius and Kateman (1993).

only selection, which is a kind of optimisation, or only recombining and mutating, which scans more different possibilities (Lucasius and Kateman 1993, 1994). Usually these algorithms ignore spatial relationships between different locations; however this can be included into the cost function like penalising sampling designs with sensors close to each other.

Genetic algorithms have been applied for many problems of spatial sampling design. Reed *et al.* (2000) could reduce the number of sensors from 30 to 15 without substantial loss in the estimate of the plume's mass by 60 generations of 1000 sets. Iterations were stopped when some locations were chosen by at least 90% of the population and the other locations by less than 10%. Wu *et al.* (2006) also wanted to determine the mass and the first two moments of a potential plume. They used two different kinds of genetic algorithms. First they ran it separately 1000 times, each for one realisation of the plume. As result they chose the locations that were most frequent among the resulting sampling designs. Then they ran it with several plume realisations; for the first iterations they took only few of them, increasing the number for later iterations. The overall effort was less than 1/6 compared with the other approach. The results of both approaches were similarly suited to determine mass and first moment of the plumes, whereas for second moments the hybrid of the 1000 individually fitted designs was superior. One advantage of this algorithm is that it returns a full set of sampling designs, instead of one single optimum. This allows the trade-offs between different designs to be seen (Cieniawski *et al.* 1995).

14.4.4 Other methods

When the evaluation of a single cost function is computationally demanding, optimisation that requires many iterations may be unfeasible. In these cases, testing some basic sampling designs or modifications of existing monitoring networks can still provide useful insights. Twenhöfel *et al.* (2007) found that calibrating dispersion parameters only worked well if at least 6 measurements within the plume were available, but more measurements did not always improve the result.

Montas *et al.* (2000) used the physical properties of a spreading plume to reduce the set of possible sensor locations. Thus it was feasible to test all the remaining ones to get the global optimum.

The algorithm of Mahar and Datta (1997) improved in turns source identification and the sampling design best suited to do that. The occurring sampling design optimisation problems could be solved completely.

Optimisation may also have to consider political and monetary constraints. Helle *et al.* (2011) optimised sampling designs for plume detection being constrained to subregions.

14.4.5 Evaluation and sensitivity

As in most cases where heuristics are used for the optimisation, we cannot be certain whether the determined optima actually are global optima. In addition, there may be other sampling designs that are of similar quality (Reed *et al.* 2000). This problem is often ignored, but there are several ways to tackle it.

Genetic algorithms always yield several different sampling designs. Thus we gain some insight about potential optima. Taking hybrids of the resulting sampling designs, the locations that occur in many of them, makes this approach robust (Wu *et al.* 2006). In principle, spatial simulated annealing also returns multiple sampling designs.

Schaetzen *et al.* (2000) used three different optimisation methods, two geometric algorithms and optimising entropy and sensitivity, to locate the sensors for pipe roughness in a water pipe network. They compared all results with expert knowledge. Finally they analysed the overlaps as well as properties that were associated with suitability for sampling like junctions or pipes of different perimeters. Altogether optimising the cost functions found the best solutions in this respect, especially for sensitivity; the other approaches, including experts, were worse.

Not only the optimisation method but also the considered plumes influence the result. Meyer *et al.* (1994) tested many different parameter configurations to determine how the cost and the geometrical properties of the optimal sampling designs changed. The set of our simulated plumes does not cover all possibilities. Therefore we estimate how good the 'optimal' sampling designs are suited to detect plumes that are not considered for the optimisation. This can also serve in finding more robust sampling designs, as some of the sensor locations may be determined by only a few plumes.

14.4.6 Use case: Combination and comparison of optimisation algorithms

We use greedy search, spatial simulated annealing, and the genetic algorithm to find optimal sensor locations for plume detection. We assume that this will give us a variety of optima that can be compared to find the most important sensor locations and to gain insight about the general structure of the problem. In addition we can evaluate the suitability of the optimisation algorithms for the given problem.

We start optimisation with greedy search, adding sensors from scratch. First we aim at a sampling design that can detect all considered plumes. Then we also look for optimal sampling designs with less sensors to compare them to sampling designs of same size from other optimisation algorithms.

We examine greedy search in more detail by checking its robustness against changes in the input plumes. We like to know, if using less plumes for the optimisation leads to different optimal sampling designs. If this is the case, we may argue that plumes missing in our simulations are unlikely to be detected by the optimal sampling designs we achieve. We try to detect plumes that are not used in the optimisation to assess this risk – knowing that plumes not considered in our simulations may be even harder to detect. We run greedy search on reduced sets of plumes. Then we sort the sensors of each sampling design according to the number of plumes that are only detected there – taking into account the same plumes as for the optimisation. Thus we can compare the 10 most important sensors of each sampling design. The resulting sampling designs are compared with the original sampling design for all plumes. However, the set of plumes may incompletely represent all possible plumes in other respects. For example the source may be of different strength or duration. Therefore we also test how well the sensors can detect 100 times weaker plumes.

For the comparison of optimisation algorithms we run spatial simulated annealing and the genetic algorithm either from random starting points or starting at the optimum of greedy search. Figure 14.4 shows the relationship of these optimisations. We run all kind of optimisations 20 times, as these two algorithms are random and each run may be different.

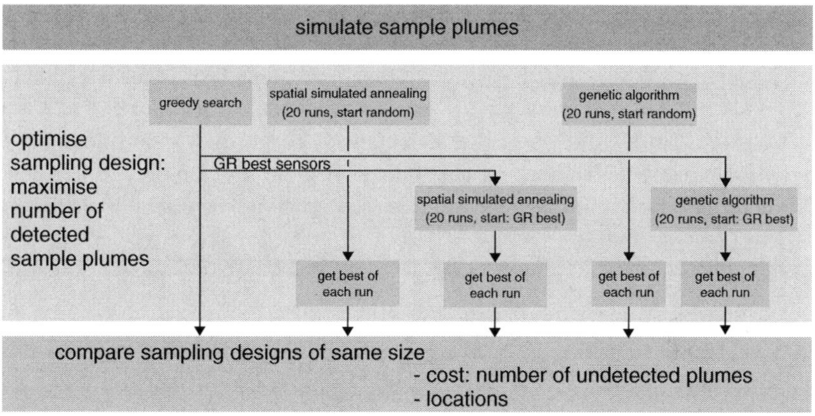

Figure 14.4 Flow chart of the use case, comparing the three optimisation algorithms and combinations of them.

Spatial simulated annealing is run for 5 and 10 sensors, respectively, and the results are compared with the best 5 or 10 sensors from greedy search – the greedy optimum of 15 sensors is less robust and therefore less suited for comparison. Optimisation is either started from random locations or from the best 5 or 10 sensors from greedy search. The genetic algorithm is started from populations of random sampling designs with on average 15 sensors or from populations whereof 10% are copies of the greedy optimum with 15 sensors. From the resulting populations we only pick the best sampling designs of each size. The number of iterations for spatial simulated annealing (15 000) and the size of the population and the number of iterations for the genetic algorithm (200 × 100) are similar to the number of evaluations of the cost function the greedy search needs. Although computation of the cost function is not the bottleneck of the algorithm in our case, this gives an idea of how much can be achieved with similar effort.

We compare the cost of the 'optima' to the cost of the sampling design of same size from greedy search. This is a benchmark for the performance of the algorithms; it shows which one is most likely to return the best result. The second focus is on the sensor locations of the different 'optima', on their similarity and location in our study area.

14.5 Results

14.5.1 Simulations

About two-thirds of the 13 times 292 simulated plumes never touch Upper Austria. Table 14.2 shows that plumes from sources in the west are most likely to reach Upper Austria. As we aim at plume detection, exceedance of the detection limit is most important. About 19% of the plumes can be detected everywhere; coverages of the other plumes are relatively uniformly distributed to all possible sizes. On average, plumes cover about 59% of Upper Austria. Plumes from nearby sources are smaller; those from the power plant Isar cover only 48% on average. The background of Figure 14.5 shows where the plumes exceed the detection limit; most plumes can be detected in the northern, less mountainous, part of Upper Austria. The properties of the plumes reflect the prevailing western winds and that plumes are narrower close to the source. As most plumes cover at least half of the region, they are relatively easy to detect. However, many events that should be detected by the sensors may have weaker sources and are therefore

Table 14.2 Sources of radioactive plumes. The last column shows how many of the 292 plumes affect Upper Austria.

Name	Long.	Lat.	Plumes in Upper Austria
Isar	12.29	48.61	185
Gundremmingen	10.40	48.52	148
Beznau	8.23	47.55	138
Neckarwestheim	9.17	49.04	110
Temelin	14.38	49.18	102
Bugey	5.27	45.80	99
Grafenrheinfeld	10.19	50.00	81
Cattenom	6.22	49.41	79
Dukovany	16.15	49.08	76
Krsko	15.52	45.94	60
Mochovce	18.46	48.26	55
Bohunice	17.69	48.50	51
Paks	18.85	46.57	42

only detectable in a smaller region. As gamma dose rate scales linearily with the source strength – if the proportion of the released nuclids is the same – we can easily determine the properties of plumes from sources with strength of 1/100: 27% of the plumes are no longer detectable when they are weaker, and coverage of the remaining ones is on average only 39%.

14.5.2 Greedy search

When we add sensors with greedy search, starting from no sensors, 15 sensors are enough to detect all plumes. Figure 14.5 shows that most of the sensors should be placed at the margins of Upper Austria where plumes enter the region. This also supports early detection. Cost decreases exponentially for each added sensor (Figure 14.6). The first sensor detects more than two-thirds of all plumes whereas adding the 10th to 15th sensor only pays off with 1 more detected plume, respectively. This means that the position of the sensors added last are determined by single plumes. We use sampling designs of less than 15 sensors for comparison as we assume them to be more robust.

Gamma dose measurements are more valuable if meteorological data from the same location is known, therefore meteorological stations are preferred locations for gamma dose sensors. If there were gamma dose rate sensors at all 36 meteorological stations, they would detect 94% of all plumes. However, we would

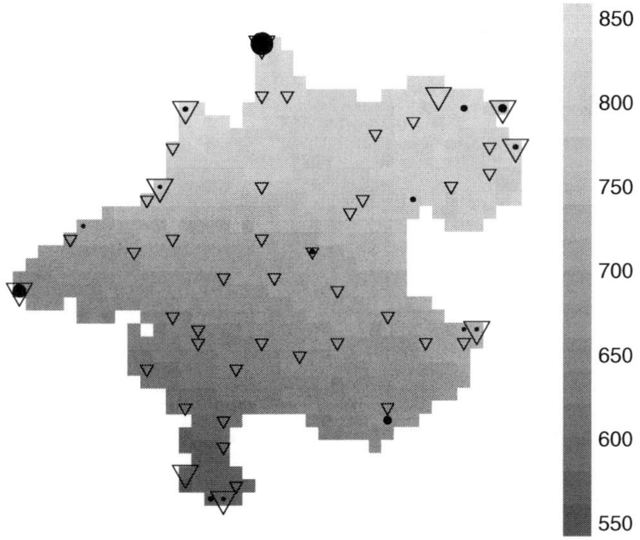

Figure 14.5 Optimal sampling designs from greedy search. The 15 dots indicate the optimum starting from scratch, their area is proportional to the plumes that are detected only here. Small triangles indicate meteorological stations; sensors there and at the big triangles can detect all plumes as well. The grey intensity of the background reflects the number of detectable plumes in each grid cell.

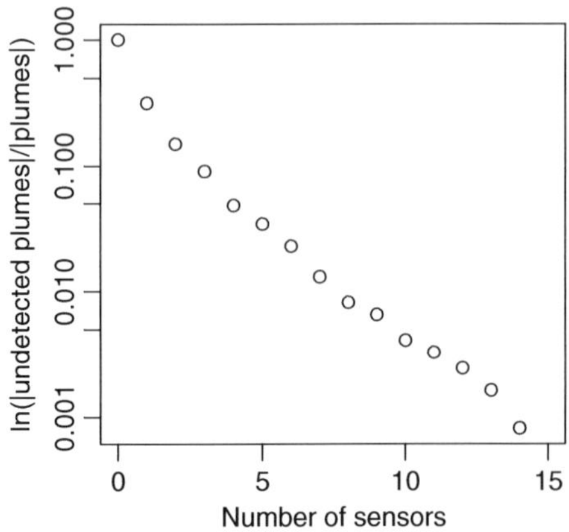

Figure 14.6 Cost (logarithmic scale) versus number of sensors for optimal sampling designs of different size from greedy search.

need 10 additional sensors to detect the remaining ones. Figure 14.5 shows that 8 of the additional sensors would coincide with those from the optimal sampling design without sensors at meteorological stations.

14.5.3 Sensitivity of greedy search to the plume simulations

First we exclude all 185 plumes from the most important source (Isar). As plumes from this source are likely to be small, they may be missed by sensors. Greedy search returns a sampling design of 16 sensors. Obviously this is not the global optimum as the 15 sensors from the sampling design above also detect all these plumes. The 16 sensors detect all the plumes from Isar even if none of them is used in the optimisation – see third column of Table 14.3. This table also shows the results of optimisations if we leave out simple or regular random subsets of the plumes, either 185 as for 'no Isar' or half of them (613). Any of the resulting sampling designs from all runs of both cases, misses more than 4% of the ignored plumes; the median loss is about 1%. The sensor locations vary

Table 14.3 Performance of sampling designs optimised ignoring some of the plumes. Regular and random are median of the 10 respective runs. Coincidence of sensors means the one with the greedy optimum.

	Ignored plumes	Plumes missed	Number of sensors	Sensors coinciding	Best 10 sensors coinciding
No Isar	185 (15%)	0 %	16	10	5
Regular	185 (15%)	1.08 %	14.5	8	5.5
Random	613 (50%)	1.06 %	12	7.5	4.5

among the results, they have about half of the sensors in common. The similarity is higher when only 185 plumes are ignored.

When we try to detect the remaining 899 plumes with 100 times weaker source term with the original 15 sensors, only 24 (3.3%) are missed.

14.5.4 Comparison of optimisation algorithms

14.5.4.1 Spatial simulated annealing versus greedy search

The performance of spatial simulated annealing is very different for 5 and for 10 sensors. For 10 sensors, the result is very poor. Runs started at random locations only find sampling designs that are worse than the one from greedy search; and when starting from this optimum, it is never improved; see fourth column of Table 14.4. On average, the sampling designs from random start detect about 4 times less plumes than the greedy optimum does – 19 plumes are missed instead of 5. Optimisation of the small sampling designs of 5 sensors works much better. The cost of the resulting sampling designs is very diverse. On average the resulting sampling designs from random start are almost as good as the one from greedy search. The worst of them miss about 1.5 times more plumes, the best ones about 8% less. Seventy-five per cent of the runs with random start are worse, 20% are better. When starting from the greedy optimum, it is improved in 35% of the runs. The resulting sensor sets detect up to 19% more plumes. Figure 14.7 shows cost development for one of the runs that could improve the initial greedy optimum. Sampling designs from random start have on average only few sensors in common with the greedy optimum – about 23% of the 5 or 10 respective sensors. In cases where the initial sensors from greedy search are improved, most of them are changed as well, only 31% are kept.

Table 14.4 Results of spatial simulated annealing compared with greedy search. Cost and number of coinciding sensors are mean values of 20 runs. Coincidence of sensors means the one with the greedy optimum of the respective size.

Initial sensors	Number of sensors	Cost	Runs better (as good)	Coinciding sensors
Random	5	0.0396	1/20 (4/20)	1.15
Greedy best	5	0.0331	7/20	3.8
Random	10	0.0157	0/20 (0/20)	2.25
Greedy best	10	0.0041	0/20	10

In general this shows that spatial simulated annealing can improve ‘optima’ of greedy search considerably in some cases, mainly for small sampling designs. When starting at random locations, the risk is high to get much worse results than from greedy search. For sampling designs of 10 sensors, it performs in all

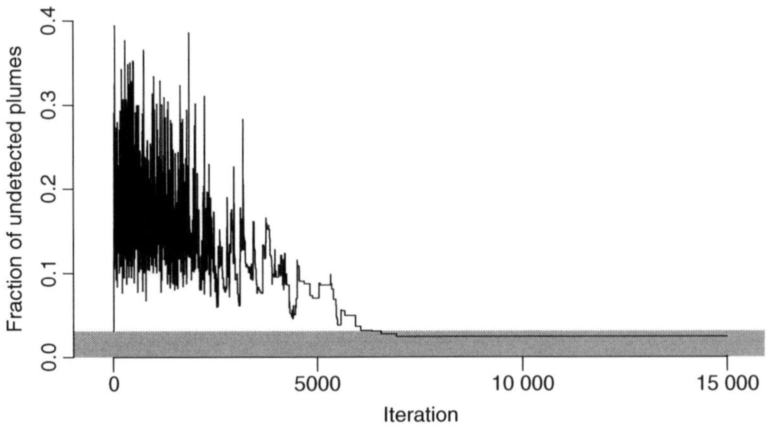

Figure 14.7 Cost curve for one run of spatial simulated annealing for 5 sensors, starting with the optimum of greedy search. Grey area: better than greedy optimum.

cases much worse than greedy search. The performance may improve for a better choice of the parameters.

14.5.4.2 Genetic algorithm versus greedy search

The genetic algorithm returns populations of sampling designs of 3–22 sensors, most of them with 10 or 11, and very few small and big ones. The sampling designs from runs starting at the greedy optimum are on average a bit bigger as they usually contain many copies of the initial sampling design of 15 sensors, see second column of Table 14.5. Figure 14.8 shows the final population from a run with random start. The size distribution is mainly due to the function used to evaluate the sampling designs during optimisation, which accounts for both plume detection and number of sensors. It could be changed to have a different or wider focus.

Table 14.5 Results of genetic algorithm compared with greedy search. Mean number of sensors is average of the complete populations of all 20 runs. Runs are marked better (as good) as greedy search if there is at least a better sampling design of any size. Better sampling designs is the number of different better sampling designs from all 20 runs.

Initial sensors	Mean number of sensors	Runs better (as good)	Better sampling designs
Random	10.43	10 (15)	33
20% greedy best	11.64	5 (20)	10

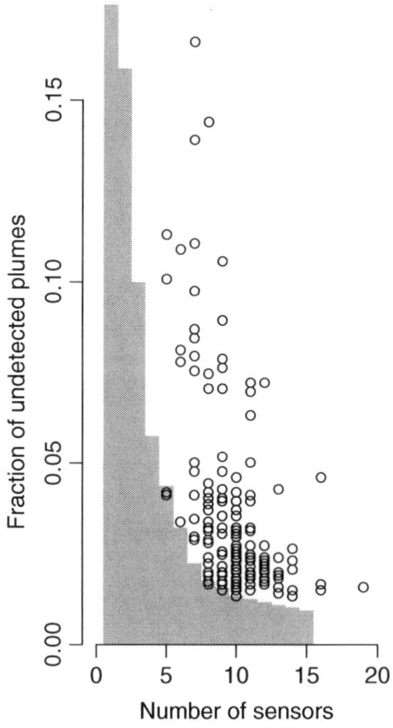

Figure 14.8 Cost and size of the 200 sampling designs of the final population from one run of genetic algorithm with random start. Grey area: better than greedy optimum.

For each run we only evaluate the best sampling designs of each size. When runs are started randomly, half of the populations contain sampling designs that are better than those from greedy search of same size. When starting at the greedy optimum, only 25% of the runs achieve such an improvement. However, for 15% of the runs with random start all results are worse than the greedy optimum. The better sampling designs all have size between 5 and 12 sensors, most have 9. The cost of these better sampling designs is on average 13% lower than for the greedy optimum, this corresponds to detecting about 1 more plume. Usually there are many duplicates of the best sampling designs. If we summarise them for all 40 runs, there are 43 different sampling designs of different sizes that are better than the ones from greedy search; 33 of them, originate from runs with random start. These optima overlap a good deal: only 44/736 of the locations on the grid are part of any of them. One sensor location is part of all the optima, and 5 more locations occur in more than 30 of them, whereas 24 locations only occur once or twice. Figure 14.9 shows all sampling designs of 9 sensors. It is obvious that the most important sensor locations are situated at the margins of Upper Austria.

These results show that the genetic algorithm is quite successful in finding sampling designs as good or even better than those from greedy search. Starting

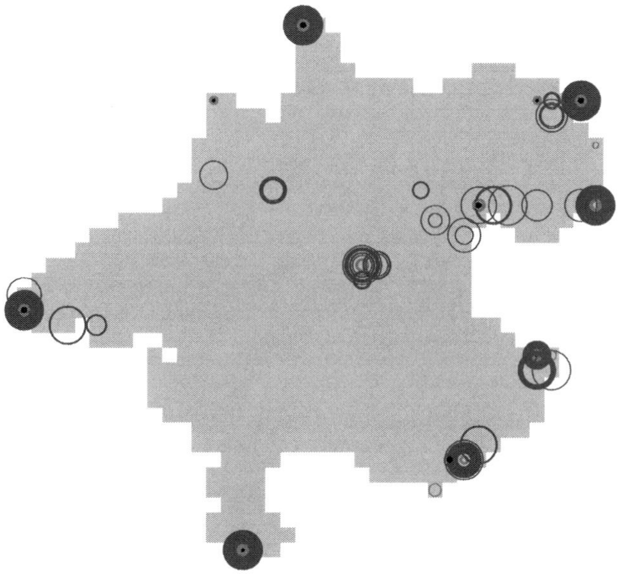

Figure 14.9 Overlay of the 25 optimal sampling designs of 9 sensors from genetic algorithm. Circles of one size represent one sampling design. Dots: best greedy. Small, light grey circles: 6 designs from runs starting at best greedy. Big, dark grey, circles: 19 designs from runs with random start.

from a completely random population increases the risk to get only worse results, but also the chance to get better ones. Most runs find sampling designs that are as good or even better than those of same size from greedy search. The very small and big sampling designs in the populations are often much worse than the greedy optimum. Several sampling designs of same size and minimal cost are found, implying that the problem may not have one but multiple global optima, which have about 7 sensors at the margins of the region and 1 in the centre in common.

14.6 Discussion

Plume simulations allow to evaluate how sensors would measure plumes. From cases where the potential plumes are quite well known, as is often the case in hydrology, we can move towards more variable plumes as they occur in the atmosphere. However, this poses new questions for the correct choice of the plume simulations. Increasing computational capacities make this approach more applicable to real world problems. It remains a challenge to allocate this capacity in the most powerful way to a good coverage of the random field by the considered realisations and their resolution in space and time, to complex plume simulation, to cost functions that are computationally demanding, or to optimisation algorithms that search the potential sampling designs thoroughly for the global optimum. A way to tackle it, is to systematically evaluate the decisions taken in all these respects.

In our use case we use simulations of medium resolution and a very simple cost function: the fraction of undetected plumes. The focus is on optimisation; we compare the results of three different algorithms. With greedy search we show that 15 sensors are sufficient to detect the 1226 potential plumes entering Upper Austria from 13 different sources. Ignoring up to half of the plumes in the optimisation results in different sampling designs, but they still detect more than 95% of the plumes they are not optimised for. Also 100 times weaker plumes are detected with high probability. The plumes seem to be not too hard to detect and similar to each other.

We optimise sampling designs of different size by greedy search. When spatial simulated annealing is started from the optimum of 5 sensors, it can improve it in 7 of 20 runs, reducing cost up to 20%. The results from random start are much worse. For sampling designs of 10 sensors no improvement can be achieved. The genetic algorithm is also started from a random population and from one containing the optimum from greedy search. It mainly returns sampling designs of medium size. On average about 1 per run is better than those from greedy search. The few small and big sampling designs it returns are very bad. Initialising this algorithm with the optima from greedy search rather reduces the chance for improvement. Thus two strategies seem successful, combining greedy search and spatial simulated annealing – mainly for small sampling designs – and genetic algorithm to get numerous 'optima' to choose from.

The 'optima' returned by the different algorithms differ: about 7 locations at the margins of the region and 1 in the centre seem to be of high importance, whereas there are many good locations for the other sensors. This implies that there may be several global optima having only some important sensors in common. Still, we do not know if any of the discovered 'optima' are really the global best ones of their size.

Acknowledgments

We are grateful to Poul Astrup (Riso National Laboratory for Sustainable Energy, Technical University Denmark) for providing the plume simulations. This work was funded by the European Commission Seventh Framework Program project DETECT, Contract No. 232662.

References

Abida R, Bocquet M, Vercauteren N and Isnard O 2008 Design of a monitoring network over France in case of a radiological accidental release. *Atmospheric Environment* **42**, 5205–5219.

Brandt J, Christensen JH, Frohn LM and Zlatev Z 2000 Numerical modelling of transport, dispersion, and deposition – validation against ETEX-1, ETEX-2 and Chernobyl. *Environmental Modelling & Software* **15**, 521–531.

Carruthers DJ, Holroyd RJ, Hunt JCR, Weng WS, Robins AG, Apsley DD, Smith FB, Thomson DJ and Hudson B 1992 UK atmospheric dispersion modelling system. In *Air Pollution Modeling and its Application IX* (eds van Dop H and Kallos G), 15–28, Plenum Press.

Cieniawski SE, Eheart JW and Ranjithan S 1995 Using genetic algorithms to solve a multiobjective groundwater monitoring problem. *Water Resources Research* **31**(2), 399–409.

Connan O, Smith K, Solier L, Maro D, Hébert D and Bacon G 2011 Mesoscale dispersion of 85-Kr in the vicinity of the AREVA La Hague reprocessing plant. *Radioprotection* **46**(6), 423–429.

Delmelle EM and Goovaerts P 2009 Second-phase sampling designs for non-stationary spatial variables. *Geoderma* **153**, 205–216.

Dyer LL and Astrup P 2010 Model evaluation of RIMPUFF within complex terrain using an 41-Ar radiological dataset. *13th Conference on Harmonisation within Atmospheric Dispersion Modelling for Regulatory Purposes, June, Paris*. www.harmo.org/Conferences/Proceedings/_Paris/publishedSections/H13-130-abst.pdf [July 2012].

Ehrhardt J and Weis A 2000 *RODOS: decision support system for off-site nuclear emergency management in Europe*. Report EUR 19144.

European Comission Joint Research Centre 2011 *European Radiological Data Exchange Platform*. http://eurdep.jrc.ec.europa.eu/Basic/Pages/Public/Home/Default.aspx [July 2012].

GADM 2012 *GADM Database of Global Administrative Areas 2.0*. http://www.gadm.org [June 2012].

Gering F and Müller H 2000 *Deposition calculation in RODOS PV 4.0*. RODOS(WG3)-TN(99)22.

Helle KB, Urso L, Astrup P, Mikkelsen T, Kaiser JC, Pebesma E, Rojas-Palma C, Holo E, Dyve JE and Raskob W 2011 Planning sensor locations for the detection of radioactive plumes for Norway and the Balkans. *Radioprotection* **46**(6) 55–61.

Heuvelink GBM, Jiang Z, De Bruin S and Twenhöfel CJW 2010 Optimization of mobile radioactivity monitoring networks. *Radioprotection* **24**(3), 365–382.

Holland JH 1975 *Adaption in Natural and Artificial Systems*. University of Michigan Press.

James BR and Gorelick SM 1994 When enough is enough: the worth of monitoring data in aquifer remediation design. *Water Resources Research* **30**(12), 3499–3513.

Kalnay E, Kanamitsu M, Kistler R, Collins W, Deaven D, Gandin L, Iredell M, Saha S, White G, Woollen J, Zhu Y, Leetmaa A, Reynolds R, Chelliah M, Ebisuzaki W, Higgins W, Janowiak J, Mo KC, Ropelewski C, Wang J, Jenne R and Joseph D 1996 The NCEP/NCAR 40-year reanalysis project. *Bulletin of the American Meteorological Society* **77**(3), 437–471.

Kim KH and Lee KK 2007 Optimization of groundwater-monitoring networks for identification of the distribution of a contaminant plume. *Stochastic Environmental Research and Risk Assessment* **21**, 785–794.

Lee YM and Ellis JH 1996 Comparison of algorithms for nonlinear integer optimisation: application to monitoring network design. *Journal of Environmental Engineering* **122**(6), 524–531.

Loaiciga HA, Charbeneau RJ, Everett LG, Fogg GE, Hobbs BF and Rouhani S 1992 Review of ground-water quality monitoring network design. *Journal of Hydraulic Engineering* **118**(1), 11–37.

Lucasius CB and Kateman G 1993 Understanding and using genetic algorithms Part 1. Concepts, properties and context. *Chemometrics and Intelligent Laboratory Systems* **19**, 1–33.

Lucasius CB and Kateman G 1994 Understanding and using genetic algorithms Part 2. Representation, configuration and hybridization. *Chemometrics and Intelligent Laboratory Systems* **25**, 99–145.

Mahar PS and Datta B 1997 Optimal monitoring network and ground-water-pollution source identification. *Journal of Water Resources Planning and Management* **123**(4), 199–207.

McPhee J and Yeh WWG 2005 Optimal experimental design for parameter estimation and contaminant plume characterization in groundwater modelling. In *Applied Optimal Designs* (eds Berger MPF and Wong WK), 219–246. John Wiley & Sons, Ltd.

Melles SJ, Heuvelink GBM, Twenhöfel CJW, Van Dijk A, Hiemstra PH, Baume O and Stöhlker U 2011 Optimizing the spatial pattern of networks for monitoring radioactive releases. *Computers and Geosciences* **37**(3), 280–288.

Meyer PD, Valocchi AJ and Eheart JW 1994 Monitoring network design to provide initial detection of groundwater contamination. *Water Resources Research* **30**(9), 2647–2659.

Mikkelsen T, Larsen S and Thykier-Nielsen S 1984 Description of the Riso puff diffusion model. *Nuclear Technology* **67**, 56–65.

Mikkelsen T, Thykier-Nielsen S and Hoe S 2007 Medium-range puff growth. *Developments in Environmental Science* **6**, 243–252.

Montas HJ, Mohtar RH, Hassan AE and AlKhal FA 2000 Heuristic space-time design of monitoring wells for contaminant plume characterization in stochastic flow fields. *Journal of Contaminant Hydrology* **43**, 271–301.

NOAA 2011 *National Oceanic and Atmospheric Administration* http://www.esrl.noaa.gov/psd/ [July 2012].

Pebesma E, Skoien J, Baume O, Chorti A, Hristopulos DT, Melles SJ and Spiliopoulos G 2012 intamapInteractive: procedures for automated interpolation – methods only to be used interactively, not included in intamap package. *R package version 1.1-1*. http://CRAN.R-project.org/web/packages/intamapInteractive/index.html [July 2012].

R Development Core Team 2011 *R: a language and environment for statistical computing* http://www.R-project.org/ [July 2012].

Reed P, Minsker B and Valocchi AJ 2000 Cost-effective long-term groundwater monitoring design using a genetic algorithm and global mass interpolation. *Water Resources Research* **36**(12), 3731–3741.

RODOS 2011 *Realtime Online Decision Support System for Nuclear Emergency Management* http://www.rodos.fzk.de/rodos.html [July 2012].

Schaetzen WBF, Walters GA and Savic DA 2000 Optimal sampling design for model calibration using shortest path, genetic and entropy algorithms. *Urban Water* **2**, 141–152.

Seinfeld JH and Pandis SN 2006 *Atmospheric Chemistry and Physics*, 2nd edn. John Wiley & Sons, Ltd.

Serón Arbeloa FJ, Pérez Caseiras C and Latorre Andrés PM 1993 Air quality monitoring: optimization of a network around a hypothetical potash plant in open countryside. *Atmospheric Environment* **27A**(5), 729–738.

Thykier-Nielsen S, Deme S and Láng E 1995 *Calculation method for gamma dose rates from Gaussian puffs*. Risoe-R-775(EN) http://www.risoe.dk/rispubl/reports_INIS/RISOR775.pdf [July 2012].

Twenhöfel CJW, Van Troost MM and Bader S 2007 *Uncertainty analysis and parameter optimisation in early phase nuclear emergency management*. RIVM Report 861004001. http://www.rivm.nl/bibliotheek/rapporten/861004001.pdf [July 2012].

Van Groenigen JW 1997 Spatial simulated annealing for optimizing sampling. In *Proc. geoENV I - Geostatistics for Environmental Applications* (eds Soares A, Gomez-Hernandez J and Froidevaux R), 351–361, Kluwer Academic Publishers.

Willighagen E 2005 genalg: R based genetic algorithm. *R package version 0.1.1*. http://CRAN.R-project.org/web/packages/genalg/index.html [July 2012].

WRF 2011 *Weather Research and Forecasting Model*. http://www.wrf-model.org/index.php [July 2012].

Wu J, Zheng C, Chien CC and Zheng L 2006 A comparative study of Monte Carlo simple genetic algorithm and noisy genetic algorithm for cost-effective sampling network design under uncertainty. *Advances in Water Resources* **29**(6), 899–911.

Index

active learning, 27, 294
  commitees, 299–300, 305, 310
  diversity, 299
  MLP, 296
  one class SVM, 297–8, 310
adaptive sampling, 233, 235, 249
additional observations, 55
air pollution, 265, 321
air quality, 131
algorithm, 198, 326
  distance, 197
anisotropy, 208, 210
area of the region, 19
atmospheric dispersion model, 321
autocorrelation, 54, 328
autoregressive model, 258
auxiliary variable, 135
average predictive variance criterion, 25
average universal kriging variance, 26

Bartlett transformation, 125
basis functions, 279
Bayesian, 25, 134
Bayesian information matrix, 79
Bayesian prior knowledge, 78
Bayesian space-time dynamical models, 24
best linear unbiased estimator (BLUE), 39
best linear unbiased predictor (BLUP), 39
bias, 131
BIC, 116
Box–Cox transformation, 81
budget, 54, 55

candidate design points, 38
candidate location, 67, 302, 311, 323
choropleth, 68
classical experimental design problem, 78
classification
  one class SVM, 292
  support vector machines, 289, 290
climatic data, 17
clustered samples, 58
collocation, 46, 48
complete symmetry, 5
composite likelihood, 13
compound design, 52
conditional covariance matrix, 232, 276
conditional entropy, 232
conditionally negative definite, 11
correlation
  ignoring, 198
correlation function, 253
cosine-sine Bessel surface harmonics, 74

Spatio-temporal design: Advances in efficient data acquisition, First Edition.
Edited by Jorge Mateu and Werner G. Müller.

cost function, 324
covariance function, 38, 210
covariance matrix, 58, 232
covariance structure, 221
cross-covariance function, 46
CUDA, 88

data
  geostatistical, 196
DEM, 213
design
  'coffee-house', 194
  compound, 197
design configuration, 66
design criterion, 14, 24, 198, 251, 253–254, 259, 324
design strategy, 255, 259, 262
design-based, 221
design, dynamic, 209, 219, 221
design, optimal, 219, 221
design, static, 209, 221
detection, 324
dimension reduction, 143, 145
discretized, 323
dispersion parameter, 272
distance, 196, 197
dynamic Gamma model, 278
dynamic model, 275
dynamic process convolution model, 257–264, 266
dynamical component, 26

EKF algorithm, 276
elevation of the district's capital, 19
EM algorithm, 107, 112–13, 126
empirical BLUP (E-BLUP), 41
Empirical Orthogonal Functions–EOFs, 273
entropy, 52, 105, 252
entropy criterion, 16, 26
entropy-based information measures, 26
environmental monitoring, 249, 265, 324
environmental network design, 24
exact optimum designs, 17
exchange algorithm, 42, 86, 278
expected utility, 274, 277, 280
extended Kalman filter (EKF), 275, 277

first-order Markov structure, 273

Gamma data model, 276
Gamma distribution, 275
Gaussian Markov random field, 140
Gaussian model, 252
Gaussian predictive process, 148
Gaussian process, 253
generalized least squares, 210
genetic algorithm, 17, 208, 329
geostatistical data, 131
Gibbs sampling, 258
GIW prior distribution, 125
grassland usage, 18
greedy algorithms, 208
greedy search, 326

Helmert contrasts, 115
heuristic, 55, 63, 68, 326
hierarchical Bayesian model, 106
hierarchical, 24, 144
homoscedasticity, 61
hybrids, 17
hyperparameter, 106

I-optimality, 78
identification number, 19
infill asymptotics, 136
information matrix, 40
informative missingness, 132
informative sampling, 132
interaction of space and time, 10
interpolation, 209, 210, 325
intrinsically stationary, 6
Integrated Nested Laplace Approximation, 146
inverse distance weighted interpolation, 325

isotropic, 38, 76
isotropy, 73

Kalman gain, 276
Karhunen–Loeve–Eigen expansion, 77
kernel, 258
kernel convolution, 147, 165, 257
knot, 143, 257
knot design, 144, 146, 149, 153, 164
knot locations, 25
knot performance, 152
knot process, 257
knot selection, 143, 150–153, 165
knot size, 152
knot-based, 146
kriging, 8, 57, 106, 196, 207–8, 211, 216, 221, 325
kriging variance, 55, 57–62, 65–8, 70
kriging, space-time, 207, 208, 211, 221
kriging, universal, 208–211, 215, 216, 218

large spatial datasets, 25
latitude, 19
learning from data, 287
  model selection, 288, 297
likelihood approximations, 145
link function, 272
local optimum, 327
longitude, 19
low rank models, 146
low-rank basis function, 27
low-rank semiparametric representation, 278

Markov chain Monte Carlo, 25, 135, 144, 277, 280
Markov random fields, 145
maximum likelihood estimator, 40
MC approximation, 277
MC-EKF dynamic design algorithm, 277
MCP, 115
mean function, 38
mean square error
  integrated, 250, 260, 265
  minimum, 254
measurement process, 271
method-of-moments, 11
missing data, 104, 107, 109, 111–13, 126, 134
mixed linear model, 77
model-based design, 26
model-based inference, 144
model-based paradigm, 2
MODIS, 209, 212–3, 215–6, 218–9
monotone data pattern, 106
Monte Carlo (MC) estimate, 274, 277, 280
Monte Carlo likelihood, 136
Monte Carlo sampling, 250
monitoring networks, 27
Moran's I, 181, 182, 191–2
mutual information, 232

network, 37, 54
nonseparable, 10
nonstationary, 55, 61, 149, 325
nonstationary covariance functions, 10
nugget effect, 57

Oberösterreich (Upper-Austria), 17
observation process, 258
optimal, 55, 116
optimal design, 37, 252
optimal spatial prediction, 23
optimisation, 326
overfitting, 289, 296, 337

pair of Fourier transforms, 7
periodicities, 56
piecewise linear function, 115
plume, 27, 321
point pattern data, 140

Poisson process, 135
polar spectral representation, 74, 77
positive-definite, 10
posterior predictive
distribution, 274
precipitation data, 104
prediction, 39, 197
spatial, 196
prediction error, 39, 57
prediction variance, 251, 254, 296
predictive distribution, 104–5, 107, 112–13, 117, 120, 122, 124–7, 250, 252–4
preferential sampling, 132
probability-based, 2
probit link, 135

radioactivity, 304, 323
rainfall, 262–4, 301
random field, 2, 38
random sampling, 170
random walk, 258
reduced rank, 143, 146, 148
region of interest, 233, 323
regression, 210, 215
kernel ridge regression, 293
supprt vector regression, 294
relative difference, 124
residual maximum likelihood
estimator, 40, 50
resolution, 323
restricted maximum likelihood
(REML), 74

sample size, 333, 338
sampling, 197
sampling design
1–adaptive, 251, 254, 256, 260, 262
fixed, 249
k–adaptive, 251, 260, 266
lattice, 252–4, 257, 260, 262
optimal, 252–4
spatial, 251
SCAD, 115
seasonality, 262
second-order stationarity, 6, 38, 73
second-phase spatial sampling, 24, 55, 59, 61, 63–4, 68
sensitivity, 221, 332
separability, 5, 48, 253
Shannon's entropy, 232
shared-variable model, 134
simulated annealing, 42, 55, 265, 286
simulation, 254, 258, 321
simultaneous addition, 65
Smith and Zhu design criterion, 74, 79, 80, 85
source term, 323
space discretization, 59
space filling, 16, 68, 251, 286
space-time interactions, 25
spatDesign toolbox, 89
spatial
dependence, 59, 196
filtering, 196
spatial basis function, 273
spatial correlation function, 107
spatial econometrics, 3
spatial mixed linear model, 76, 77
spatial network design, 23
spatial prediction, 252, 265
spatial process, 148
spatial simulated annealing, 17, 26, 208–9, 211, 221, 328
spatial variation, 55, 58, 61
spatially anisotropic, 10
spatially referenced, 143
spatio-temporal covariance models, 10
spatio-temporal dependence
structure, 235
spatio-temporal dynamic process, 27
spatio-temporal kriging, 8
spatio-temporal monitoring network
design, 21
spatio-temporal prediction, 10

spatio-temporal process, 233–6, 238, 253, 257, 273, 275
spatio-temporal sampling, 231
spectral density function, 8, 81
spectral distribution function, 81
staircase structure, 106
state space model, 253
stationarity, 6, 252–3
stationary, 61
stationary covariance function, 8
stochastic process, 147, 250

temporal linear model, 108
temporally asymmetrical, 10
thin-plate spline approach, 107
time series, 208, 212–3
trans-Gaussian kriging, 81, 97
trend, 208–210, 215–6, 218, 221
type-II maximum likelihood estimation, 126

Upper Austria, 300
  rainfall data, 262–264, 301
  temperature data, 300
utility function, 274

variogram, 57–60, 196–7, 215, 216
  empirical, 196
  estimation, 197
  model, 196

weather, 321
wind power, 307

Statistics in Practice

Human and Biological Sciences

Berger – Selection Bias and Covariate Imbalances in Randomized Clinical Trials
Berger and Wong – An Introduction to Optimal Designs for Social and Biomedical Research
Brown and Prescott – Applied Mixed Models in Medicine, Second Edition
Carstensen – Comparing Clinical Measurement Methods
Chevret (Ed) – Statistical Methods for Dose-Finding Experiments
Ellenberg, Fleming and DeMets – Data Monitoring Committees in Clinical Trials: A Practical Perspective
Hauschke, Steinijans & Pigeot – Bioequivalence Studies in Drug Development: Methods and Applications
Källén – Understanding Biostatistics
Lawson, Browne and Vidal Rodeiro – Disease Mapping with Win-BUGS and MLwiN
Lesaffre, Feine, Leroux & Declerck – Statistical and Methodological Aspects of Oral Health Research
Lui – Statistical Estimation of Epidemiological Risk
Marubini and Valsecchi – Analysing Survival Data from Clinical Trials and Observation Studies
Millar – Maximum Likelihood Estimation and Inference: With Examples in R, SAS and ADMB
Molenberghs and Kenward – Missing Data in Clinical Studies
O'Hagan, Buck, Daneshkhah, Eiser, Garthwaite, Jenkinson, Oakley & Rakow – Uncertain Judgements: Eliciting Expert's Probabilities
Parmigiani – Modeling in Medical Decision Making: A Bayesian Approach
Pintilie – Competing Risks: A Practical Perspective
Senn – Cross-over Trials in Clinical Research, Second Edition
Senn – Statistical Issues in Drug Development, Second Edition
Spiegelhalter, Abrams and Myles – Bayesian Approaches to Clinical Trials and Health-Care Evaluation
Walters – Quality of Life Outcomes in Clinical Trials and Health-Care Evaluation
Welton, Sutton, Cooper and Ades – Evidence Synthesis for Decision Making in Healthcare
Whitehead – Design and Analysis of Sequential Clinical Trials, Revised Second Edition
Whitehead – Meta-Analysis of Controlled Clinical Trials
Willan and Briggs – Statistical Analysis of Cost Effectiveness Data
Winkel and Zhang – Statistical Development of Quality in Medicine

Earth and Environmental Sciences

Buck, Cavanagh and Litton – Bayesian Approach to Interpreting Archaeological Data
Chandler and Scott – Statistical Methods for Trend Detection and Analysis in the Environmental Statistics
Glasbey and Horgan – Image Analysis in the Biological Sciences
Haas – Improving Natural Resource Management: Ecological and Political Models
Helsel – Nondetects and Data Analysis: Statistics for Censored Environmental Data
Illian, Penttinen, Stoyan, H and Stoyan D-Statistical Analysis and Modelling of Spatial Point Patterns
Mateu and Müller (Eds) – Spatio-temporal design: Advances in efficient data acquisition
McBride – Using Statistical Methods for Water Quality Management
Webster and Oliver – Geostatistics for Environmental Scientists, Second Edition
Wymer (Ed) – Statistical Framework for Recreational Water Quality Criteria and Monitoring

Industry, Commerce and Finance

Aitken – Statistics and the Evaluation of Evidence for Forensic Scientists, Second Edition
Balding – Weight-of-evidence for Forensic DNA Profiles
Brandimarte – Numerical Methods in Finance and Economics: A MATLAB-Based Introduction, Second Edition
Brandimarte and Zotteri – Introduction to Distribution Logistics
Chan – Simulation Techniques in Financial Risk Management
Coleman, Greenfield, Stewardson and Montgomery (Eds) – Statistical Practice in Business and Industry
Frisen (Ed) – Financial Surveillance
Fung and Hu – Statistical DNA Forensics
Gusti Ngurah Agung – Time Series Data Analysis Using EViews
Kenett (Eds) – Operational Risk Management: A Practical Approach to Intelligent Data Analysis
Kenett (Eds) – Modern Analysis of Customer Surveys: With Applications using R
Kruger and Xie – Statistical Monitoring of Complex Multivariate Processes: With Applications in Industrial Process Control
Jank and Shmueli (Ed.) – Statistical Methods in e-Commerce Research
Lehtonen and Pahkinen – Practical Methods for Design and Analysis of Complex Surveys, Second Edition
Ohser and Mücklich – Statistical Analysis of Microstructures in Materials Science
Pourret, Naim & Marcot (Eds) – Bayesian Networks: A Practical Guide to Applications
Taroni, Aitken, Garbolino and Biedermann – Bayesian Networks and Probabilistic Inference in Forensic Science
Taroni, Bozza, Biedermann, Garbolino and Aitken – Data Analysis in Forensic Science